CELL and MUSCLE MOTILITY

MOTILITY

Volume **3**

Cell and Muscle Motility

CELL and MUSCLE MOTILITY

MOTILITY
Volume 3

Edited by
Robert M. Dowben
and
Jerry W. Shay

University of Texas Health Science Center
Dallas, Texas

PLENUM PRESS • NEW YORK AND LONDON

The Library of Congress cataloged the first volume of this title as follows:

Main entry under title:
Cell and muscle motility.

Includes bibliographies and indexes.
1. Muscle contraction. 2. Cells—Motility. I. Dowben, Robert M. II. Shay, Jerry
W. [DNLM: 1. Cytology—Periodical. 2. Muscles—Periodical. 3. Movement—Periodical. W1 CE127].

| QP321.C365 | 599.01′852 | 81-13827 |

ISBN-13: 978-1-4615-9298-3   e-ISBN-13: 978-1-4615-9296-9
DOI: 10.1007/978-1-4615-9296-9

© 1983 Plenum Press, New York
Softcover reprint of the hardcover 1st edition 1983
A Division of Plenum Publishing Corporation
233 Spring Street, New York, N.Y. 10013

Contributors

Peter Cooke, Department of Physiology, University of Connecticut Health Center, Farmington, Connecticut 06032

Setsuro Ebashi, Department of Pharmacology, Faculty of Medicine, University of Tokyo, Bunkyo-ku, Tokyo, Japan

M. S. Forbes, Department of Physiology, University of Virginia School of Medicine, Charlottesville, Virginia 22908

S. M. Heywood, Department of Biology, University of Connecticut, Storrs, Connecticut 06268

Maureen G. Price, Department of Anatomy, University of Pennsylvania Medical School, Philadelphia, Pennsylvania 19104; present address: Cardiovascular Research Laboratory, Department of Physiology, University of California, Los Angeles, California 90024

Marschall S. Runge, Department of Molecular Biology, Vanderbilt University, Nashville, Tennessee 37235

Joseph W. Sanger, Department of Anatomy, University of Pennsylvania Medical School, Philadelphia, Pennsylvania 19104

Robert J. Schwartz, Department of Cell Biology, Baylor College of Medicine, Texas Medical Center, Houston, Texas 77030

E. Siegel, Department of Biology, University of Connecticut, Storrs, Connecticut 06268

N. Sperelakis, Department of Physiology, University of Virginia School of Medicine, Charlottesville, Virginia 22908

Edwin M. Stone, Department of Cell Biology, Baylor College of Medicine, Texas Medical Center, Houston, Texas 77030

M. C. Thibault, Department of Biology, University of Connecticut, Storrs, Connecticut 06268

Robert W. Tucker, The Johns Hopkins Oncology Center, Baltimore, Maryland 21205

Robley C. Williams, Jr., Department of Molecular Biology, Vanderbilt University, Nashville, Tennessee 37235

Preface

Motility is a fundamental property of living systems, from the cytoplasmic streaming of unicellular organisms to the most highly differentiated and developed contractile system of higher organisms, striated muscle. Although scientific investigations into the mechanisms of motility have a long and interesting history, the knowledge of molecular processes, especially in the area of regulation of control of motility, has been developing at an ever more rapid pace with the utilization of multidisciplinary approaches from physiology, cell biology, genetics, biochemistry, pharmacology, and biophysics.

In Volume 3, *Cell and Muscle Motility* continues the same philosophy as that of the preceding volumes. The essays are meant to focus on topics of current interest, to be critical rather than exhaustive, and to indicate the current trends of research efforts. The series is intended to foster an interchange of concepts among various workers in a variety of disciplines and to serve as a reference for students and workers who wish to familiarize themselves with the most current progress in motility.

Robert M. Dowben
Jerry W. Shay

Dallas

Contents

Chapter 5
The Membrane System and Cytoskeletal Elements of Mammalian Myocardial Cells
M. S. Forbes and N. Sperelakis

Chapter 6
Control of Gene Expression in Muscle Development
S. M. Heywood, M. C. Thibault, and E. Siegel

Chapter 7
Cloning of Contractile Protein
Robert J. Schwartz and Edwin M. Stone

Chapter 8
Role of Microtubules and Centrioles in Growth Regulation of Mammalian Cells
Robert W. Tucker

1

Intermediate Filaments in Striated Muscle

A REVIEW OF STRUCTURAL STUDIES IN EMBRYONIC AND ADULT SKELETAL AND CARDIAC MUSCLE

Maureen G. Price and Joseph W. Sanger

1. Introduction

In this chapter we will review the results and conclusions of ultrastructural and immunofluorescent studies of intermediate filament proteins in striated muscle. The majority of presently available publications on the structure of striated muscle are concerned either with the contractile elements of the myofibrils or with the membranous systems. We consider the evidence that striated muscle contains filaments that have average diameters intermediate between those of actin and myosin and that are neither thick (myosin) filaments nor thin (actin) filaments. The intermediate filaments have also been designated by the term 10-nm filaments on the basis of their average diameter, but it is important to remember that their diameters range from 8 to 13 nm.

Intermediate filaments were first observed in embryonic muscle cells, where they are prominent cytoplasmic components (Rash *et al.*, 1967, 1970a; Ishikawa *et al.*, 1968). The discovery of intermediate filaments in the cytoplasm of the mononucleated cells present in cultures of embryonic skeletal muscle (Ishikawa *et al.*, 1968), as well as in epithelial cells, chondrogenic cells, and nerve cells (Ishikawa *et al.*, 1969), indicated that intermediate filaments may be a common subcellular feature.

Maureen G. Price and Joseph W. Sanger • Department of Anatomy, University of Pennsylvania Medical School, Philadelphia, Pennsylvania 19104. Present address for Dr. Price: Cardiovascular Research Laboratory, Department of Physiology, University of California, Los Angeles, California 90024.

A series of studies on intermediate filaments in many kinds of cells was prompted by the initial reports of cytoplasmic filaments that are morphologically distinct from actin and myosin filaments. Intermediate filaments have been localized in nearly every cell type that has been examined by electron microscopy or immunofluorescence using antibodies to the protein components of the filaments (for reviews, see Gaskin and Shelanski, 1976; Lasek and Hoffman, 1976; Goldman *et al.*, 1979; Lazarides, 1980; Small and Sobieszek, 1980). Although the intermediate filaments of different cell types exhibit similar diameters and solubility properties, they are biochemically and immunologically distinct. The intermediate filaments have been classified into five groups (Lazarides, 1980): (1) neurofilaments, (2) glial filaments, (3) keratin filaments, (4) vimentin-containing filaments from cells of mesenchymal origin, and (5) desmin filaments from muscle cells. Classification cannot be based strictly on the tissue of origin, as there is increasing evidence for the presence of two and possibly three types of intermediate filaments within single cells (Starger *et al.*, 1978; Bennett *et al.*, 1979; Franke *et al.*, 1979; Gard *et al.*, 1979; Granger and Lazarides, 1979; Tuszynski *et al.*, 1979; Zackroff and Goldman, 1979; Gard and Lazarides, 1980; Osborn *et al.*, 1980).

Following the designation of a third type of filament, the intermediate filament, in the cytoplasm of developing skeletal and cardiac cells and in numerous other kinds of cells, the biochemical characterization of the protein subunits of the filaments was begun. It was determined that embryonic striated muscle contains two intermediate filament proteins. One of these is desmin, the major intermediate filament protein of smooth muscle (Cooke, 1976; Small and Sobieszek, 1977; Johnson and Yun, 1980; Hubbard and Lazarides, 1979; Huiatt *et al.*, 1980). Desmin has been discovered in skeletal muscle (Lazarides and Hubbard, 1976; Izant and Lazarides, 1977; Lazarides and Balzer, 1978; Granger and Lazarides, 1978; Bennett *et al.*, 1979; O'Shea *et al.*, 1979; Gard and Lazarides, 1980; Young *et al.*, 1980; O'Shea *et al.*, 1981), cardiac muscle (Lazarides and Hubbard, 1976; Bennett *et al.*, 1978; Lazarides, 1978; Campbell *et al.*, 1979; Fuseler *et al.*, 1981), and Purkinje cells (Eriksson *et al.*, 1978; Stigbrand *et al.*, 1978; Eriksson *et al.*, 1979). Vimentin, the major component of the intermediate filaments in fibroblastic cells (Bennett *et al.*, 1978; Franke *et al.*, 1978), has also been determined to be a component of the intermediate filaments of striated muscle, on the basis of immunofluorescence and two-dimensional gel elecrophoresis (Bennett *et al.*, 1978; Bennett *et al.*, 1979; Granger and Lazarides, 1979; Gard and Lazarides, 1980).

It is ironic that although intermediate filaments were first observed in developing striated muscle, it proved particularly difficult to determine the fate of these filaments after myogenesis. A major impediment to observing intermediate filaments in mature muscle was the inability of many investigators to look beyond the immediate attraction of the thin and thick filaments in the highly organized sarcomeres. In addition, it is likely that filaments other than actin and myosin filaments would be inconspicuous in or among the tightly packed myofibrils that are closely encased in the elaborate sarcoplasmic reticulum. The myofibrils of cardiac muscle cells are not as densely packed as those in skeletal muscle cells; this fact allowed earlier detection of intermediate filaments in mature cardiac muscle (Ferrans and Roberts, 1973). Detection of intermediate filaments in either type of striated muscle was hampered by the lack of good fixation methods prior to 1963, when glutaraldehyde was introduced as a reliable preservative (Sabatini, 1963). The improved methods for electron microscopy, coupled with the use of fluorescently labeled antibodies for localization of proteins at the level of the light microscope, have resulted in a better understanding of the fate of intermediate filaments during myogenesis.

2. The Search for the Third Type of Filament in Striated Muscle: A Historical Overview

The search for a third type of filament, distinct from actin and myosin, in adult striated muscle was originated when Hanson and Huxley observed that the myofbrillar Z-discs remained in linear register following extraction of most of the myosin filaments (Hanson and Huxley, 1953, 1955) or of the bulk of both myosin and actin filaments (Hanson, 1956). Densitometric scans of the region between the tips of actin filaments in myosin-depleted myofibrils revealed the presence of proteinaceous material (Huxley and Hanson, 1957). To explain the linear continuity of the myofibrillar Z-discs in the absence of the bulk of the myosin and actin filaments, they proposed that a third type of filament connects the Z-discs along the longitudinal axis of the myofibrils. Since the Z-discs of extracted myofibrils are often separated by a distance several times the normal sarcomere length, it was proposed that the linking filaments are extensible or stretchable; hence the name S-filaments.

In the model of 1955, Hanson and Huxley proposed that the S-filaments are thinner than actin filaments and that they extend through the H-zone of each sarcomere to connect the actin filaments of opposite I-bands. Finding that the myofibrillar Z-discs remained in a linear array in the virtual absence of both the myosin and actin filaments (Hanson, 1956) necessitated revising the model so that the S-filaments directly linked the Z-discs along the longitudinal axis (see Hoyle, in Ernst and Straub, 1968).

A number of ultrastructural studies were sparked by the proposal of a third type of intrafibrillar filament with a diameter less than actin. A common approach was to examine skeletal muscle that was fixed after being severely stretched to minimize or even eliminate the overlap of the myosin and actin filaments (Sjostrand, 1962; Carlsen *et al.*, 1965; McNeill and Hoyle, 1967; Garamvolgyi, 1969; Locker and Leet, 1975). Those searching for the third intrafibrillar filament gave particular attention to the gap between the A-bands and I-bands of highly stretched muscles. The other approach taken in the search for a third filament type was to use electron microscopy on the material remaining in muscle fibers or myofibrils after treatment with either myosin-extracting solutions (Garamvolgyi, 1962; Walcott and Ridgeway, 1967; Guba *et al.*, 1968) or after treatment with both myosin-extracting and actin-extracting solutions (Walcott and Ridgeway, 1967; Guba *et al.*, 1968; Locker and Leet, 1976).

The results and conclusions of the investigation that were designed to elucidate the nature of the third type of muscle filament can be placed into several categories. Some investigators concluded that longitudinal filaments with diameters between 3 and 6 nm emerged from the actin filaments and connected them to the myosin filaments (Sjostrand, 1962; Carlsen *et al.*, 1965; McNeill and Hoyle, 1967). Others concluded that very thin filaments connect the myosin filaments to the Z-discs in insect flight muscle (Garamvolgyi, 1963, 1969) and ox muscle (Locker and Leet, 1975, 1976). A third conclusion was that filaments measuring 2.5–4 nm in diameter are primary connections between Z-discs, parallel with the actin and myosin filaments (Walcott and Ridgeway, 1967; Guba *et al.*, 1968).

The conflicting conclusions reached by the various investigators are not equally well supported by the evidence. All of the investigations are weakened by reliance on longitudinal sections that are either so thick that the question of superimposition of filaments in different planes arises, or else so thin that the full length of few, if any, of the putative connecting filaments is included. Several reviewers of the evidence for thin intrafibrillar connecting filaments have cautioned that further research must be done before drawing any firm conclusions (Peachey, 1968; Sandow, 1970).

Further evidence for the presence of thin intrafibrillar filaments linking the myosin filaments to the Z-discs in insect muscle has accumulated since these cautionary statements were made. In accordance with the results of Garamvolgyi (1963, 1969), others can clearly discern filamentous connections (C-filaments) between the myosin filaments and the Z-discs in thin sections of insect muscle fixed at various sarcomere lengths (Trombitas and Tigyi-Sebes, 1977). Moreover, myosin filaments that are isolated from insect flight muscle are linked in pairs and triplets by very thin filaments (Guba, in Ernst and Straub, 1968; Trombitas and Tigyi-Sebes, 1979). This phenomenon of inter-filamentous connections appears to be specific to certain invertebrate muscles since no similarly convincing evidence for such connections has come from research on vertebrate muscle.

Recognizing the difficulty of identifying and tracing filaments in longitudinal sections of muscle, some investigators tried to quantitate filament types in cross-sections of different muscles. The results were not directly comparable as different methods were used by each investigator. The most accurate method of comparing numbers of filaments would be to count the filaments within individual myofibrils rather than quantitating filaments per unit area, since muscles may shrink or swell during preparation for electron microscopy. One group (Ullrick *et al.*, 1977) recently used this method for comparative quantitation of the thick and thin filaments in cross-sections through the A-band and the I–Z junction of myofibrils in striated muscles from two vertebrates and one invertebrate (frog sartorious, chameleon tongue, and water bug flight muscle). They found that the total number of filaments in the I–Z junction of insect myofibrils corresponds well with the sum of the predicted number of actin filaments (3 times the number of myosin filaments) plus the number of myosin filaments counted in the A-band. This result provides further evidence for direct filamentous extensions of the myosin filaments to the Z-discs in insect muscle. Conversely, the quantitative results rule out the possibility that the myosin filaments in vertebrate myofibrils are linked to the Z-discs.

The possibility that the Z-discs of vertebrate and invertebrate skeletal myofibrils are directly connected by longitudinal filaments cannot be ruled out by the quantitative study, because one cannot predict how many non-myofibrillar filaments would be associated with each sarcomere. There could be small numbers of longitudinal filaments extending between the Z-discs. In this regard, one may note that in the I–Z junction of vertebrate myofibrils there are between 20 and 80 more filaments per sarcomere than can be accounted for by the calculated number of actin filaments (Ullrick *et al.*, 1977). The question of longitudinal filamentous linkages between the Z-discs will be considered further in Section 4.1.4.

In the course of ultrastructural investigations of insect muscle, Garamvolgyi (1962, 1963, 1965) provided evidence for interfibrillar connections in addition to the intrafibrillar filamentous connections described above. Garamvolgyi obtained electron micrographs of bee flight muscle showing filamentous bundles extending obliquely between the Z-discs of adjacent myofibrils. The Z-bridges, as he called them, also link myofibrils to the sarcolemma (1965).

The observation of interfibrillar filamentous connections in several vertebrate muscles (Sandborn *et al.*, 1967; Page, 1969) suggests that transverse filamentous linkages between the Z-discs of neighboring myofibrils may be a general feature of skeletal muscle throughout the animal kingdom. Nevertheless, it should be pointed out that in 1970 the evidence for transverse filamentous links between the Z-discs was far more convincing for insect muscle than for vertebrate muscle. Garamvolgyi (1962) isolated sheets or planar arrays of Z-discs from bee flight muscle by dissolving the contractile filaments in dilute acid. When the isolated sheets were viewed *en face*, it was obvious that each Z-disc within a plane was linked to its neighbors by filaments radiating from it.

The primary emphasis of the search for the third filament was to find a new filament type within the myofibril, linking the contractile filaments or linking the Z-discs. A major shift in emphasis occurred following the observation of intermediate filaments between the assembling myofibrils of embryonic striated muscle, a finding that suggested that these filaments could become associated with the myofibrils to form extramyofibrillar linking elements. Garamvolgyi's (1962) ultrastructural documentation of interfibrillar Z-bridges was the harbinger of the modern view of the distribution and function of the intermediate filaments in striated muscle.

3. Localization of Intermediate Filaments in Embryonic Striated Muscle

3.1. Skeletal Myogenic Cells

Skeletal myogenic cells have been the subject of numerous morphological investigations since the 19th century, when Schwann (1847) postulated that multinucleated myotubes arise by the coalescence of many mononucleated cells. With the advent of electron microscopy, the filamentous components of skeletal myogenic cells could be examined (for historical references, see Hay, 1963; Dessouky and Hibbs, 1965). Since it was known that actin and myosin filament comprised the filamentous elements of mature skeletal muscle (Huxley and Hanson, 1957; Huxley, 1963), it was expected that these would also be present in embryonic muscle. Therefore, most investigators of developing muscle describe two classes of cytoplasmic filaments in addition to the microtubules. These two classes are thin actin filaments with an average diameter of 7 nm, and thicker myosin filaments that have diameters of between 10 and 18 nm, depending on the method of fixation (Hay, 1963; Dessouky and Hibbs, 1965; Allen and Pepe, 1965; Pryzbylski and Blumberg, 1966; Firket, 1967; Fischman, 1967, 1970; Shimada *et al.*, 1967; Etlinger and Fischman, 1972; Hilfer *et al.*, 1973). The two types of filaments can be divided into two categories, those that are free in the cytoplasm and those that are assembling into the hexagonal arrays that represent incipient myofibrils.

In retrospect, one finds that the early reports on the ultrastructure of developing skeletal muscle included a few indications that some of the filaments resembled neither actin nor myosin filaments. Dessouky and Hibbs (1965) noted the presence of many free filaments in the cytoplasm of myogenic cells. Allen and Pepe (1965) reported that these cells contain scattered filaments with diameters approximating those of actin filaments, but at least 50% longer than the 1-μm-long actin filaments that they isolated from adult skeletal muscle. Allen and Pepe assumed from the diameter of the scattered filaments that they were actin and that actin is synthesized before myosin in developing muscle cells.

Several researchers described classes of filaments that are morphologically different from mature actin and myosin filaments. Heuson-Steinnon (1965) categorized two additional classes of filaments, one of which is a class of smooth randomly dispersed filaments with diameters between 7 and 8 nm. She observed that these smooth curved filaments are the earliest filamentous components of myoblasts and proposed that they are precursors of myosin filaments. A similar conclusion was reached by Firket (1967) to account for the identity of the smooth filaments with average diameters of 9.5 nm, which undulate through the cytoplasm of chicken myogenic cells.

Ishikawa *et al.* (1968) were the first to identify the wavy cytoplasmic filaments in skeletal myogenic cells as being members of class of filaments distinct from actin and myosin, on the basis of elegant experiments using heavy meromyosin. Heavy meromyosin is a proteolytic fragment of myosin (Szent-Gyorgyi, 1953) that binds selectively to actin to form polarized arrowhead complexes along the filaments (Huxley, 1963). Ishikawa *et al.* (1969) found that the morphology of the intermediate filaments in a number of different types of cells was unaltered by treatment with heavy meromyosin, whereas the actin filaments were decorated by characteristic arrowheadlike structures. This exciting result represents the initial definitive evidence that intermediate filaments are composed of a protein other than actin. This paper was also an important milestone in documenting the presence of actin filaments in a wide variety of nonmuscle cells. The use of heavy meromyosin decoration clearly enables one to sort out two groups of filaments with diameters less than those of myosin filaments. There are still several papers each year in which the investigators identify intermediate filaments as actin filaments. It is important that heavy meromyosin decoration be used regularly in morphological investigations to distinguish between actin and intermediate filaments.

The majority of the filaments that do not bind heavy meromyosin cells have diameters between 8 nm and 11 nm, with the highest concentration at 10 nm (Ishikawa *et al.*, 1969). Not only do the filaments have diameters intermediate between those of actin and myosin, but they have other distinctive morphological characteristics. The intermediate filaments are curved rather than straight, so that they appear to be more flexible than actin or myosin filaments. Most of the intermediate filaments are longer than 1 μm (Ishikawa *et al.*, 1968; Kelly, 1969; Fischman, 1970), while the actin filaments present in skeletal muscle are 1 μm long. The length of some intermediate filaments exceeds that of myosin filaments (1.6 μm); they can be traced for more than 2.5 μm. The substructure of the intermediate filaments differs from the double helical configuration diagnostic of actin filaments (Hanson and Lowy, 1963). In addition, the intermediate filaments lack the cross-bridges that are a distinguishing feature of myosin filaments.

With the knowledge that intermediate filaments are representative of a unique class of filaments instead of being precursors of either myosin or actin filaments, it was necessary to reexamine skeletal myogenic cells to determine the localization of these filaments during muscle development. It was found that the majority of the filaments in presumptive myoblasts are intermediate filaments (Ishikawa *et al.*, 1968; Kelly, 1969; Shimada, 1971; Holtzer *et al.*, 1972). Intermediate filaments occupy a greater percentage of a unit of area in early myotubes than they do in myoblasts that are in the process of fusing with other cells (Shimada, 1971). This indicates that immediately following fusion, there is an increase in either the number of intermediate filaments or in the length of existing intermediate filaments, or both.

As myofibrillogenesis proceeds, the number of intermediate filaments in the skeletal myotubes decreases (Kelly, 1969; Holtzer *et al.*, 1972; Ishikawa, 1974). The fate of the intermediate filaments has not yet been conclusively determined. Nevertheless, the combined results of electron microscopic and immunofluorescent studies using antibodies to several protein subunits of intermediate filaments provide some insights into this problem. The results of these two types of investigations will be discussed in sequence.

In the initial report of intermediate filaments in embryonic skeletal muscle, the authors (Ishikawa *et al.*, 1968) concluded that the intermediate filaments coursed freely through the cytoplasm without making contact with any subcellular organelles or with the sarcolemma. In a more recent study (Bennett *et al.*, 1979), this group maintains that although some intermediate filaments can be found near the myofibrils of chicken myogenic cells that have grown for several weeks *in vitro,* they do not appear to be associated with the myofibrils.

This conclusion is at variance with that drawn by Kelly (1969) and Price and Sanger (1981). In an ultrastructural study of the skeletal myogenesis in salamanders, Kelly (1969) found that while the intermediate filaments of presumptive myoblasts appear to course randomly through the cytoplasm, a significant proportion of them run transverse to the long axis in myotubes containing immature myofibrils. The intermediate filaments are particularly abundant near the Z-discs of the immature myofibrils, where some of them loosely encircle the myofibrils or insert into the narrowing interfibrillary spaces (Fig. 1). In longitudinal sections some of the intermediate filaments appear to form transverse linkages between the Z-discs of adjacent myofibrils. Kelly concluded that during maturation of skeletal muscle some of the intermediate filaments become deposited close to the Z-discs, while still remaining separate from the myofibrils. He hypothesized that the intermediate filaments may act as a structural framework during early myogenesis.

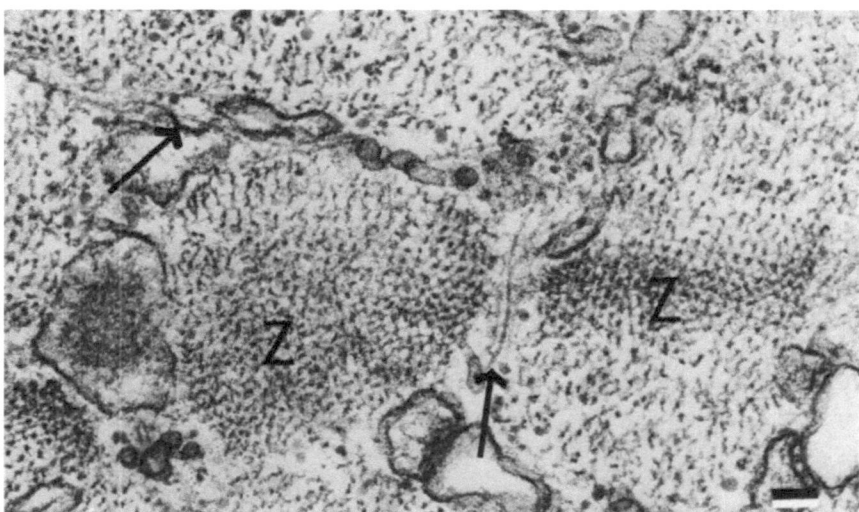

Figure 1.   Cross-section of an older myotube from a salamander embryo, cut at the level of the Z-discs (Z) of several myofibrils. Intermediate filaments (arrows) that are oriented transverse to the long axis of the myofibrils course between the myofibrils and between the myofibrils and the vesicles of sarcoplasmic reticulum. Bar: 100 nm. From Kelly (1969).

In order to gain more information concerning the orientation and localization of intermediate filaments in skeletal myogenic cells, Price and Sanger (1980, 1981) examined thin sections of breast muscle taken from chick embryos at five different stages of development. Intermediate filaments are a prominent component of stage 29 (6 days) cells that contain rudimentary myofibrils. The intermediate filaments occur as individual elements oriented at various angles with respect to the developing myofibrils. Those filaments in the immediate vicinity of the myofibrils appear to encircle them. Some of the intermediate filaments appear to approach and contact the sarcolemma, especially where there are subsarcolemmal densities.

Between 6 and 10 (stage 36) days of incubation, intermediate filaments are redistributed. Rather than being randomly distributed, most of the intermediate filaments in the vicinity of the myofibrils are oriented parallel to them (Fig. 2). A number of these intermediate filaments appear to be associated with the Z-discs or with sarcoplasmic reticulum and T-tubules at the level of the Z-discs (Figs. 2 and 3). In agreement with Kelly, transverse intermediate filaments are found, extending between the Z-discs of adjacent myofibrils (Figs. 2 and 4). Both the transverse and the longitudinal intermediate filaments appear to insert into the amorphous material at the periphery of the developing Z-discs. Some intermediate filaments appear to contact the nuclear envelope and the sarcolemma and may link the myofibrils to these membranes at the level of the Z-discs (Fig. 4). The disposition of intermediate filaments in breast muscle taken from chick embryos between 10 and 17 days (stages 36 to 43) is similar. Between stage 43 and hatching, there is a sharp increase in the number of myofibrils and in the complexity of the sarcoplasmic reticulum within the embryonic breast muscle, so that muscles of late embryos are virtually identical to adult breast muscle.

Morphological details that are difficult to perceive in plastic-embedded specimens are easily resolved in negatively stained preparations. Intermediate filaments isolated from embryonic skeletal muscle by homogenization of the tissue have been examined by negative staining (Ishikawa, 1974; Breckler and Lazarides, 1982). The isolated intermediate filaments are curvilinear structures with an average diameter of 10 nm, consisting of four to six protofilaments twisted in a parallel array. Their substructure is remarkably similar to that of the desmin-containing intermediate filaments of smooth muscle (Rice and Brady, 1972; Cooke, 1976; Small and Sobieszek, 1977; Hubbard and Lazarides, 1979).

3.2. Cardiac Myogenic Cells

While excellent ultrastructural investigations of the developing myocardium have been done (for references, see Lindner, 1960; Wainrach and Sotelo, 1961; Manasek, 1968; Markwald, 1973; Kelly and Chacko, 1976; Manasek, 1979), only Rash and his colleagues (1967, 1970a,b) considered the topic of intermediate filaments.

In a preliminary report, Rash and his collaborators (1967) gave the first

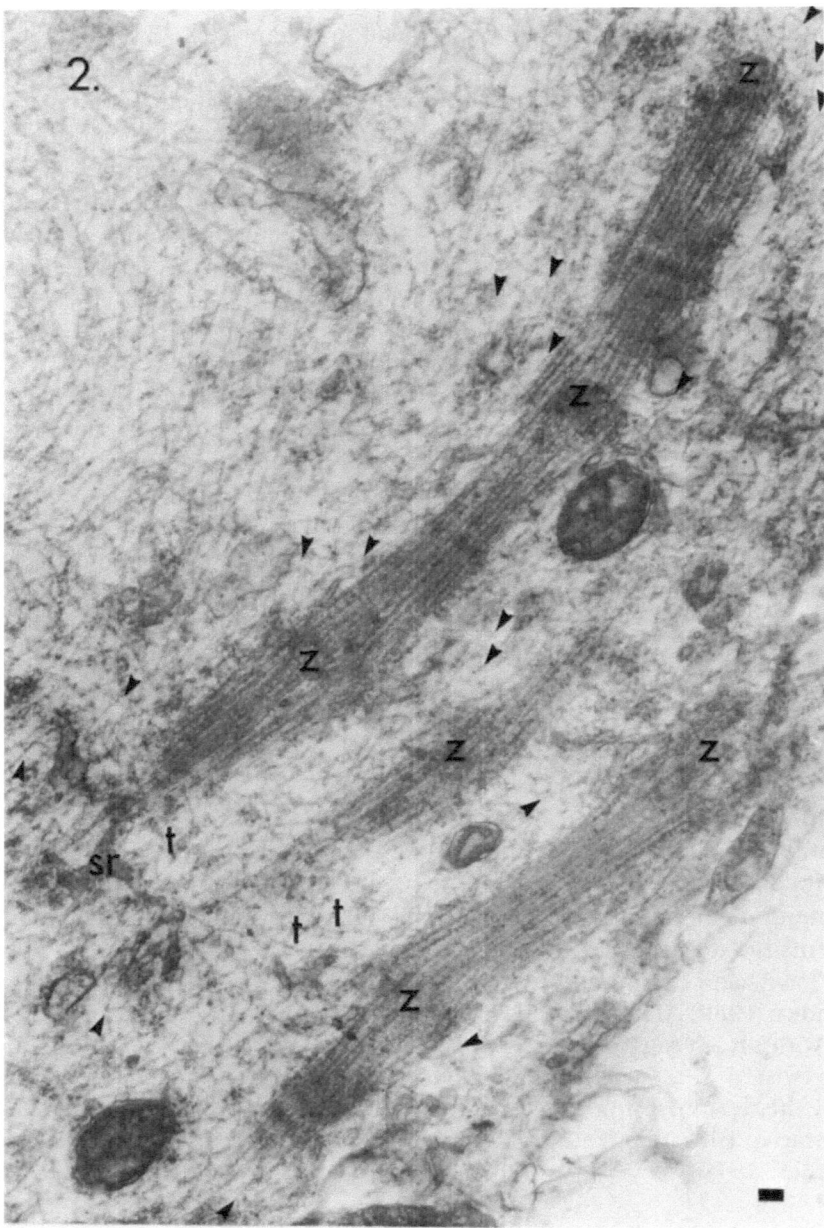

Figure 2. Many of the intermediate filaments (arrow heads) in the vicinity of the myofibrils in embryonic chicken muscle are oriented parallel to the myofibrils. Some of the intermediate filaments appear to merge with the dense material at the periphery of the developing Z-discs (Z) or to terminate at the vesicles of sarcoplasmic reticulum (sr) near the Z-discs. Transverse intermediate filaments (t) extend between the Z-discs of adjacent myofibrils. Bar: 100 nm.

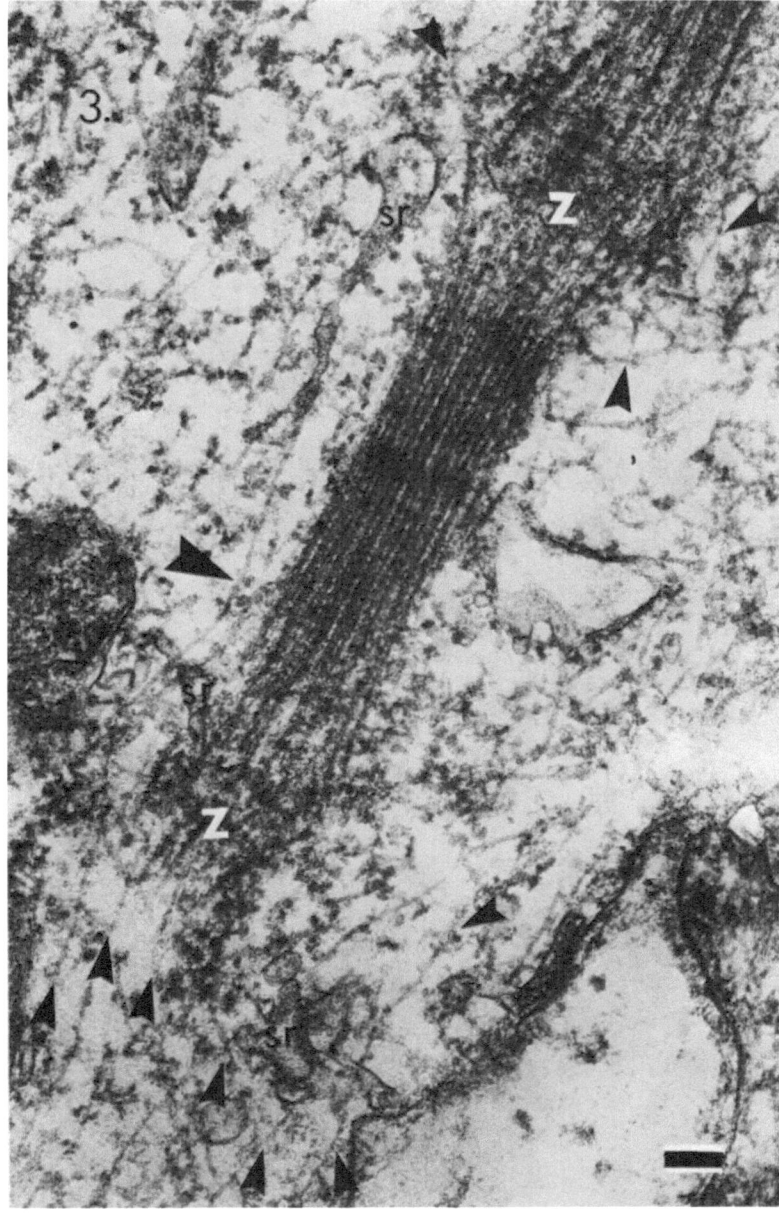

Figure 3. Section through a developing myofibril in a young skeletal myogenic cell from a chick embryo, demonstrating intermediate filaments (arrowheads) coursing parallel to the myofibril. The intermediate filaments follow sinuous courses, sometimes appearing to attach to the sarcoplasmic reticulum (sr) at the level of the Z-discs or at the periphery of the Z-discs (Z). The intermediate filament marked with a large arrowhead is over 1 μm in length. Bar: 100 nm.

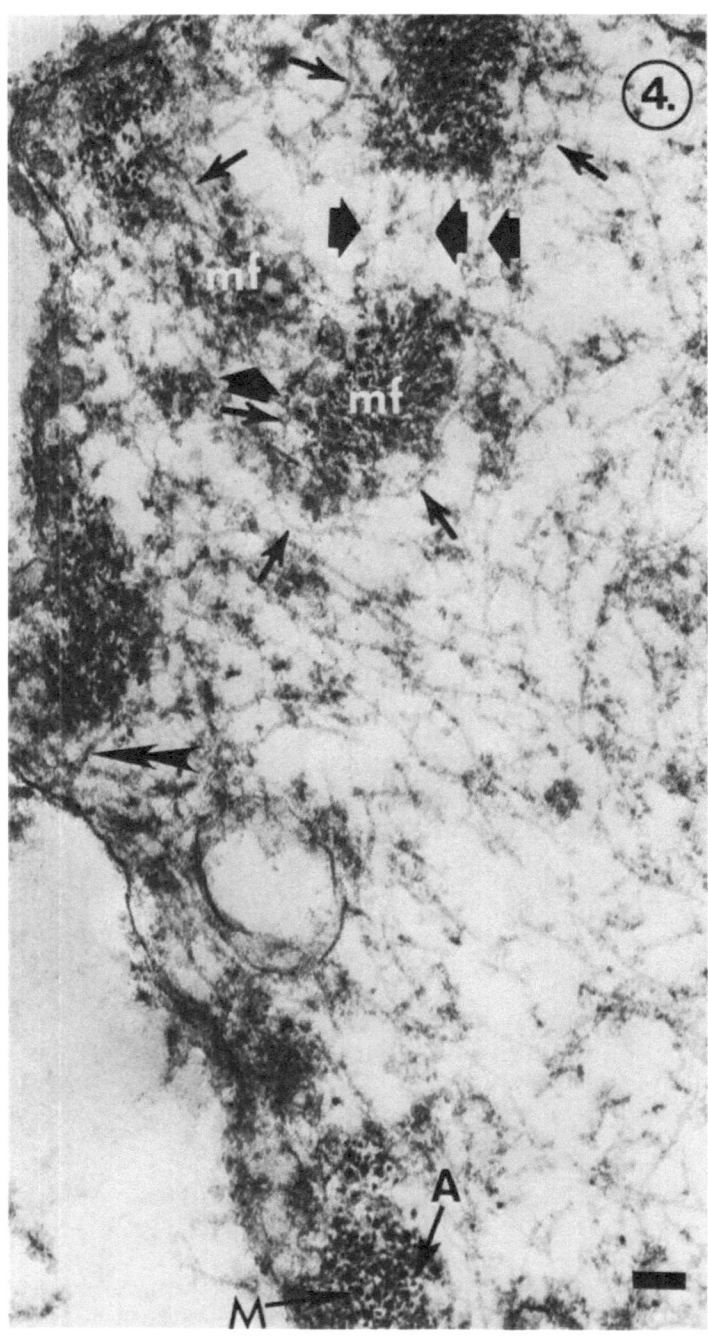

Figure 4.   A transverse section through a myogenic cell from a chick embryo. The developing myobrils are surrounded by intermediate filaments (small arrows). Some of the intermediate filaments appear to link pairs of adjacent myofibrils (large arrows) or to connect a myofibril to the sarcolemma (double arrowhead). Bar: 10 nm.

descriptions of a third class of filaments in developing cardiac cells. The class consists of filaments with heterogeneous diameters, ranging from 8 nm to 14 nm, with the majority measuring 9.5–11 nm in diameter. The intermediate filaments are observed in chick precardiac mesenchyme, presumptive cardiac myoblasts, and cardiac myocytes. Masses of intertwined intermediate filaments are seen in myocytes containing small aggregates of actin and myosin filaments. As occurs in skeletal myogenesis, the number of intermediate filaments decreases as myofibrils accumulate (Rash *et al.*, 1970b). The majority of the intermediate filaments in older myocytes course parallel to the developing myofibrils (Fig. 5). As in skeletal myogenic cells, some of the intermediate filaments surround the myofibrils. Some of the intermediate filaments approach the subsarcolemmal dense patches, the Z-discs, and the desmosomes from oblique angles and appear to insert into these structures. The authors found no evidence of transverse intermediate filaments extending between the Z-discs of adjacent myofibrils, as some have observed in developing skeletal muscle (Kelly, 1969; Price and Sanger, 1980, 1981). They concluded that

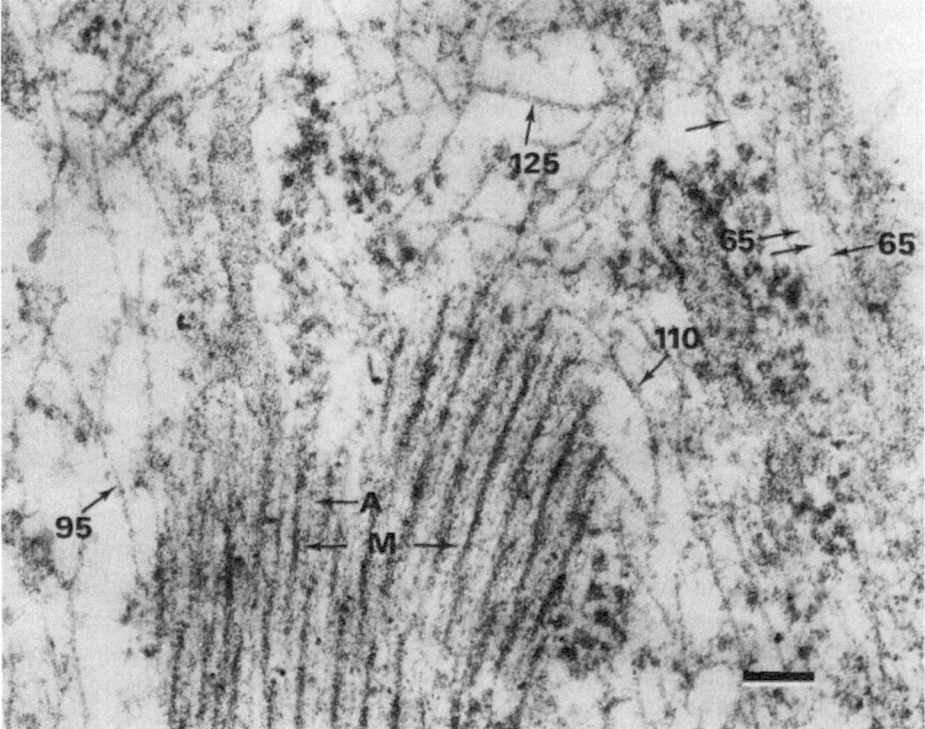

Figure 5.   Periphery of an embryonic cardiomyocyte from a 14-somite chick. The intermediate filaments, some of which are labeled (95, 110, 126 Å), closely surround the myofibrils. Actin (A) and myosin (M) filaments, and the thin filaments (65 Å) of the subsarcolemma are indicated for comparison with the intermediate filaments. Bar: 100nm. From J. E. Rash *et al.* (1970a).

the intermediate filaments may serve as a cytoskeleton of embryonic cardiac cells and may perhaps play a role in the longitudinal alignment of the myofibrils.

Using embryonic cardiac muscle, Rash *et al.* (1970a) provided the initial biochemical demonstration that the protein constituents of intermediate filaments are not actin or myosin. In comparison to actin and myosin filaments, the intermediate filaments are extremely insoluble. Following treatment of embryonic cardiac tissue with solutions that solubilize the majority of actin and myosin filaments, such as 0.6 M KI and osmium, the intermediate filaments were identical to those in untreated tissue.

3.3. Immunofluorescent Localization of Vimentin and Desmin

The major components of the intermediate filaments in striated myogenic cells are vimentin, the major intermediate filament protein from fibroblastic cells, and desmin, the primary component of intermediate filaments of smooth muscle. Presumptive myoblasts synthesize vimentin as the major intermediate protein (Gard and Lazarides, 1980), so that antibodies to vimentin stain filamentous bundles in presumptive myoblasts and young skeletal myotubes (Fellini *et al.*, 1978; Bennett *et al.*, 1979; Gard *et al.*, 1979; Gard and Lazarides, 1980). Desmin can be detected in an occasional postmitotic myoblast, suggesting that desmin synthesis begins in myogenic cells prior to fusion (Bennett *et al.*, 1978). There is a surge of synthesis of desmin following the fusion of myoblasts, after which the rate of desmin synthesis exceeds the rate of vimentin synthesis, resulting in a net increase of desmin compared to vimentin (Gard and Lazarides, 1980). The filamentous immunofluorescent patterns of anti-desmin and anti-vimentin antibodies in young myotubes are exactly superimposable (Bennett *et al.*, 1979; Gard and Lazarides, 1980).

The intermediate filament proteins are gradually redistributed during myogenesis. The two intermediate filament proteins are translocated in a temporal sequence. Within a week of culturing, when striated myofibrils are observed within the myotubes, and antibodies to α-actinin bind to definitive Z-discs, anti-vimentin antibodies bind to the Z-discs of about one third of the myotubes in addition to binding the cytoplasmic filaments in each myotube (Bennett *et al.*, 1979; Gard and Lazarides, 1980). In the next few days in culture, desmin is discerned in the region of the Z-discs. The filamentous staining pattern achieved with antibodies to either desmin or vimentin diminishes as the staining at the level of the Z-discs increases, suggesting that the intermediate filament proteins are redistributed to the Z-discs (Gard and Lazarides, 1980). It is not yet known whether the cytoplasmic intermediate filaments present in young myotubes coalesce at the myofibrillar Z-discs or whether newly synthesized desmin and vimentin condense in those regions.

The immunofluorescent staining patterns of desmin and vimentin in longitudinal views of older skeletal myotubes in culture are almost indistinguishable from the pattern produced by antibodies specific for α-actinin

(Gard and Lazarides, 1980). There is agreement that desmin persists in the region of the Z-discs throughout myogenesis (Lazarides and Hubbard, 1976; Granger and Lazarides, 1978, 1979; Bennett *et al.*, 1979; Gard and Lazarides, 1980). Some investigators (Bennett *et al.*, 1979) find that their vimentin-specific antibodies no longer stain older myotubes, while others demonstrate anti-vimentin staining at the Z-discs of older (37 days *in vitro*) myotubes (Gard and Lazarides, 1980; Granger and Lazarides, 1979).

Desmin and vimentin have been localized in primary cultures of cardiac cells. Antibodies to vimentin stain a fine meshwork of filaments in every cell in the cultures (Bennett *et al.*, 1978), whereas desmin-specific antibodies produce a similar pattern in only half the cells (Bennett *et al.*, 1978; Lazarides, 1978). Since the number of cells that bind desmin antibodies corresponds to the number of contracting cells within the population, the authors infer that the stained cells are functional cardiac cells. Desmin is present in a fine filamentous network that extends throughout the cytoplasm (Bennett *et al.*, 1978; Lazarides, 1978; Campbell *et al.*, 1979). In some cardiac cells, the desmin filaments are arranged in bundles of various sizes, with the thickest ones concentrated in the perinuclear region (Lazarides, 1978).

It appears that during cardiac myogenesis, intermediate filaments proteins are progressively redistributed as they are in skeletal myogenic cells. Since embryonic cardiac cells are notoriously difficult to keep in culture due to the overgrowth of fibroblasts, few investigators have been able to directly examine the progressive redistribution of intermediate filament proteins during myocardial development. Until recently, it has been necessary to deduce that intermediate filament proteins are redistributed during cardiac myogenesis, because desmin and vimentin antibodies stain a filamentous network in embryonic cardiac cells while desmin antibodies stain discrete nonfilamentous structures in newborn or adult cardiac cells (Lazarides and Hubbard, 1976; Campbell *et al.*, 1979) (see Section 4.2).

Fuseler and Shay (1982) have recently provided direct immunofluorescent evidence supporting the hypothesis that desmin is redistributed during cardiac myogenesis. Cardiac ventricular cells were taken from rat embryos at several stages and from newborn rats, and grown in culture for almost a week before being stained with antibodies to desmin. Desmin is localized primarily in a filamentous meshwork in ventricular cells from young embryos. In older cardiac cells, desmin is localized both in the filamentous network as well as in spike-shaped aggregations near the developing Z-discs. The spike-like aggregates at the level of the Z-discs may represent desmin in transition from the cytoplasm to the Z-discs, because in fully mature cells, there is bright immunofluorescence in the regions of the Z-discs. The filamentous staining pattern is progressively diminished and the antibodies are gradually confined to phase-dense structures. In the most mature cardiac cells, some of which are conjoined by intercalated discs, the majority of desmin is distributed in the regions of the Z-discs, in the intercalated discs, and in plaques on the membranes of laterally associated cells.

4. Localization and Characterization of Intermediate Filaments in Adult Striated Muscle

4.1. Ultrastructural Studies of Intermediate Filaments in Striated Muscle

4.1.1. Cardiac Muscle

Taken together, the results of the ultrastructural and immunofluorescent studies of developing cardiac muscle strongly suggest that during myocardial differentiation, desmin-containing intermediate filaments become associated with Z-discs, intercalated discs, and desmosomes. The limit of resolution of the fluorescent microscope precludes determining from immunofluorescent micrographs whether the desmin in mature cardiac cells is in filamentous polymers or in condensed aggregates.

There is ultrastructural evidence for the presence of intermediate filaments in the heart tissue of some representatives of the phlyla Amphibia and Mammalia. Intermediate filaments have not been described in the reports on the cardiac ultrastructure of many different invertebrates (Page and Fozzard, 1973); they certainly are not observed in the tubular hearts of arthropods or in the chambered hearts of molluscs (Sanger, 1979).

The original observations of nonmyofibrillar filaments in cardiac tissue were made in electron microscopic studies of the frog lymphatic heart (Lindner and Schaumburg, 1968) and the papillary muscle and annulus fibrosus of monkey (rhesus) heart (Viragh and Challice, 1969). The average diameters of the nonmyofibrillar filaments observed in these two tissues are 6 nm. In both tissues, the nonmyofibrillar filaments are associated with the Z-discs and with various membrane systems.

The nonmyofibrillar filaments course parallel to the long axis of the cells, between the myofibrils, and transverse to the myofibrils. The longitudinal nonmyofibrillar elements are individual filaments that appear to be associated with some of the Z-discs (Lindner and Schaumburg, 1968). Many of the nonmyofibrillar filaments are arranged in bundles of less than 20 filaments within a given section, extending at right and oblique angles from the Z-discs of the sparsely distributed myofibrils (Lindner and Schaumburg, 1968; Viragh and Challice, 1969). These filament bundles pass around a Z-disc, linking it to a Z-disc in an adjacent myofibril or to a nearby subsarcolemmal density. Apparent linkages of nonmyofibrillar filaments with the sarcolemma are seen at densities associated with junctional complexes (Viragh and Challice, 1969) and the sarcoplasmic reticulum, especially in regions where vesicles of sarcoplasmic reticulum are attached to the invaginated sarcolemma (Lindner and Scnaumburg, 1968). In the atrioventricular annulus fibrosus of monkey heart, the filament bundles approach the intercalated discs from perpendicular and oblique angles and appear to insert into membrane-associated densities, frequently linking pairs of membrane-bound densities at the intercalated discs (Viragh and Challice, 1969).

The electron microscopic studies reported above were carried out before intermediate filaments were widely accepted as a third class of filaments; nevertheless, Lindner and Schaumburg (1968) and Viragh and Challice (1969) recognized that the cytoplasmic filaments of cardiac muscle represent a third class of filament performing cytoskeletal and integrative functions. Later electron microscopic investigations of mammalian myocardium provided support for their observations and conclusions, except that the diameter of the mammalian cytoplasmic filaments is larger, approximately 10 nm (Ferrans and Roberts, 1973; Junker and Sommer, 1977; Thornell *et al.*, 1978; Forbes and Sperelakis, 1980).

Intermediate filaments are closely associated with the Z-discs of the myofibrils in myocardium from human and canine hearts (Ferrans and Roberts, 1973), opossum heart (Junker and Sommer, 1977; Sommer and Johnson, 1979), bovine heart (Thornell *et al.*, 1978) and murine heart (Forbes and Sperelakis, 1980). In each of these tissues, transverse bundles of intermediate filaments that extend laterally from the Z-discs cross the interfibrillar space to link two or three adjacent Z-discs (Fig. 6). The transverse bundles, which are composed of less than 20 filaments (per section) as in the previously described cases, are seldom found beyond the I–Z junction. From cross-sections and grazing longitudinal sections, it is obvious that the intermediate filament bundles surround the Z-discs (Fig. 7). The transverse intermediate filaments were also found in an avian heart, which lacks T-tubules, in preparations of isolated myofibrils that were decorated with anti-desmin antibodies (Richardson *et al.*, 1981). The anti-desmin antibodies, indirectly labeled with peroxidase, bind to intermediate filaments which project from the periphery of the Z-discs.

Transverse intermediate filaments stretch from the Z-discs of peripheral myofibrils to the sarcolemma (Ferrans and Roberts, 1973; Junker and Sommer, 1977; Forbes and Sperelakis, 1980); this observation led to the speculation that a network of transverse filaments may extend the width of a cardiac muscle at the level of the Z-discs, linking the Z-discs together and binding the most peripheral myofibrils to the sarcolemma. Apparent contact with the sarcolemma usually occurs in the regions where there are subsarcolemmal dense patches resembling the material of the Z-disc (Ferrans and Roberts, 1973).

In one study, it was noted that the transverse bundles of intermediate filaments curve between the Z-discs and neighboring mitochondria (Forbes and Sperelakis, 1980); the authors conjecture that the filament bundles may segregate the mitochondria from the Z-discs. Alternatively, the filaments may attach the mitochondria to the Z-discs. In the perinuclear region (Fig. 8), bundles of intermediate filaments extend laterally from the Z-discs to contact the outer nuclear membrane (Ferrans and Roberts, 1973; Forbes and Sperelakis, 1980).

In addition to the transverse intermediate filaments, there are longitudinally oriented intermediate filaments (Fig. 9) that merge with the amorphous density at the periphery of the Z-disc (Thornell *et al.*, 1978). In the region of the intercalated discs, longitudinal and oblique intermediate filaments approach the associated membrane densities and appear to insert into them

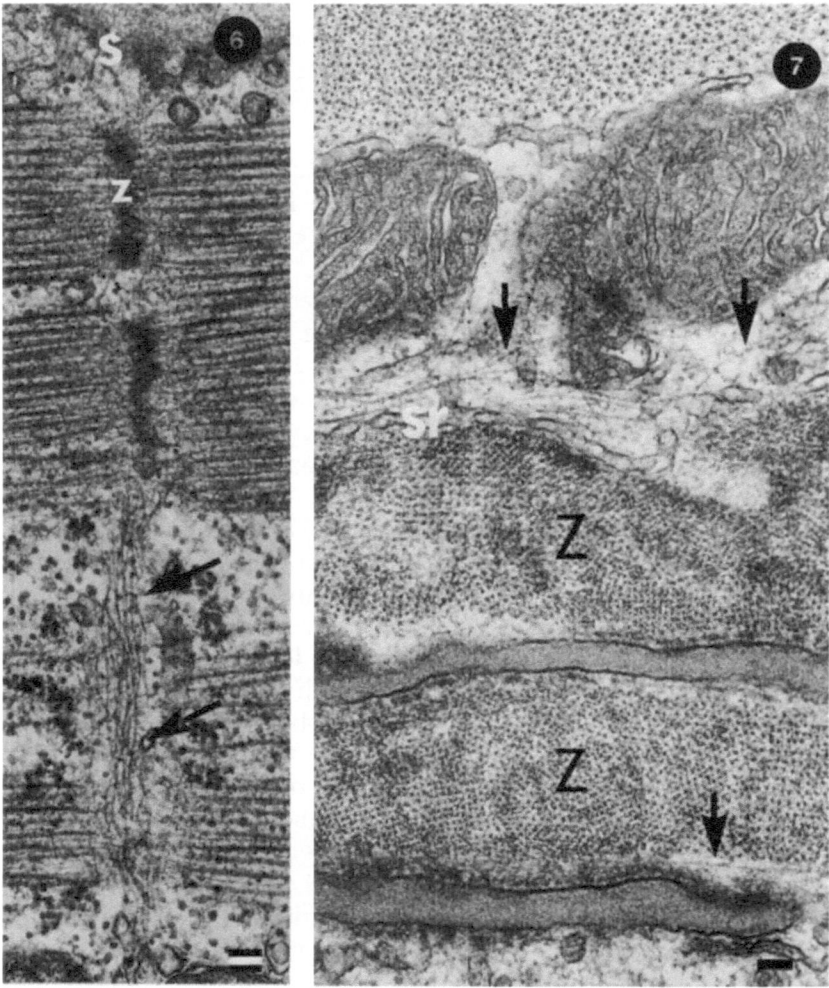

Figure 6.   Longitudinal section of an opussum cardiac muscle, showing intermediate filaments (arrows) extending transversely across the cell. The intermediate filaments are arranged in bundles that extend from the sarcolemma to a Z-disc (Z) and continue to cross the cell, linking all the myofibrils at the level of the Z-discs. Bar: 100 nm. From Junker and Sommer (1977).

Figure 7.   Cross section of an opussum cardiac muscle which demonstrates that the transverse intermediate filaments (arrows) encircle the myofibrils at the level of the Z-discs (Z). Some of the intermediate filaments appear to be attached to vesicles of the sarcoplasmic reticulum (sr). Bar: 100 nm. From Junker and Somer (1977).

(Fig. 9). This nonmyofibrillar filamentous network that appears to insert into the intercalated discs was described earlier by Fawcett and McNutt (1969), who postulated that the filaments are extraneous actin filaments. Small bundles of intermediate filaments loop between pairs of membrane-associated densities in the intercalated discs (Ferrans and Roberts, 1973; Thornell *et al.*, 1978; Forbes and Sperelakis, 1980). Bundles of the intermediate filaments are closely applied to intra- and intercellular desmosomal plaques (Fig. 10) (Forbes and Sperelakis, 1980) (also see Chapter 5, this volume).

4.1.2. Cardiac Impulse-Conducting System

Regular sequential contraction of the atria and ventricles of the mammalian heart requires the propagation of impulses from the sinoatrial region via a conducting system composed of modified myocardium. The conducting system consists of the sinoatrial node or pacemaker, the atrioventricular node, and the atrioventricular bundle and its two branches that terminate in the ventricles. The distal portions of the two branches of the atrioventricular bundle contain Purkinje cells. The general morphological details distinguishing the cells of the conducting system from normal myocardium are a relatively sparser population of myofibrils and a relatively greater population of cytoplasmic filaments. Biochemical and immunological characterization of the 7-nm cytoplasmic filaments of Purkinje cells demonstrates that they are intermediate filaments containing a protein similar to desmin from smooth muscle (Eriksson *et al.*, 1978; Eriksson and Thornell, 1979; Eriksson *et al.*, 1979). It remains to be determined whether the cytoplasmic filaments in the other cells of the conducting system are also composed largely of desmin.

The cytoplasmic filaments are similarly distributed in the cells of the conducting system despite there being differences in the number and order of myofibrils in the various cells (Viragh and Challice, 1969). As in the force-

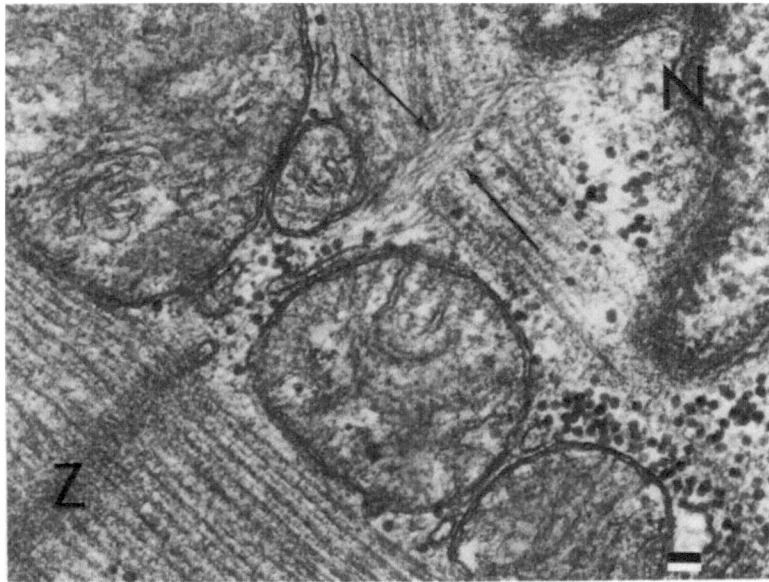

Figure 8.   A transverse bundle of intermediate filaments (arrows) in the perinuclear region of a normal canine myocardium extends between the adjacent Z-discs (Z) of two myofibrils and terminates near the nuclear membrane (N). Bar: 100 nm. From Ferrans and Roberts (1973).

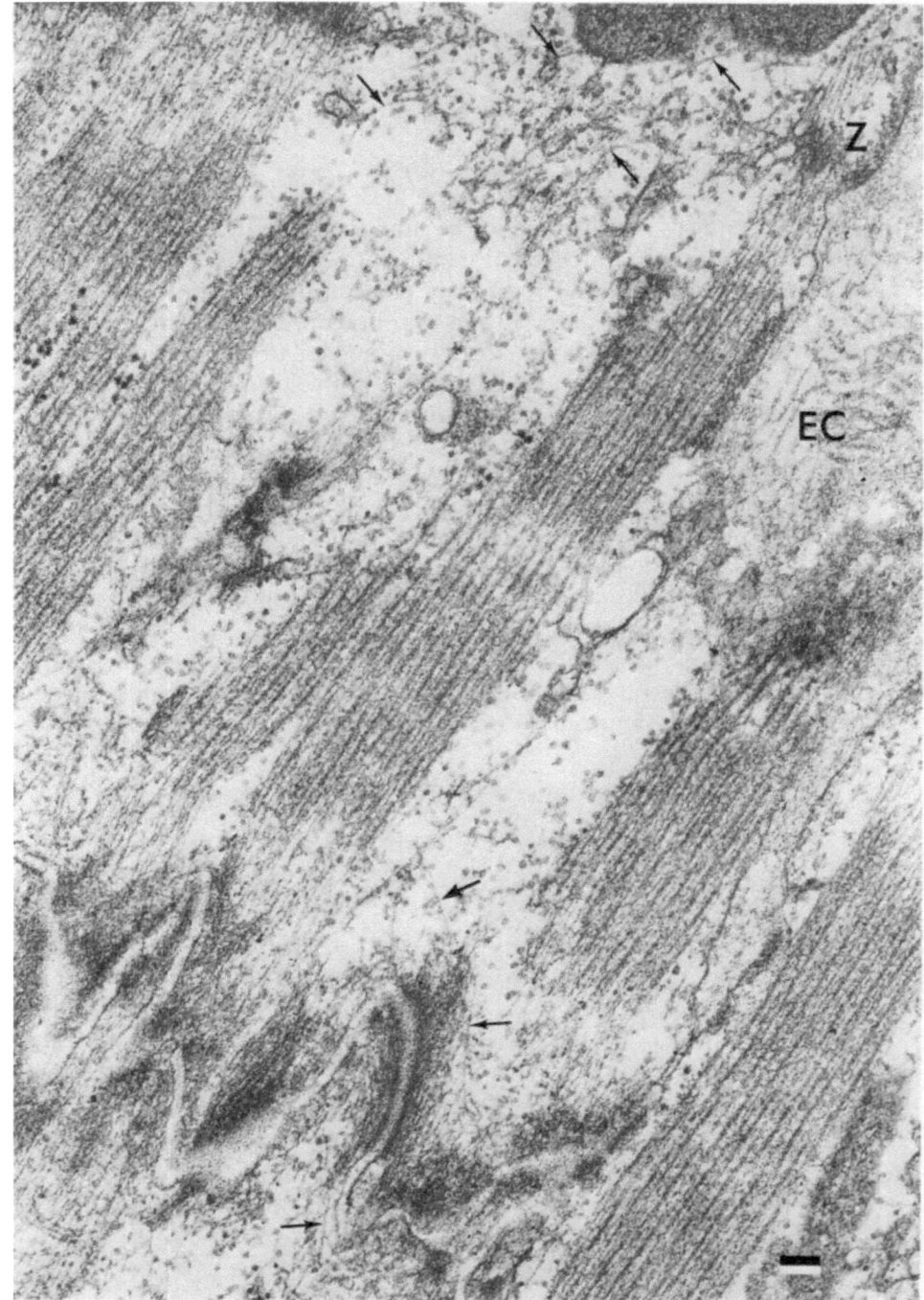

Figure 9. A section of normal mammalian ventricle demonstrating intermediate filaments (arrows) in the vicinity of the intercalated disc (lower left) and the Z-discs (Z). Note that bundles of intermediate filaments loop between the dense plaques on the intercalcated disc. The extracellular space (EC) is indicated. Bar: 100 nm. From Thornell (1978).

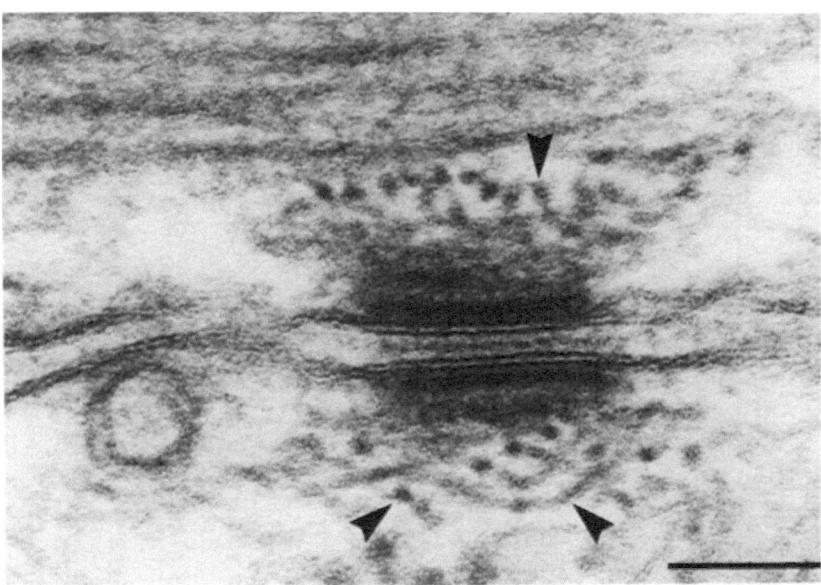

Figure 10. Intermediate filaments (arrowheads) at an internal, or intracellular, desmosome in a mammalian cardiac cell. The intermediate filaments are oriented randomly with respect to the desmosome, so they are viewed in longitudinal and cross sections. The intermediate filaments cannot be distinquished in the dense plaques on either side of the desmosome. Part of a myofibril is visible at the top. Bar: 100 nm. From Forbes and Sperelakis (1980).

generating myocardial cells, the intermediate filaments of conducting cells are associated with the myofibrillar Z-discs, with modified intercalated discs and with desmosomes (Viragh and Challice, 1969; Oliphant and Loewen, 1976; Thornell *et al.*, 1976, 1978; Eriksson and Thornell, 1979). Transverse bundles of intermediate filaments surround the Z-discs and extend between neighboring discs. Small groups of intermediate filaments appear to terminate in the dense material associated with desmosomes; adjacent desmosomes are conjoined by loops of intermediate filaments (Fig. 11).

In the vicinity of myofibrils, the majority of the intermediate filaments are often arranged in bundles coursing parallel to the myofibrils, whereas they are dispersed throughout the rest of the cytoplasm (Eriksson and Thornell, 1979). The central cytoplasm of the cells in the bifurcation of the atrioventricular bundle (Viragh and Challice, 1969) and the Purkinje cells of some species contain whorls of intermediate filaments that make no conspicious connections with membranes or organelles (Oliphant and Loewen, 1976; Eriksson and Thornell, 1979).

4.1.3. Cardiomyopathic Tissues

The intermediate filaments of cardiomyopathic tissue are between 7 and 11 nm in diameter and are found in all the same locations as the intermediate

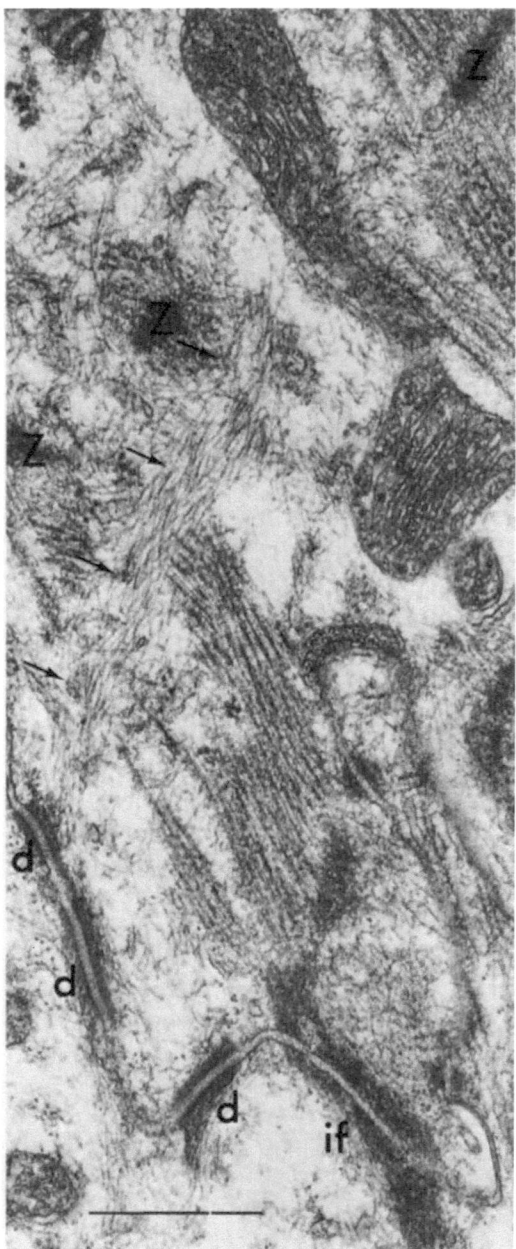

Figure 11.   A tangential section of two Purkinje cells joined by a series of desmosomes (d), to which intermediate filaments (if) are attached. The bundle of intermediate filaments (arrows) that crosses the upper cell at approximately the level of the Z-discs (Z) of three adjacent myofibrils may serve to interconnect the myofibrils. This bundle appears to terminate at a desmosome. Bar: 500 nm. From Eriksson and Thornell (1979).

filaments of normal myocardium. The major difference between the intermediate filaments of normal and diseased myocardium is their prevalence. In addition to there being more intermediate filaments in hypertrophied myocardium, there are coincidentally fewer myofibrils, many of which are degenerating; therefore observation of the intermediate filaments in diseased myocardium is facilitated (Fig. 12).

Most of the ultrastructural investigations of intermediate filaments in cardiomyopathy were performed on human heart muscle that had undergone hypertrophy due to various causes, including congenital septal defects, obstructive and nonobstructive stenosis (Ferrans and Roberts, 1973), mitral valvular disease (Theidemann and Ferrans, 1976), and auriculo-ventricular block (Porte *et al.*, 1980). There is one case of experimentally induced hypertrophy of intermediate filaments in rat myocardium, due to exposure to an androgenic steroid with anabolic effects (Behrendt, 1977).

The intermediate filaments in each cardiomyopathic tissue are disposed similarly; many of the intermediate filaments are arranged in bundles at the level of the Z-discs, extending at right or oblique angles between adjacent Z-discs as well as between the peripheral Z-discs and the subsarcolemmal densities. The filamentous connections between myofibrils appear to be quite

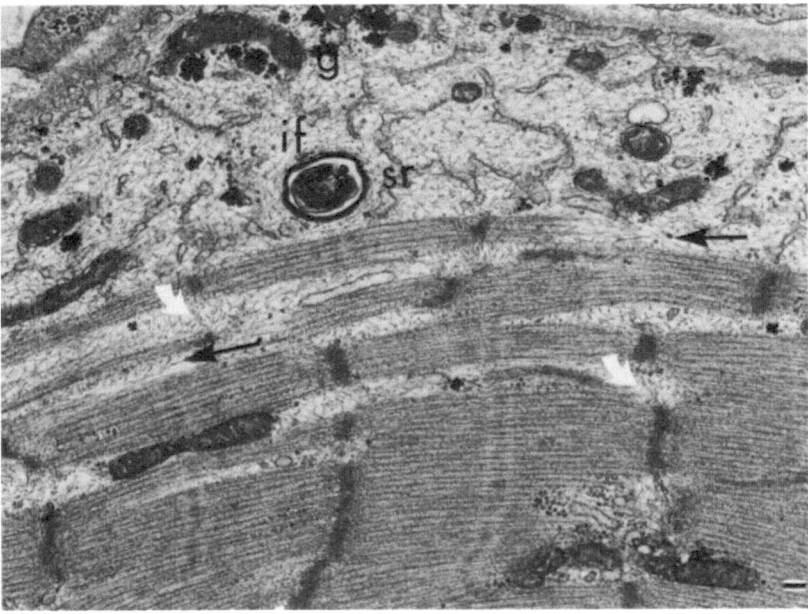

Figure 12. Pathological atrial tissue, with large numbers of intermediate filaments (if). There is a focal accumulation of intermediate filaments in the subsarcolemmal region that is coincidentally devoid of myofibrils. This region also includes greatly elevated amounts of sarcoplasmic reticulum (sr) and glycogen granules (g). There are transverse intermediate filaments (white curved arrows) between adjacent Z-discs as well as intermediate filaments (black arrows) coursing parallel to the myofibrils. Bar: 100 nm. From Theidemann and Ferrans (1976).

strong, since they exist even between the severely dissarranged myofibrils found in the ventricles of patients with obstructive cardiomyopathy (Ferrans and Roberts, 1973). Some intermediate filaments project from the Z-discs of myofibrils in the perinuclear region and converge upon the outer nuclear membrane. Behrendt (1977) has proposed that the bundles of intermediate filaments form a horizontal network that integrates the Z-discs within a plane of the fiber and further links them to the nuclear and plasma membranes. Many of the intermediate filaments in the human cardiomyopathic tissues run parallel to the myofibrils (Ferrans and Roberts, 1973; Theidemann and Ferrans, 1976).

Intermediate filaments are attached to the membrane-bound dense plaques on the intercalated discs, and appear to connect the desmosomes of hypertrophied myocardial cells. Associations between intermediate filaments and the T-tubules are more prominent in hypertrophied cardiac muscle than in normal myocardium (Ferrans and Roberts, 1973; Theidemann and Ferrans, 1976; Behrendt, 1977). Bundles of intermediate filaments can extend from a given T-tubule across an adjacent myofibril at the level of the I–Z junction (see Behrendt, 1977). As in normal myocardium, the intermediate filaments appear to attach the T-tubules and the sarcoplasmic reticulum to the myofibrils.

Hypertrophic cardiomyopathic cells of humans (Fig. 12) contain massive focal accumulations of intermediate filaments (Ferrans and Roberts, 1973; Theidemann and Ferrans, 1976; Porte et al., 1980), reminiscent of the whorls of filaments in the central cytoplasm of Purkinje cells. Aggregates of glycogen are interspersed among the filaments. The fact that these regions lack intact myofibrils, while containing both dense bodies resembling degenerated Z-discs and abnormally complex formations of sarcoplasmic reticulum, has contributed to the hypothesis of a causal relationship between degeneration of myofibrillar filaments and proliferation of sarcotubular elements (Maron and Ferrans, 1974; Theidemann and Ferrans, 1976).

4.1.4. Skeletal Muscle

The ultrastructural studies of stretched or extracted skeletal fibers from vertebrate muscle that were performed prior to the discovery of intermediate filaments led to conflicting conclusions that were inadequately supported by the evidence. Two major difficulties were faced by those who relied on thin sections to determine whether a third type of myofibrillar filament extends along the longitudinal axis of vertebrate skeletal myofibrils. The first problem is that the contractile filaments and membrane elements are extremely tightly packed, rendering other filaments inconspicuous. Secondly, if the noncontractile longitudinal filaments proposed by Hanson and Huxley (1955) are extensible and collapsible, it is likely that they follow sinuous courses between the Z-discs so that longitudinal sections of skeletal muscle would not contain their entire lengths.

Dos Remedios and Gilmour (1978) sought to circumvent these technical problems by examining isolated myofibrils that had been placed on grids and then treated with a myosin-extracting solution or with KI, which solubilizes most of the myosin and actin filaments. By electron microscopy of the negatively stained remnants of myofibrils, filaments between 2 and 7 nm in diameter are observed between the tufts of actin filaments that remain after extraction of myosin. Similar filaments are present in the 1-μm gap between the residual actin filaments of KI-extracted myofibrils. The longitudinal filaments could not be traced directly from one Z-disc to the next, because they intermingle with the residual actin filaments attached to the Z-discs. The results suggest that longitudinal filaments that are relatively insoluble, compared to actin and myosin, connect the Z-discs of vertebrate myofibrils.

This conclusion has been supported and extended by results obtained from electron microscopy of vertebrate skeletal myofibrils that were extracted in solutions before being placed on grids for negative staining (Price and Sanger, 1979a). The experimental protocol was designed to eliminate the possibility that myofibrillar remnants that have already bound to the grid remain in place when the myofibrils are extracted directly on the grid. Isolated myofibrils were treated with detergent to remove membrane components, then repeatedly extracted with a solution that solubilizes myosin filaments. The myofibrils were pelleted by centrifugation before addition of fresh extraction solution, and after four or five extractions, some of the remaining material was treated with heavy meromyosin. The myofibrillar elements that resisted repeated myosin extraction are linear arrays of Z-discs with attached actin filaments, connected by long smooth filaments with average diameters of 10.5 nm (Fig. 13). The smooth longitudinal filaments span distances of 1–7 μm between opposite I-bands. Heavy meromyosin binds to the actin filaments, but not to the 10-nm filaments, indicating that the two filament types are biochemically heterogeneous. The longitudinal filaments were designated as intermediate filaments, on the basis of their average diameter and the lack of binding of heavy meromyosin.

In order to determine if the intermediate filaments directly connect successive Z-discs rather than linking the actin filaments of opposite I-bands, isolated myofibrils were extracted with a KI solution (Price and Sanger, 1979b). The extracted myofibrils consist of a series of Z-discs linked by intermediate filaments spanning the distances of 1.5–4.5 μm between the Z-discs. Even though some short segments of actin filaments are attached to the Z-discs of extracted myofibrils, the longitudinal intermediate filaments can be traced between Z-discs. Occasionally, an intermediate filament appears to link three successive Z-discs. The tips of the intermediate filaments merge with the dense amorphous material of the Z-discs. The number of longitudinal intermediate filaments in a sarcomere ranges from 8 to 50; these figures are well within the range of the excess of longitudinal filaments over the calculated number of actin filaments in several vertebrate sarcomeres (Ullrick *et al.*, 1977).

The longitudinal filamentous connections between Z-discs are strong enough to withstand repeated centrifugation and resuspension. The inter-

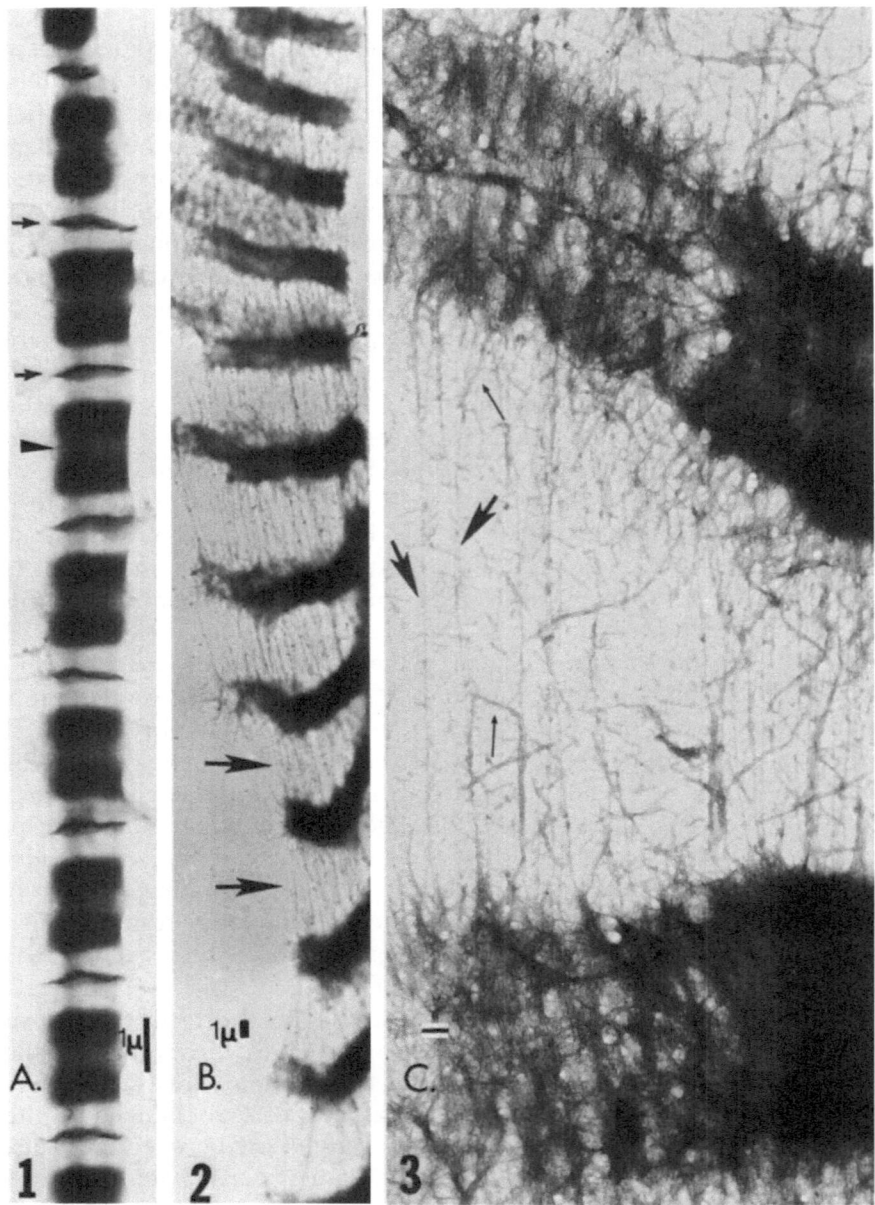

Figure 13. (A) Control myofibril from a pectoralis muscle of an adult chicken, negatively stained on a grid. Small arrows point to Z-discs and the large arrow points to an H-zone. Bar: 1 μm. (B) Myosin-extracted myofibril. The Z-discs of this myofibril, which was extracted in solution, are separated by about three times the normal sarcomere distance. The Z-discs, with actin filaments attached to either side, are connected by longitudinal filaments (arrows). Bar: 1 μm. (C) A myosin-extracted myofibril, treated with heavy meromyosin (HMM). The long interconnecting filaments (large arrows) do not bind HMM. They remain as smooth filaments of 10 nm in diameter. The leftmost connecting filament is over 3 μm in length. The actin filaments do bind HMM (small arrows). Bar: 500 nm. From Price and Sanger (1979).

mediate filaments are extensible to accomodate the Z-discs of extracted myofibrils being separated by two to four times the length of a normal sarcomere; individual intermediate filaments of between 2.2 and 3 μm are found. Wang and Ramirez-Mitchell (1979) have presented preliminary evidence corroborating the finding of small numbers of intermediate filaments connecting the Z-discs of vertebrate skeletal myofibrils that were extracted with KI. It is proposed that the longitudinal intermediate filaments are anchored at the periphery of the Z-discs, so that they extend along the outside of myofibrils (Price and Sanger, 1979a; Wang and Ramirez-Mitchèll, 1979). This model is compatible with the honey-comb pattern obtained by staining planar sheets of Z-discs with antibodies to intermediate filament proteins (Granger and Lazarides, 1978, 1979, 1980).

It appears that two sets of intermediate filaments persist in adult skeletal muscle; in addition to the longitudinal intermediate filaments that are closely associated with the periphery of the Z-discs, there are transverse intermediate filaments at the level of the Z-discs (Nunzi and Franzini-Armstrong, 1980; Richardson *et al.*, 1981) (Fig. 14). Short intermediate filaments that project radially from the Z-discs form transverse connections between the peripheral regions of the Z-discs and the sarcoplasmic reticulum in a muscle with the T-tubules at the A–I junctions (Fig. 15). Some of the transverse intermediate filaments are long wavy filaments like those in embryonic muscle. Only the short transverse filaments which link myofibrils at the level of the Z-discs were seen in the preparations of Richardson *et al.* (1981), which were groups of myofibrils isolated from avian thigh muscles and decorated with antibodies to desmin.

To demonstrate the transverse intermediate filaments in skeletal muscles with the T-tubules located at the level of the Z-discs, Nunzi and Franzini-Armstrong (1980) used muscles in which the spaces between the myofibrils and the membranes were greatly increased by soaking the muscles in a hypotonic solution. This clever technique allows the clear demonstration of transverse filaments linking the Z-discs to the sarcoplasmic reticulum and T-tubules (Fig. 16). In both types of skeletal muscle, the subsarcolemmal myofibrils are joined to the sarcolemma by long intermediate filaments. The transverse intermediate filaments extending between the Z-discs of vertebrate skeletal muscle resemble the Z-bridges first demonstrated by Garamvolgyi (1963) in sheets of Z-discs from insect flight muscle, indicating that these filamentous connections are a general feature of skeletal muscle.

The ultrastructural evidence indicates that some of the intermediate filaments that become associated with the Z-discs during myogenesis of vertebrate skeletal muscle persist in the adult tissue. In both embryonic skeletal muscle and adult skeletal muscle, there are modest numbers of longitudinal and transverse intermediate filaments. Both sets of intermediate filaments are characterized by their apparent attachment to the periphery of the Z-discs. The persistence of the transverse intermediate filaments in drastically swollen muscle indicates that these filaments are as extensible and as firmly anchored as are the longitudinal intermediate filaments observed in myofibrils that were extracted in solution by Price and Sanger (1979a).

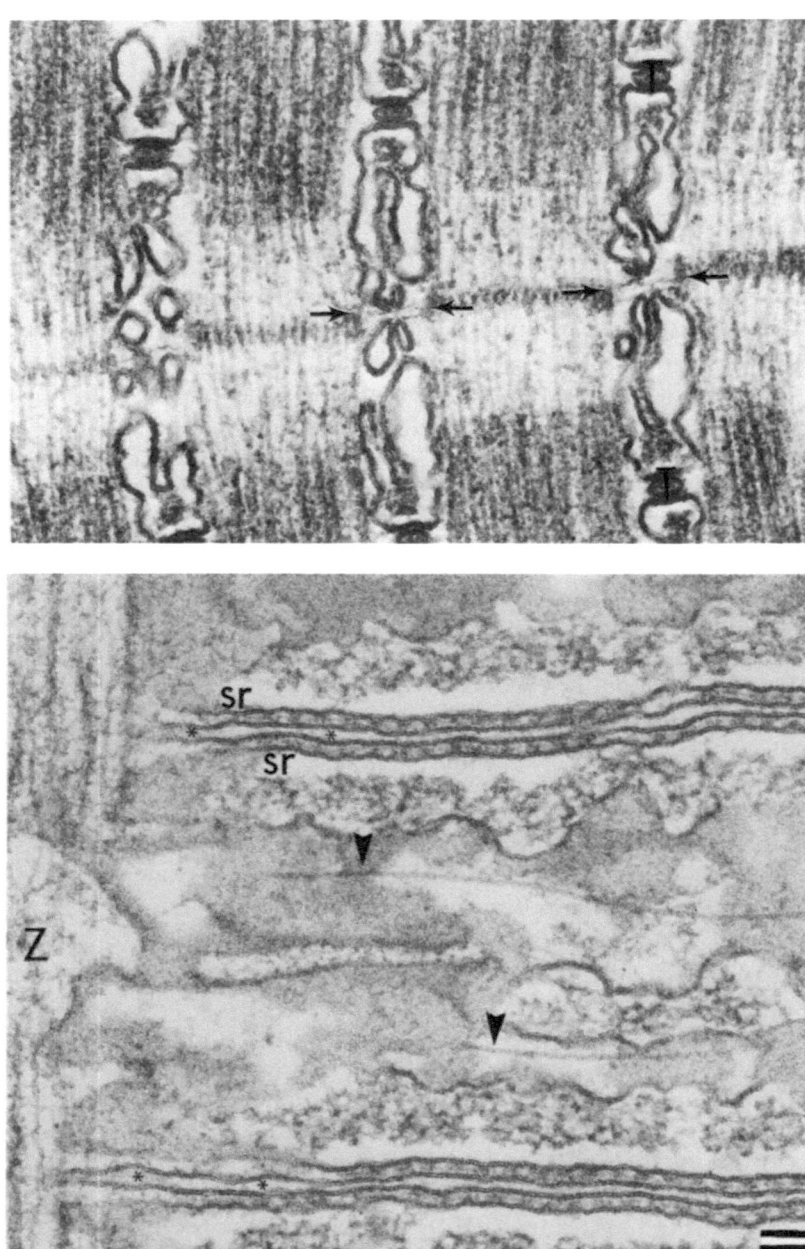

Figure 14.   Upper panel is longitudinal section of the striated muscle in the toadfish swim bladder. The triads (T) are at the level of the A–I junction. Intermediate filaments (arrows) connect the Z-discs to the sarcoplasmic reticulum. Lower panel is a section taken at right angles to the one above. This section includes longitudinal views of the intermediate filaments (arrowheads) at the level of the Z-discs (Z), indicating that the intermediate filaments radiate from the periphery of the Z-discs. The spaces between the T-tubules (*) and the sarcoplasmic reticulum (sr) are filled with dense material. Bar: 100 nm. From Nunzi and Franzini-Armstrong (1980).

4.2. Immunofluorescence of Intermediate Filament Proteins

The immunofluorescent staining patterns obtained by staining adult skeletal muscle and adult cardiac muscle bear many similarities. In both types of mature striated muscle, desmin is localized in the region of the Z-discs and in subsarcolemmal patches adherent to the membranes of apposed cells (Lazarides and Hubbard, 1976; Bennett *et al.*, 1978; Campbell *et al.*, 1979; Fuseler *et al.*, 1981). In mature cardiac myocytes, desmin is also found in the region of the intercalated discs (Lazarides and Hubbard, 1976; Campbell *et al.*, 1979; Fuseler *et al.*, 1981).

Although individual intermediate filaments would not be detected by the immunofluorescent technique, longitudinal filamentous elements are visible in some of the micrographs of skeletal muscle stained with anti-desmin (Lazarides and Hubbard, 1976; Lazarides, 1978b).

The precise localization of desmin within the Z-discs of skeletal muscle was determined by immunofluorescent staining of planar arrays of Z-discs that can be viewed *en face* (Lazarides and Granger, 1978; Granger and Lazarides, 1978). Antibodies to actin, α-actinin, and the intermediate filament proteins desmin and vimentin bind to specific regions of the Z-discs, indicating that molecular domains exist in the Z-discs (Granger and Lazarides, 1978, 1979). The immunofluorescent staining patterns of α-actinin and the intermediate filament proteins are complementary, with α-actinin in the central region of the Z-discs and the intermediate filament proteins confined to a collar around the periphery of each Z-discs. While intermediate filaments *per se* are not seen in the Z-discs (Granger and Lazarides, 1978), it is assumed that the intermediate filament proteins at the periphery of each Z-discs serve to bind together the Z-discs in a plane. Several other proteins have recently been localized to the peripheral domain of the skeletal Z-discs; these are synemin, a protein associated with intermediate filaments (Granger and Lazarides, 1980), and filamin, an actin-binding protein (Gomer and Lazarides, 1981).

5. Role of Intermediate Filaments in Embryonic and Adult Striated Muscle

5.1. Transverse Intermediate Filaments

During the development of skeletal and cardiac muscle, many intermediate filaments become associated with Z-discs, intercalated discs, desmosomal plaques, and patches of dense material applied to various membrane systems (Fig. 17). In muscle cells with rudimentary assemblies of contractile filaments, the intermediate filaments are scattered throughout the cytoplasm. In the ensuing period when myofibrils are rapidly assembled, the pattern of the intermediate filaments seems less random, with the majority of the intermediate filaments oriented parallel or transverse to the myofibrils. At this time, the sarcoplasmic reticulum develops around the myofibrils and, in most skeletal and cardiac myogenic cells, portions of the sarcolemma invaginate to form the

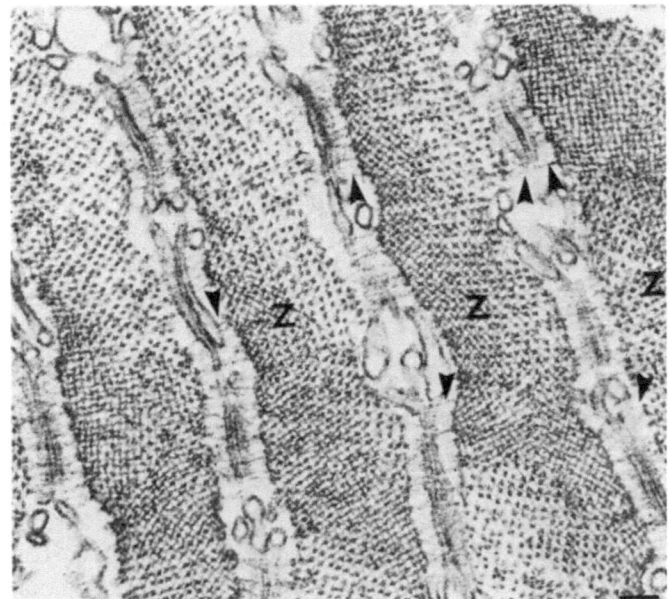

Figure 15. Cross section of a toadfish swim bladder, showing intermediate filaments (arrows) radiating from the periphery of the Z-discs (Z) to the sarcoplasmic reticulum that surrounds the myofibrils. Bar: 200 nm. From Nunzi and Franzini-Armstrong (1980).

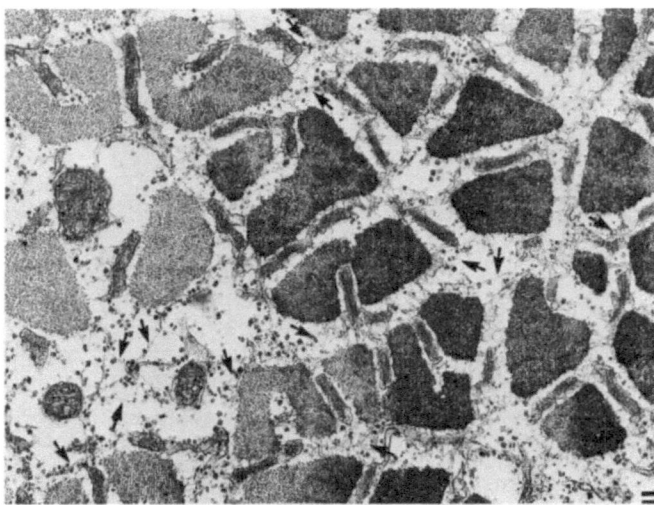

Figure 16. Cross section at the level of the Z-discs of frog striated muscle that was exposed to hypotonic shock prior to fixation. The Z-discs are recognized by the basket-weave pattern. Intermediate filaments (arrows) extend from the Z-discs to the swollen sarcoplasmic reticulum and T-tubules. Some of the intermediate filaments link myofibrils to one another. Bar: 200 nm. From Nunzi and Franzini-Armstrong (1980).

Figure 17. A schematic model of the arrangement of intermediate filaments (white lines) in a striated muscle. There are both longitudinal and transverse intermediate filaments, which appear to be anchored at the periphery of the Z-discs (Z). The wavy longitudinal filaments link the Z-discs within a myofibril, while the transverse intermediate filaments connect adjacent myofibrils at the level of the Z-discs. The transverse intermediate filaments radiate from the periphery of the Z-discs to the T-tubules (T) and to the sarcoplasmic reticulum (SR). In the subsarcolemmal region, the transverse intermediate filaments extend to the sarcolemma. Drawing by John D. Woolsey of the Department of Biology, University of Pennsylvania.

T-tubules. The myofibrils become aligned with one another so that the Z-discs assume lateral registry across the cells. The pattern of desmin and vimentin distribution, as visualized by the immunofluorescent technique, is gradually and dramatically altered during this phase, from a profuse filamentous pattern to a pattern of regular striations corresponding to the distribution of the Z-discs. By time that well-developed myofibrils and membrane systems are present, transverse intermediate filaments link neighboring myofibrils at the level of the Z-discs, and project from the Z-discs to the T-tubules, sarcoplasmic reticulum, nuclear membranes, and the sarcolema.

Considering the sequence of coordinated events has led some to speculate that the intermediate filaments play a role in the lateral alignment of myofibrils (see Lazarides, 1981). The transverse intermediate filaments that are anchored at the Z-discs may guide the invaginating sarcolemma to the myofibrils (Granger and Lazarides, 1978; Lazarides, 1980). The mechanism for guidance may be entirely based on the intermediate filaments having a strong affinity for a protein common to the peripheral region of the Z-discs and to the subsarcolemmal dense patches on the invaginating sarcolemma.

The transverse filamentous connections among the Z-discs within a plane and those between the periphery of the Z-discs and the nuclear and plasma membranes persist in the adult striated muscle. In the interfibrillary spaces, the transverse intermediate filaments seem to contact the membranes of the sarcoplasmic reticulum and the T-tubules. By physically integrating the myofibrils and the membranes within cross-sectional planes, the transverse intermediate filaments may play a role in coordinating the contractile activity of the myriad myofibrils within a cell. The transverse intermediate filaments may mediate the maintenance of the invariant relationship of the membrane systems to the myofibrils throughout the contraction–relaxation cycle, especially in muscles lacking a T-tubular system (Junker and Sommer, 1977). In the same way, the transverse intermediate filaments extending from the Z-discs to the nuclear membrane may secure the nucleus in place during cellular contraction (Ferrans and Roberts, 1973; Forbes and Sperelakis, 1980).

As a cytoskeleton within striated muscle cells, the transverse intermediate filaments may provide tensile strength and convey tension force to the sarcolemma. This speculation is based on the direct relationship between the quantity of intermediate filaments in a cell and the mechanical stress to which the cell is subjected, so that the noncontractile cells of the cardiac impulse-conducting system contain more intermediate filaments than do the myocardial cells, while skeletal muscle cells have the fewest number of intermediate filaments. The intermediate filaments that loop between nearby subsarcolemmal dense patches in the intercalated disc may provide more elastic recoil for the recovery of the shape of this convoluted region following contraction.

5.2. Longitudinal Intermediate Filaments

The longitudinal intermediate filaments (Fig. 17) found between the myofibrils of embryonic striated muscle may serve as more than a passive cytoskeleton. While providing a structural support for the developing myofibrils, they may also form corridors which would facilitate transport throughout the elongated cells. Such a function is proposed for the intermediate filaments of nerve cells (Lasek and Hoffman, 1976). The longitudinal intermediate filaments may play a role in aligning the developing myofibrils in parallel with the long axis of the cells.

Longitudinal intermediate filaments become associated with the periphery of the maturing Z-discs at the time that the rudimentary sarcoplasmic reticulum loosely surrounds the myofibrils. These intermediate filaments extend along the outside of the adult myofibrils, connecting successive Z-discs. The function of these filaments, which seem to be extensible elements, may be to prevent overstretching of the sarcomeres as well as to contribute to the elastic recoil of skeletal muscle.

6. Future Directions for Research

6.1. Biochemistry of Intermediate Proteins from Striated Muscle

Smooth muscle cells are the model system for the study of desmin-containing intermediate filaments, since in these cells desmin comprises 10% of the total cellular protein and at least 50% of the residue that is insoluble in high salt solutions. In smooth muscle, all of the dense bodies are linked together by firmly attached intermediate filaments (Campbell *et al.*, 1970; Somlyo *et al.*, 1971; Cooke and Chase, 1971; Uehara *et al.*, 1971; Rice and Brady, 1972; Cooke, 1976; Small and Sobieszek, 1977).

The explanation for the association of intermediate filaments with certain subcellular organelles is not yet available. There is some indication that the dense bodies of smooth muscle are biochemically similar to the Z-discs, intercalated discs, membrane-bound dense patches and desmosomes with which the intermediate filaments of striated muscle are associated. These structures are all relatively insoluble organelles that contain α-actinin (Rash *et al.*, 1969; Schollmeyer *et al.*, 1976). In addition, with the exception of the desmosomal plaques, all the structures associated with intermediate filaments have actin filaments attached to them. A more complete biochemical characterization of the organelles to which intermediate filaments bind is needed for an understanding of morphogenesis at the molecular level.

It has been shown that purified desmin from smooth muscle polymerizes *in vitro* to form filaments with diameters ranging from 10 to 14 nm (Cooke, 1976; Small and Sobieszek, 1977; Bennett *et al.*, 1978; Hubbard and Lazarides, 1979; Huiatt *et al.*, 1980; Johnson and Yun, 1980). It proved more difficult to isolate desmin from striated muscle, where it comprises only a fraction of a percent of the total protein. Purification of desmin from skeletal muscle has been accomplished, and the purified protein forms filaments identical to those formed by smooth muscle desmin (O'Shea *et al.*, 1981). This finding was expected, since it has been demonstrated that, within a species, the desmins present in smooth, cardiac, and skeletal muscle have identical molecular weights and isoelectric points (Izant and Lazarides, 1977; Lazarides and Balzer, 1978) and are recognized by the same antibodies (Lazarides and Hubbard, 1976). Predictably, the desmins yield similar peptide maps (Bennett *et al.*, 1978; O'Shea *et al.*, 1979; Gard *et al.*, 1979) and have similar amino acid compositions (O'Shea *et al.*, 1979; Johnson and Yun, 1980; Young *et al.*, 1980).

The intermediate filament system of embryonic striated muscle is fundamentally different from that of smooth muscle in containing both desmin and vimentin. Further research on striated muscle desmin is needed to determine whether desmin and vimentin copolymerize into the same filaments. It is unlikely that the intermediate filaments present in muscle cells are composed of pure desmin or vimentin. There is increasing evidence that certain high molecular weight proteins are closely associated with the intermediate filaments of some muscle cells (Granger and Lazarides, 1980; Breckler and Lazarides, 1982). It will be interesting to determine the effect of the filament-associated proteins on the polymerization of desmin and vimentin and to discover whether different filament-binding proteins confer specific physical properties to the intermediate filaments.

The biochemical studies of desmin that have been performed to date have employed desmin isolated under denaturing conditions. It has been assumed that the purified desmin is native since it polymerizes to form filaments, but this may not be the case. Several enzymes, such as M-creatinine phosphokinase (Eckert *et al.*, 1980; Fuseler *et al.*, 1982) and three different types of kinases that phosphorylate desmin and vimentin (O'Connor *et al.*, 1981) are associated with intermediate filaments. The functional importance of the phosphorylation of intermediate filaments and the co-localization of enzymes with the filaments remains to be elucidated. Perhaps a biological assay for the native form of muscle intermediate filament proteins can be based on their ability to bind enzymes or accept phosphate groups.

6.2. Immunoelectron Microscopy

For a detailed understanding of the disposition of desmin and vimentin in striated muscle cells, antibody-binding must be determined at the level of resolution of the electron microscope, in studies similar to those of Richardson *et al.* (1981) using antibodies to desmin. The subcellular components of striated muscle are so dense that antibodies penetrate the cells poorly, and unbound antibodies tend to be trapped, leading to nonspecific binding; these are some of the reasons for the use of isolated myofibrils in the immunoelectron microscopic study of Richardson and colleagues (1981). These problems, as well as the difficulty of visualizing the intermediate filaments in the narrow spaces between the sarcoplasmic reticulum and the myofibrils, may be overcome by swelling the muscle in a hypotonic solution, as Nunzi and Franzini-Armstrong (1980) did.

The accuracy of protein localization at the ultrastructural level may be significantly improved by avoiding the use of ferritin-labeled antibodies, which tend to bind nonspecifically. The ultrastructural distribution of antibodies coupled to gold particles or to iron dextran is believed to be more representative of the distribution of the antigen. It would be a good idea to use monospecific antibodies that are directly coupled to electron-dense particles. Immunoelectron microscopic techniques may be adapted to determine the localization of proteins in the filaments and aggregates that form *in vitro* from extracts of striated muscle.

ACKNOWLEDGMENTS. We would like to thank Drs. Gabriel de la Haba, James Lash, and Frank Pepe, as well as Barbara Langer, for helpful discussions of many of the ideas presented in this review. John Woolsey executed the schematic drawing. Robert Smith helped immensely by editing and providing generous technical assistance with the microprocessor. M. G. P. was a fellow in Cancer Research supported by Grant DRG-442-F of the Damon Runyon-Walter Winchell Cancer Fund. J. W. S. was supported by NIH grants GM 25653 to his laboratory and HL 15835 to the Pennsylvania Muscle Institute.

References

Allen, E. R., and Pepe, F. A., 1965, Ultrastructure of developing muscle cells, *Am. J. Anat.* **116:**115–148.

Behrendt, H., 1977, Effect of anabolic steroids on rat heart muscle cells. I. Intermediate filaments, *Cell Tiss. Res.* **180:**303–315.

Bennett, G. S., Fellini, J. A., and Holtzer, H., 1978, Immunofluorescent visualization of 100 Å filaments in different cultured chick embryo cell types, *Differentiation* **12:**71–81.

Bennett, G. S., Fellini, S. A., Toyama, Y., and Holtzer, H., 1979, Redistribution of intermediate filament subunits during skeletal myogenesis and maturation in vitro, *J. Cell Biol.* **82:**577–584.

Breckler, J., and Lazarides, E., 1982, Isolation of a new high molecular weight protein associated with desmin and vimentin filaments from avian embryonic skeletal muscle, *J. Cell Biol.* **92:**795–806.

Campbell, G. R., Uehara, Y., Mark, G., and Burnstock, G., 1970, Fine structure of smooth muscle cells grown in tissue culture, *J. Cell Biol.* **49:**21–34.

Carlsen, F., Fuchs, F., and Knappeis, G. P., 1965, Contractility and ultrastructure in glycerol-extracted muscle fibers. II. Ultrastructure in resting and shortened fibers, *J. Cell Biol.* **27:**35–46.

Chamley-Campbell, J., Campbell, G. R., Groschel-Stewart, U., Small, J. V., and Anderson, P., 1979, Antibody staining of 10-nm (100 Å) filaments in cultured smooth, cardiac and skeletal muscle cells. *J. Cell Sci.* **37:**303–322.

Cooke, P. H., 1976, A filamentous cytoskeleton in vertebrate smooth muscle fibers, *J. Cell Biol.* **68:**539–556.

Cooke, P. H., and Chase, R. H., 1971, A potassium chloride-insoluble myofilament in vertebrate smooth muscle cells, *Exp. Cell Res.* **66:**417–425.

Dessouky, S. A., and Hibbs, R. G., 1965, An electron microscopic study of the development of the somatic muscle of the chick embryo, *Am. J. Anat.* **116:**523–566.

dos Remedios, C. G., and Gilmour, G., 1978, Is there a third type of filament in striated muscle? *J. Biochem.* **84:**235–238.

Eckert, B. P., Koons, S. J., Schantz, A. W., and Zobel, C. R., 1980, Association of creatinine phosphokinase with the cytoskeleton of cultured mammalian cells, *J. Cell Biol.* **86:**1–5.

Eriksson, A., and Thornell, L.-E., 1979, Intermediate (skeletin) filaments in heart Purkinje fibers. A correlative morphological and biochemical identification with evidence of a cytoskeletal function, *J. Cell Biol.* **80:**231–347.

Eriksson, A., Thornell, L.-E., and Stigbrand, T., 1978, Cytoskeletal filaments of heart conducting system localized by antibody against a 55,000 dalton protein, *Experimentia* **34:**792–794.

Eriksson, A., Thornell, L.-E., and Stigbrand, T., 1979, Skeletin immunoreactivity in heart Purkinje fibers from several species, *J. Histochem. Cytochem* **27:**1604–1609.

Ernst, E., and Straub, F. B., 1968, *Symposium on Muscle.* Akademiai Kiado, Budapest.

Etlinger, J. P. and Fischman, D. A., 1972. M and Z Band components and the assembly of myofibrils, *Cold Spring Harbor Symp. Quant. Biol.* **37:**511–522.

Fawcett, S. W. and McNutt, N. J., 1969, The ultrastructure of the cat myocardium. I. Ventricular papillary muscle, *J. Cell Biol.* **42:**1.

Fellini, S. A., Bennett, G. S., Toyama, Y., and Holtzer, H., 1978, Biochemical and immunological heterogeneity of 100 Å filament subunits from different chick cell types, *Differentiation* **12:**59–69.

Ferrans, V. J., and Roberts, W. C., 1973, Intermyofibrillar and nuclear-myofibrillar connections in human and canine myocardium: An ultrastructural study, *J. Mol. Cell. Cardiol.* **5**:247–257.

Firket, H., 1967, Ultrastructural aspects of myofibril formation in cultured skeletal muscle, *Z. Zellforsch.* **78**:313–327.

Fischman, D. A., 1967, An electron microscope study of myofibril formation in embryonic skeletal muscle, *J. Cell Biol.* **32**:557–574.

Fischman, D. A., 1972, Development of striated muscle, in *The Structure and Function of Muscle*, 2nd edition, Vol. 1 (G. H. Bourne, ed.), pp. 75–149, Academic Press, New York.

Fischman, D. A., 1970, The synthesis and assembly of myofibrils in embryonic muscle, *Curr. Top. Devel. Biol.* **5**:235–280.

Forbes, M. S., and Sperelakis, N., 1980, Structures located at the levels of the Z bands in mouse ventricular myocardial cells, *Tissue and Cell* **12**:467–489.

Franke, W. W., Schmid, E., Breitkreutz, D., Luder, M., Boukamp, P., Fusenig, N. E., Osburn, M., and Weber, K., 1979, Simultaneous expression of two different types of intermediate sized filaments in mouse keratinocytes proliferating in vitro. *Differentiation* **14**:35–50.

Fuseler, J. W., and Shay, J. W., 1982, The association of desmin with the developing myofibrils of cultured embryonic rat heart myocytes, *Dev. Biol.* **91**:448–457.

Fuseler, J. W., Eckert, B. S., Koon, S. J., and Shay, J. W., 1982, The association of creatine phosphokinase with the mitotic spindle, in: *Cell and Muscle Motility*, Vol. II (R. M. Dowben and J. W. Shay, eds.), pp. 103–119, Plenum Press, New York.

Garamvolgyi, N., 1962, Interfibrillare Z-Verbin-dungen des quergestreiften Muskel, *Acta Physiol. Hung.* **22**:235–241.

Garamvolgyi, N., 1963, Observations preliminaires sur la structure de la strie Z dans le muscle alaire de l'abeille, *J. Microsc.* **2**:107–113.

Garamvolgyi, N., 1965, Inter-Z bridges in the flight muscle of the bee. *J. Ultrastruct. Res.* **13**:435–443.

Garamvolgyi, N., 1969, The structural basis of the elastic properties in the flight muscle of the bee, *J. Ultrastruct. Res.* **27**:462–471.

Gard, D. L., and Lazarides, E., 1980, The synthesis and distribution of desmin and vimentin during myogenesis in vitro, *Cell* **19**:263–275.

Gard, D. L., Bell, P. B., and Lazarides, E., 1979, Coexistence of desmin and the fibroblastic intermediate filament subunit in muscle and non-muscle cells: identification and comparative peptide analysis, *Proc. Natl. Acad. Sci.* **76**:3894–3898.

Gaskin, F., and Shelanski, M. L., 1976, Microtubules and intermediate filaments, *Essays in Biochem.* **12**:115–146.

Goldman, R. D., Milsted, A., Schloss, J. A., Starger, J., and Yerna, M.-J., 1979, Cytoplasmic fibers in mammalian cells: cytoskeletal and contractile elements, *Ann. Rev. Physiol.* **41**:713–722.

Gomer, R. G., and Lazarides, E., 1981, The synthesis and deployment of filamin in chicken skeletal muscle, *Cell* **23**:524–532.

Granger, B. L., and Lazarides, E., 1978, The existence of an insoluble Z disc scaffold in chicken skeletal muscle, *Cell* **15**:1255–1268.

Granger, B. L., and Lazarides, E., 1979, Desmin and vimentin coexist at the periphery of the myofibril Z disc, *Cell* **18**:1053–1063.

Granger, B. L., and Lazarides, E., 1980, Synemin: A new high molecular weight protein associated with desmin and vimentin filaments in muscle, *Cell* **22**:727–738.

Guba, F., Harsonyi, V., and Vajda, E., 1968, Ultrastructure of myofibrils after selective protein extraction, *Acta Biochim. Biophys. Acad. Sci. Hung.* **3**:433–440.

Hanson, J., 1956, Studies on the cross-striation of indirect flight myofibrils of the blowfly Calliphora, *J. Biophys. Biochem. Cytol.* **2**:691–710.

Hanson, J., and Huxley, H. E., 1953, Structural basis of the cross-striations in muscle, *Nature* **172**:530–532.

Hanson, J., and Huxley, H. E., 1955, The structural basis of contraction in striated muscle, *Symp. Soc. Exp. Biol.* **9**:228–264.

Hanson, J., and Lowry, J., 1963, The structure of F-actin and of actin filaments isolated from muscle, *J. Mol. Biol.* **6**:46–60.

Hay, E. D., 1963, The fine structure of differentiating muscle in the salamander tail, *Z. Zellforsch.* **59:**6–34.

Heuson-Steinnon, J. A., 1965, Morphogenèse de la cellule musculaire striée, etudiée au microscope électronique, *J. Microscop.* **4:**657–678.

Hilfer, S. R., Searls, R. L., and Fonte, V. G., 1973, An ultrastructural study of early myogenesis in the chick wing bud, *Dev. Biol.* **30:**374–391.

Holtzer, H., Sanger, J. W., Ishikawa, H., and Strahs, K., 1972, Selected topics in skeletal myogenesis, *Cold Spring Harbor Symp. Quant. Biol.* **37:**549–566.

Hubbard, B. D., and Lazarides, E., 1979, Copurification of actin and desmin from chicken smooth muscle and their copolymerization in vitro to intermediate filaments, *J. Cell Biol.* **88:**166–182.

Huiatt, T. W., Robson, R. M., Arakawa, N., and Stromer, M. H., 1980, Desmin from avian smooth muscle. Purification and partial characterization, *J. Biol. Chem.* **255:**6981–6939.

Huxley, H. E., 1963, Electrin microscope studies on the structure of natural and synthetic filaments from striated muscle, *J. Mol. Biol.* **7:**281–307.

Huxley, H. E., and Hanson, J., 1957, Quantitative studies on the structure of cross-striated myofibrils. I. Investigations by interference microscopy, *Biochim. Biophys. Acta* **23:**229–249.

Ishikawa, H., 1974, Arrowhead complexes in a variety of cell types, in: *Exploratory Concepts in Muscular Dystrophy II,* (A. T. Milhorat, ed.), pp. 37–50, Excerpta Medica, Amsterdam.

Ishikawa, H., Bischoff, R., and Holtzer, H., 1968, Mitosis and intermediate-sized filaments in developing skeletal muscle, *J. Cell Biol.* **38:**538–555.

Ishikawa, H., Bischoff, R., and Holtzer, H., 1969, Formation of arrow-head complexes with heavy meromyosin in a variety of cell types, *J. Cell Biol.* **38:**538–555.

Izant, J. G., and Lazarides, E., 1977, Invariance and heterogeneity in the major structural and regulatory proteins of chick muscle cells revealed by two-dimensional electrophoresis, *Proc. Nat. Acad. Sci.* **74:**1450–1454.

Johnson, P., and Yun, J. S., 1980, Intermediate filaments of bovine pulmonary artery smooth muscle: Isolation and polypeptide composition, *Int. J. Biochem.* **11:**143–154.

Junker, J., and Sommer, J. S., 1977, Anchorfibers and the topography of junctional SR. 35th Annual EMSA Meeting, pp. 582–583.

Kelly, D. A., 1969, Myofibrillogenesis and Z-band differentiation, *Anat. Rec.* **163:**403–426.

Kelly, A. M., and Chacko, S., 1976, Myofibril organization and mitosis in cultured cardiac muscle cells, *Dev. Biol.* **43:**421–430.

Lasek, R. J., and Hoffman, P. N., 1976, The neuronal cytoskeleton, axonal transport and axonal growth, in: *Cell Motility,* Vol. 3, Book C, Cold Spring Harbor Conference on Cell Proliferation (R. D. Goldman, T. D. Pollard, and J. L. Rosenbaum, eds.), pp. 1021–1050, Cold Spring Harbor Laboratory, Cold Spring Harbor, New York.

Lazarides, E., 1978, The distribution of desmin (100 Å) filaments in primary cultures of embryonic chick cardiac cells. *Exp. Cell Res.* **112:**265–273.

Lazarides, E., 1980, Intermediate filaments as mechanical integrators of cellular space, *Nature (Lond.)* **283:**249–256.

Lazarides, E., 1981, Molecular morphogenesis of the Z-disc in muscle cells, in: *International Cell Biology, 1980–1981* (H. G. Schweiger, ed.), pp. 392–398, Springer-Verlag, Berlin.

Lazarides, E., and Balzer, D. R., 1978, Specificity of desmin to avian and mammalian muscle cells, *Cell* **14:**429–438.

Lazarides, E., and Granger, B. L., 1978, Fluorescent localization of membranes sites in glycerinated skeletal muscle fibers and the relationship of these sites to the protein composition of the Z-discs, *Proc. Natl. Acad. Sci.* **75:**3683–3687.

Lazarides, E., and Hubbard, B. D., 1976, Immunological characterization of the subunit of 100 Å filaments from muscle cells, *Proc. Natl. Acad. Sci.* **73:**4344–4348.

Lindner, F., 1960, Myofibrils in the early development of chick embryo hearts as observed with the electron microscope, *Anat. Rec.* **136:**234.

Lindner, E., and Schaumburg, G., 1968, Zytoplasmatische filamente in den quergestreiften muskelzellen des kaudalen lymphherzens von Rana temporaria L. I. Untersuchungen am lymphherzen, *Z. Zellforsch. Mikrosk. Anat.* **84:**549–562.

Locker, R. H., and Leet, N. G., 1975, Histology of highly stretched beef muscle. I. The fine structure of grossly stretched single fibers, *J. Ultrastruct. Res.* **52**:64–75.

Locker, R. H., and Leet, N. G., 1976, Histology of highly stretched beef muscle. II. Further evidence on the location and nature of gap filaments, *J. Ultrastruct. Res.* **55**:157–172.

Manasek, F. J., 1968, Embryonic development of the heart. I. A light and electron microscopic study of myocardial development in the early chick embryo, *J. Morph.* **125**:329–366.

Manasek, F. J., 1979, Organization, interactions, and environment of heart cells during myocardial ontogeny, in: *Handbook of Physiology*, Section 2; *The Cardiovascular System*, Vol. I, *The Heart*, (R. M. Berne, ed.), pp. 29–43, American Physiological Society, Bethesda, Maryland.

Markwald, R. R., 1973, Distribution and relationship to precursor Z material to organizing myofibrillar bundles in embryonic rat and hamster ventricular myocytes, *J. Mol. Cell Cardiol.* **5**:341–350.

Maron, B. J., and Ferrans, V. J., 1974, Aggregates of tubules in human cardiac muscle cells, *J. Mol. Cell. Cardiol.* **6**:249–264.

McNeill, P. A., and Hoyle, G., 1967, Evidence for superthin filaments, *Am. Zoologist.* **7**:483–498.

Nunzi, M. G., and Franzini-Armstrong, C., 1980, Trabecular network in adult skeletal muscle, *J. Ultrastructure Res.* **73**:21–26.

O'Connor, C. M., Gard, D. L., and Lazarides, E., 1981, Phosphorylation of intermediate filament proteins by cAMP-dependant protein kinases. *Cell* **23**:135–143.

O'Shea, J. M., Robson, R. M., Huiatt, T. W., Hartzer, M. K., and Stromer, M. H., 1979, Purified desmin from adult mammalian skeletal muscle: A peptide mapping comparison with desmins from adult mammalian and avian smooth muscle, *Biochem. Biophys. Res. Commun.* **89**:972–980.

O'Shea, J. M., Robson, R. M., Hartzer, M. K., Huiatt, T. W., Rathbun, W. E., and Stromer, M. H., 1981, Purification of desmin from adult mammalian skeletal muscle, *Biochem. J.* **195**:345–356.

Oliphant, L. W., and Loewen, R. D., 1976, Filament systems in Purkinje cells of the sheep heart: Possible alterations of myofibrillogeneisis, *J. Mol. Cell. Cardiol.* **8**:69–688.

Osborn, M., Franke, W. W., and Weber, K., 1980, Direct demonstration of the presence of two immunologically distinct intermediate-sized filament systems in the same cell by double immunofluorescence microscopy, *Exp. Cell Res.* **125**:37–46.

Page, S. G., 1969, Structures and some contractile properties of fast and slow muscles of the chicken, *J. Physiol.* **205**:131–145.

Page, E., and Fozzard, H. W., 1973, Capacitive, resistive, and syncytial properties of heart muscle—ultrastructural and physiological consideration, in: *The Structure and Function of Muscle*, 2nd edition, Vol. 1 (G. H. Bourne, ed.), pp. 91–158, Academic Press, New York.

Peachey, L. P., 1968, Muscle, *Ann. Rev. Physiology* **30**:401–440.

Porte, A., Stoeckel, M. E., Sacrez, A., and Batzenschlager, A., 1980, Unusual familial cardiomyopathy with storage of intermediate filaments in the cardiac muscular cells, *Virchows Arch. A. Path. Anat. Histol.* **386**:43–58.

Price, M., and Sanger, J. W., 1979a, Intermediate filaments connect Z-discs in adult chicken muscle, *J. Exp. Zool.* **208**:263–269.

Price, M., and Sanger, J. W., 1979b, Intermediate filaments in chicken skeletal myofibrils: localization and reconstitution, *J. Cell Biol.* **83**:318a.

Price, M., and Sanger, J. W., 1980, Intermediate filaments are redistributed during myogenesis and become associated with Z-discs and membranes, *J. Cell Biol.* **87**:182a.

Price, M., and Sanger, J. W., 1981, The distribution of intermediate filaments during skeletal myogenesis, in preparation.

Przybylski, R. J., and Blumberg, J. M., 1966, Ultrastructural aspects of myogenesis in the chick, *Lab. Invest.* **15**:836–863.

Rash, J. E., Shay, J. W., and Biesele, J. J., 1967, Ultrastructure of differentiating muscle cells after actin and tropomyosin extractions, *J. Cell Biol.* **35**:110a.

Rash, J. E., Shay, J. W., and Biesele, J. J., 1969, A third class of filaments in early cardiac myocytes, *J. Cell Biol.* **43**:112a.

Rash, J. E., Shay, J. W., and Biesele, J. J., 1970a, Preliminary biochemical investigations of the intermediate filaments, *J. Ultrastruct. Res.* **33**:399–407.

Rash, J. E., Biesele, J. J., and Gey, G. O., 1970b, Three classes of filaments in cardiac differention, *J. Ultrastruct. Res.* **33**:408–435.

Rice, R. V., and Brady, A. C., 1972, Biochemical and ultrastructural studies on vertebrate smooth muscle, *Cold Spring Harbor Symp. Quant. Biol.* **37**:429–442.

Richardson, F. L., Stromer, M. H., Huiatt, T. W., and Robson, R. M., 1981, Immunoelectron and immunofluorescence localization of desmin in mature avian muscles, *Eur. J. Cell Biol.* **26**:91–101.

Sabatini, D. D., Bensch, K., and Barnett, R. J., 1963, Cytochemistry and electron microscopy. The preservation of cellular ultrastructure and enzymatic activity by aldehyde fixation, *J. Cell Biol.* **17**:19.

Sandborn, E. B., Cote, M. G., Roberge, J., and Bois, P., 1967, Microtubules et filaments cytoplasmiques dans le muscle mammiferes, *J. Microscopie* **6**:169–178.

Sandow, A., 1970, Skeletal muscle, *Ann. Rev. Physiol.* **32**:87–138.

Sanger, J. W., 1979, Cardiac fine structures in selected arthropods and molluscs, *Amer. Zool.* **19**:9–27.

Schollmeyer, J. E., Furcht, L. T., Goll, D. E., Robson, R. M., and Stromer, M. H., 1976, Localization of contractile proteins in smooth muscle cells and in normal and transformed cells, in: *Cell Motility*, Vol. 3, Book A, Cold Spring Harbor Conference on Cell Proliferation (R. D. Goldman, T. D. Pollard, and J. L. Rosenbaum, eds.), pp. 364–388, Cold Spring Harbor Laboratory, Cold Spring Harbor, New York.

Schwann, T., 1847, Microscopial researches into the accordance in the structure and growth of animals and plants. Translated from the German by H. Smith. C. and J. Adlard, printers, London.

Shelanski, M. L., Yen, S.-H., and Lee, V. M., 1976, Neurofilaments and glial filaments, in: *Cell Motility*, Vol. 3, Book C, Cold Spring Harbor Conference on Cell Proliferation (R. D. Goldman, T. D. Pollard, and J. L. Rosenbaum, eds.), pp. 1007–1020, Cold Spring Harbor Laboratory, Cold Spring, New York.

Shimada, Y., 1971, Electron microscope observations on the fusion of chick myoblasts in vitro, *J. Cell Biol.* **48**:128–142.

Shimada, Y., Fischman, D. A., and Moscona, M. M., 1967, The fine structure of embryonic chick skeletal muscle cells differentiated in vitro, *J. Cell Biol.* **35**:445–453.

Sjostrand, F. S., 1962, The connections between A- and I- band filaments in striated frog muscle, *J. Ultrastruct. Res.* **7**:225–246.

Small, J. V., and Sobieszek, A., 1977, Studies on the function and composition of the 10 nm (100 A) filaments of vertebrate smooth muscle, *J. Cell Sci.* **23**:243–268.

Small, J. V., and Sobieszek, A., 1980, The contractile apparatus of smooth muscle, *Int. Rev. Cytol.* **64**:241–306.

Somlyo, A. P., Somlyo, A. V., Devine, C. E., and Rice, R. V., 1971, Aggregation of the thick filaments into ribbons in mammalian smooth muscle. *Nature* **231**:243–246.

Sommer, J. R., and Johnson, E. A., 1979, Ultrastructure of cardiac muscle, in: *Handbook of Physiology*, Section 2, *The Cardiovascular System*, Vol. I, *The Heart*, (R. M. Berne, N. Sperelakis, and S. R. Geiger, eds.), pp. 113–186, American Physiological Society, Bethesda, Maryland.

Starger, J. M., Brown, W. E., Goldman, A. E., and Goldman, R. D., 1978, Biochemical and immunological analysis of rapidly purified 10-nm filaments from baby hamster kidney (BHK-21) cells, *J. Cell Biol.* **78**:93–109.

Stigbrand, T., Eriksson, A., and Thornell, L.-E., 1979, Isolation and partial characterization of intermediate filament protein (skeletin) from cow heart Purkinje fibres, *Biochim. Biophys. Acta* **577**:52–60.

Szent-Gyorgyi, A. G., 1953, Meromyosins, the subunits of myosin, *Arch. Biochem. Biophys.* **42**:305–320.

Thiedemann, K. U., and Ferrans, V. J., 1976, Ultrastructure of sarcoplasmic reticulum in atrial myocardium of patients with mitral valvular disease, *Am. J. Pathol.* **83**:1–37.

Thornell, L.-E., Sjostrom, M., and Andersson, K.-E., 1976, The relationship between mechanical stress and myofibrillar organization on heart Purkinje fibres, *J. Mol. Cell. Cardiol.* **8:**689–695.

Thornell, L.-E., Eriksson, A., Stigbrand, T., and Sjostrom, M., 1978, Structural proteins in cow Purkinje and ordinary ventricular fibres—a marked difference, *J. Mol. Cell. Cardiol.* **10:**605–616.

Trombitas, K., and Tigyi-Sebes, A., 1977, Fine structure and mechanical properties of insect muscle, in: *Insect Flight Muscle*, (R. T. Tregear, ed.), pp. 79–90, North Holland Publ. Co. Amsterdam.

Trombitas, K., and Tigyi-Sebes, A., 1979, The continuity of thick filaments between sarcomeres in honeybee flight muscle, *Nature* **281:**319–320.

Tuszynski, G. P., Frank, E. D., Damsky, C. H., Buck, C. A., and Warren, L., 1979, The detection of smooth muscle desmin-like protein in BHK-21/C-13 fibroblasts, *J. Biol Chem.* **254:**6138–6143.

Uehara, Y., Campbell, G. R., and Burnstock, G., 1971, Cytoplasmic filaments in developing and adult vertebrate smooth muscle, *J. Cell Biol.* **50:**484–497.

Ullrick, W. C., Toselli, P. A., Chase, D., and Dasse, K., 1977, Are there extensions of thick filaments to the Z-line in vertebrate and invertebrate striated muscle? *J. Ultrastruc. Res.* **60:**263–271.

Viragh, S. Z., and Challice, C. E., 1969, Variations in filamentous and fibrillar organization, and associated sarcolemmal structures, in cells of the normal mammalian heart, *J. Ultrastruct. Res.* **28:** 321–334.

Wainrack, J., and Sotelo, J. R., 1961, Electron microscope study of the developing chick embryo heart, *Z. Zellforsch.* **55:**622.

Walcott, B., and Ridgeway, E. B., 1967, The ultrastructure of myosin-extracted striated muscle fibers, *Am. Zoologist.* **7:**499–504.

Wang, K., and Ramirez-Mitchell, R., 1979a, Titin: a possible candidate as components of putative longitudinal filaments in striated muscle, *J. Cell Biol.* **83:**389a.

Young, O. A., Graafhuis, A. E., and Parey, L. L., 1980, Post-mortem changes in cytoskeletal proteins of muscle, *Meat Science* **5:**41–55.

Zackroff, R. V., and Goldman, R. D., 1979, In vitro assembly of intermediate filaments from baby hamster kidney (BHK-21) cells, *Proc. Natl. Acad. Sci.* **76:**6226–6230.

2

Biochemistry and Structure of Mammalian Neurofilaments

Robley C. Williams, Jr., and Marschall S. Runge

1. Introduction

Neurofilaments are the apparently unique intermediate filaments of neurons, different in their composition and structure from the intermediate filaments of other types of cells. This chapter summarizes current knowledge of the biochemistry and structure of neurofilaments. Special emphasis is placed on studies *in vitro* and on the properties of mammalian neurofilaments. Several unresolved questions are discussed that appear to arise naturally from the work done thus far.

Definition

Intermediate filaments in different types of cells appear similar to each other when observed with the electron microscope. Until recently, it was believed by many investigators that they would also prove to be similar to each other in protein composition, in structure, and in function, as microtubules from different sources have proven to be. Instead, they have turned out to be quite diverse. In review articles, Lazarides (1980) distinguishes between five groups of intermediate filaments and Goldman *et al.* (1979) point out the differences in protein composition between intermediate filaments. Recent studies by immunohistochemical methods (Bennett *et al.*, 1981) show that neurons (at least in the chicken) contain no vimentin, desmin, cytokeratins, or

Robley C. Williams, Jr., and Marschall S. Runge • Department of Molecular Biology, Vanderbilt University, Nashville, Tennessee 37235.

glial-fibrillary-acidic protein. This result, and other evidence discussed below, leads to the conclusion that neurofilaments must be made of proteins not closely related to the subunits of the other major types of intermediate filaments. It is therefore reasonable to study them separately from the other intermediate filaments.

In thin sections of neurites and axons of the neurons of both vertebrate and invertebrate species, neurofilaments appear as darkly staining structures, sometimes interspersed with microtubules and sometimes as the sole filamentous element. In axons of rat anterior horn cells (Wuerker, 1970) they appear to be multiply stranded structures of 8 nm diameter at their narrowest and 11 nm at their widest. As discussed below, they are composed primarily of three polypeptides, having molecular weights of approximately 210,000, 160,000 and 70,000 and known as the "neurofilament triplet" of proteins.

When a macromolecule or a small organelle is isolated by biochemical techniques, the question arises of how one knows that one has isolated the "right" material. In the case of an enzyme, for instance, one employs its function, catalysis of a particular reaction, to define it. In the case of microtubules, one employs their unique structure as a means of identification. Neurofilaments, however, share a common morphology with other intermediate filaments. Their functions are still essentially unknown. Therefore, for the purposes of isolation and biochemistry, they can be taken to be defined by the *combination* of their morphology and their origin in neurons.

2. Isolation and Identification in Vitro

2.1. Historical Aspects

The isolation of neurofilaments has presented a difficult problem. Neurons occur in close association with cells of other types, all of which have intermediate filaments of morphology similar to that of neurofilaments. The early literature (for examples, see Shelanski *et al.*, 1971; Davison and Winslow, 1974; Yen *et al.*, 1976; Liem *et al.*, 1977; Davison and Hong, 1977; Iqbal *et al.*, 1977, 1978; J. E. Goldman *et al.*, 1978) mistakenly suggested that mammalian neurofilaments are composed of a single kind of subunit having a molecular weight variously estimated to be between 47,000 and 54,000. These conclusions apparently were drawn as a result of differences in solubility of the two types of filaments in buffers of low ionic strength (Liem *et al.*, 1978), as follows. In an effort to isolate filaments from starting material rich in neurons, homogenized brain tissue was subjected to an axonal flotation step, followed by demyelination in .01 M phosphate, and then by centrifugations in sucrose solutions buffered by .01 M Tris. Unfortunately, it was not realized that prolonged exposure to low ionic strength solubilizes neurofilaments but not glial filaments (Schlaepfer, 1971; Schlaepfer, 1977a; Schlaepfer, 1978; Schlaepfer and Lynch, 1977; Liem *et al.*, 1978). Thus, although the axonal flotation step apparently yielded relatively pure neuronal material, the demyelination step and subsequent centrifugations dissolved the desired neurofilaments and yielded a small amount of contaminating glial filaments, highly enriched because of their insolubility. The consequence of the misidentification of the protein composition of the neurofilament has been a certain confusion in the literature, especially surrounding the question of whether antisera to neurofilaments would bind to filaments of other cell types (Blose *et al.*, 1977; Yen *et al.*, 1976), and over the question of whether the proteins of neurofilaments might be identical to those of the intermediate filaments of astroglia (J. E. Goldman *et al.*, 1978).

2.2. Current Isolation and Identification

A series of investigations by Schlaepfer and his co-workers, together with extensive studies by Lasek and his co-workers, have established that mammalian neurofilaments can be defined *in vitro* by their morphology and by their characteristic composition of a triplet of proteins, of molecular weights near 70,000, 160,000, and 210,000. Micko and Schlaepfer (1978) showed, in fact, that the major nonmyelin polypeptides of peripheral nerves are tubulin and the neurofilament triplet. No polypeptide of a molecular weight in the range of 47,000–54,000 is present in their preparations. By isolating neurofilaments, in buffers of approximately 0.1 M ionic strength, from excised and desheathed peripheral nerves of the rat, and from anterior and posterior nerve roots of the rat, Schlaepfer and Freeman (1978) were able to prepare fractions of neurofilament that would be expected to have only small amounts of any contaminating structures of glial origin. These filaments consisted largely of polypeptides of apparent molecular weights corresponding to the neurofilament triplet, in addition to smaller amounts of tubulin and other minor proteins. Proteins in the molecular weight range 50,000–54,000 were lacking in this material from the peripheral nervous system. (Filaments composed of the same triplet of proteins were isolated by these workers from the spinal cord and medulla oblongata of rat, but in that case large amounts of a protein with a molecular weight of 51,000 were present as well.) Schlaepfer and Micko (1978) found that the disappearance of the neurofilament triplet proteins coincides with the disappearance of morphological neurofilaments in peripheral axons that have been transected, and later showed that these coordinated changes result from the penetration of Ca^{2+} into the axoplasm (Schlaepfer and Micko, 1979). Thus, the crucial connection between the structural element and its constituent polypeptides was made, first by isolation from a relatively homogeneous population of neurons, and second by showing the coordinate disappearance of the structure and the proteins during degeneration. These findings, and a number of supporting studies, are reviewed by Schlaepfer (1979).

Studies of axonal transport showed (Hoffman and Lasek, 1975; Lasek and Hoffman, 1976) that the chief proteins carried distally in the slow component are tubulin and the neurofilament triplet. Radioactive amino acids were injected into the L5–L6 ventral horn region of the spinal cords of rats. At each of several intervals after injection, the sciatic nerve of a rat was excised, sectioned, and counted for radioactivity. The polypeptide composition of each 3-mm segment was examined by SDS-gel electrophoresis. Although numerous polypeptides were seen in the stained gels, in the slowest moving component radioactivity was confined primarily to the neurofilament triplet and to tubulin. The proteins of the triplet appeared to move at a common rate (as did much of the tubulin). Similar results have been obtained in similar experiments with the retinal ganglion cells of the rabbit (Willard and Hulebak, 1977) and with numerous other neurons. In these experiments, the axon acts, in a sense, as a chromatographic column. The various rates at which different components move serve to separate from each other proteins that are not parts of a common structure. Although co-migration of proteins does not prove in any case that they are parts of the same structure, it creates a strong presumption that they are. Further evidence of the correlation between the morphological neurofilament and the triplet of proteins is provided by the effects of β,β'-iminodipropionitrile. This compound inhibits the slow component of axonal transport selectively (Griffin *et al.*, 1978) and produces large neurofilament-filled swellings in the most proximal part of the axon (Chou and Hartmann, 1964, 1965; Chou and Klein, 1972). The proteins whose blocked transport coincides with the blocked transport of neurofilaments are the neurofilament triplet and probably tubulin (Griffin *et al.*, 1978).

2.3. Preparation of Neurofilaments for Biochemistry

A number of isolation procedures have been devised to yield microtubules in the quantities required for successful biochemical studies. As is true in the early stages of isolation of any organelle, some of the properties of the resulting filaments may depend upon their source and the procedure used to isolate them. Schlaepfer (1977a,b) prepared from peripheral nerve of rat small quantities (a few milligrams) of neurofilaments. In this method, dissected, desheathed nerve is disrupted by osmotic shock and extracted by stirring in a hypotonic buffer. A neurofilament-enriched supernatant is then obtained by a single high-speed centrifugation step and has been used (Schlaepfer, 1977b; Schlaepfer, 1978; Schlaepfer and Freeman, 1978) without further purification. A second isolation procedure (Runge *et al.*, 1981a) allows preparation of approximately 80 mg of neurofilaments from bovine brain. Tissue is disrupted by blade homogenization in a buffer of about 0.15 M ionic strength. Neurofilaments appear in the supernatant after a high-speed centrifugation of the tissue extract. They are then separated from free proteins and smaller structures by gel filtration on a column of Bio-Gel A-150m (the filaments emerge in the void volume) and freed of residual membranous contaminants by centrifugation through a sucrose cushion. A third procedure (Shecket and Lasek, 1980) yields several milligrams of neurofilaments from peripheral nerve and spinal cord of guinea pig. Tissue extracts are prepared by vigorous homogenization in a hypotonic buffer, followed by centrifugation. Some unwanted proteins in the supernatant of the centrifugation step are precipitated with ammonium sulfate, and many of the rest by gel filtration on Sepharose 4B. The filaments are then precipitated with ethanol and resuspended as a final step. A fourth preparation (Delacourte *et al.*, 1980) provides 100 mg of neurofilaments from bovine spinal cord via a procedure resembling the preparation of microtubules (Shelanski *et al.*, 1973). Tissue is homogenized in a buffer of 0.15 M ionic strength, and a high-speed supernatant is prepared. Glycerol (20%) is added to this supernatant and it is incubated at 37°, then centrifuged at 30°. The pellets are resuspended in buffer at 4° and centrifuged at 4°. The pellet from this cold centrifugation consists largely of neurofilaments. This preparation appears to work partly because neurofilaments are dragged into the pellet during the warm centrifugation by their association with microtubules present in the solution (see below). A fifth isolation technique (Czosnek *et al.*, 1980) has been used to prepare several milligrams of neurofilaments (apparently mixed with glial filaments) from rabbit spinal cord by means of homogenization in 0.5 M and 1.1 M sucrose, followed by a series of sedimentation steps in sucrose. The filaments that result from these schemes differ somewhat in protein composition and microscopic appearance. The Schlaepfer and Delacourte filaments are relatively smooth in profile and even in diameter. Both the Runge and Shecket filaments appear somewhat more uneven in diameter. The kinds and amounts of nontriplet proteins present in the preparations also differ. In particular, neurofilaments prepared from brain seem to have associated with them a much larger amount of tubulin than do filaments prepared by the same methods from peripheral nerve (Delacourte *et al.*, 1980; E. Aamodt and R. C. Williams, Jr., unpublished observations). After accounting for these minor differences, one can assert that there are several feasible methods of preparing neurofilaments for further biochemical work.

3. Protein Composition

3.1. Properties of the Neurofilament Triplet

The proteins of the neurofilament triplet are easily separated by SDS-gel electrophoresis, and most studies of the individual proteins have been done with material isolated in that way. Attempts to separate the proteins by chromatographic and gel filtration techniques have proved relatively unrewarding thus far. A series of studies of the gel filtration behavior of neurofilament proteins in 6 M guanidine HCl and in 8 M urea (M. R. Lifscis and R. C. Williams, Jr., unpublished) has led to the conclusion that the components of the triplet form an associated complex (or several different complexes) in these solvents. Liem and Turey (1980) have separated the components of the neurofilament triplet from contaminating tubulin and glial filament protein by chromatography on hydroxylapatite in 8 M urea, but the proteins of the triplet are not fully resolved from each other by this method. Moon *et al.* (1981) have obtained a nearly complete resolution of the high-molecular-weight triplet protein from the other two by ion-exchange chromatography in a solvent of 8 M urea, pyridine, formic acid, and mercaptoethanol. They concluded that substantial aggregation of the triplet proteins was present in this solvent. The neurofilament proteins are thus highly unusual in their tendency to remain associated in common denaturing solvents. Full characterization of the proteins of the triplet awaits their separation in quantity, and that separation awaits the discovery of effective dissociating conditions.

An estimate of the molar amounts of the proteins in the neurofilament has appeared. From the integration of scans of SDS gels stained with Fast Green, Shecket and Lasek (1980) estimated that in the isolated neurofilaments of the guinea pig, the molar ratio of small to intermediate to large proteins is 6:2:1. This estimate, while useful, is subject to possible systematic errors arising from the nonlinear relationship between stain absorbance and the action of a calcium-activated protease during isolation.

Apparent differences in the molecular weights of homologous neurofilament proteins are observed when filaments from different mammalian species are subjected to SDS-gel electrophoresis (Chiu et al., 1980; Davison and Jones, 1980; Davison, 1981; Runge et al., 1981a). The differences are observed regardless of the preparative technique employed, and are as large as 10% of the total molecular weight in some cases. Apparently, precisely equal molecular weights are not requisite for homologous function in these proteins, but the biological significance of the size difference is still obscure.

Substantial immunological cross-reactivity has been reported between the three neurofilament subunits (Liem et al., 1978; Willard and Simon, 1981; Yen and Fields, 1981), leading one to speculate that they have regions of sequence homology and perhaps a common evolutionary origin. Peptide mapping of each of the three subunits (Davison and Jones, 1981; Mori et al., 1979) and studies of their biosynthesis (Czosnek et al., 1980) indicate that they are not derived from one large precursor, but are uniquely and independently synthesized. The possibility remains, however, that the three genes arose from a common ancestral gene. Further assessment of possible common features would be greatly aided by sequence determination.

An interesting minor point is that the large triplet polypeptide appears to exist in two genetically determined forms of slightly different molecular weight, at least in the rabbit (Willard, 1976) and guinea pig (Shecket and Lasek, 1980). This dimorphism has been used a probe to investigate the likely invariance of expression of neurofilament protein genes in different regions of the nervous system (Czosnek et al., 1981).

3.2. Minor Proteins

Proteins other than the neurofilament triplet are persistently present in neurofilament preparations. Tubulin is found in substantial quantities in neurofilaments prepared from brain (Delacourte et al., 1980; Runge et al., 1981a), although it is present in vastly reduced amount in neurofilaments prepared by a similar technique from spinal cord or peripheral nerve. Proteins in the molecular weight range 57,000–67,000 often appear in small but significant amounts (Runge *et al.*, 1981a; Shecket and Lasek, 1980; Black and Lasek, 1980; Delacourte *et al.*, 1980). Two particular polypeptides, having molecular weights of 62,000 and 64,000, are carried in the slow component of axonal transport and have been tentatively identified (Black and Lasek, 1980) as belonging to the group of "tau" proteins usually associated with microtubules (Weingarten *et al.*, 1975). It appears that none of these proteins is essential to the structure of the neurofilament, since their amounts relative to the triplet proteins are highly variable. Their association with neurofilaments may not be entirely fortuitous, however, and they may serve "bridging" or "capping" functions *in vivo*.

4. Structure and Attempted Reversible Disassembly

4.1. Substructure

Details of the substructure of neurofilaments are still largely obscure. Early micrographs of cross-sections of neurofilaments showed two or four densely stained dots (Wuerker, 1970), suggesting that the filaments are composed of a number of protofilaments. Electron micrographs of partially disrupted and negatively stained filaments (Schlaepfer, 1977a) show "frayed" regions in which a distinct appearance of a cable of protofilaments is seen. It appears not to be possible yet to count the number of protofilaments, although they must number fewer than 10.

A recently proposed structural model was developed by means of labeling with antibodies to each of the proteins of the triplet (Willard and Simon, 1981). According to this model, the filament has a central core composed of an unspecified number of protofilaments, each made up of the small protein (molecular weight 73,000, in their study). The large (molecular weight 195,000) protein is helically disposed on the outside of this core, in an apparently regular, but discontinuous, periodic array. The protein of intermediate size (molecular weight 145,000) is also peripherally located, and the suggestion is made that it and the large protein may alternate to form a continuous helix. This model is compatible with the known properties of neurofilaments. It is to be noted, however, that we have only one tentative estimate of the mass ratios of the three proteins of the triplet, and no estimate of the mass per unit length of the neurofilaments, to delimit the models that are proposed. Achievement of more precise models will have to await these and similar measurements.

4.2. Assembly and Disassembly

Although successful disassembly and reassembly of other intermediate filaments has been achieved, most notably in the cases of vimentin (Renner *et al.*, 1981), of desmin (Small and Sobiszek, 1977; Cooke, 1976; Huiatt *et al.*,

1980), and of keratins (Steinert *et al.*, 1976; Milstone, 1981), only partial success has thus far been achieved with neurofilaments. In experiments of Moon *et al.* (1981), assembled filaments were obtained after dialysis of a sample of neurofilament proteins from a urea–pyridine–formic-acid solvent into dilute aqueous buffer. The experiment did not show a complete disassembly–assembly cycle, however, since the "solubilized" mixture of proteins is present in an aggregated state, probably as short pieces of filament. Apparent temperature-dependent disassembly and assembly is implied by the preparation of Delacourte *et al.* (1980), but the products of disassembly are again much larger than individual polypeptide chains. It is clear that many of the structural questions raised above are potentially answerable by assembly–disassembly experiments. In particular, if the small protein of the triplet forms an appropriate filament when purified, the structural model of Willard and Simon (1981) would be greatly strengthened.

5. Enzymic Activities

The presence of enzymic activities in neurofilament preparations has been reported by a number of laboratories. It is difficult to state firmly whether or not these activities are intrinsic to the structure of the neurofilament, but for the most part their properties and the reactions catalyzed by these enzymes suggest that they are. It has been shown that protein kinase activities are present in neurofilament preparations from bovine and rat brain (Runge *et al.*, 1979b; Runge *et al.*, 1981b) and from guinea pig peripheral nerve (Shecket and Lasek, 1979). Squid axoplasm also contains a protein kinase activity (Pant *et al.*, 1978, 1979). In many ways the protein kinase activities from these different sources show similarities. They all catalyze the phosphorylation of their constituent proteins, are all cAMP-independent, and all show similar effects in the presence of various cations.

The mammalian neurofilament-associated protein kinase activities demonstrated by Runge *et al.* (1979b, 1981b) and those demonstrated by Shecket and Lasek (1979, 1982) phosphorylate each of the neurofilament triplet proteins, show no enhancement of activity in the presence of cAMP, and are not affected by the presence of the Walsh protein kinase inhibitor (Walsh *et al.*, 1971; Runge *et al.*, 1981b). It is of interest that the protein kinase activity associated with bovine brain neurofilaments is stimulated in the presence of MAP-2 [MAP-2 is a microtubule associated protein which may function as a cross-bridging element between microtubules and neurofilaments (Kim *et al.*, 1979)]. This cAMP-independent protein kinase is almost certainly distinct from a cAMP-dependent protein kinase found in cycled microtubule preparations. The microtubule-associated kinase preferentially phosphorylates the medium-sized neurofilament protein when it is combined with neurofilaments, (Shelanski *et al.*, 1981) in contrast to the uniform phosphorylation of the triplet catalyzed by the neurofilament-associated enzyme.

Shelanski *et al.* (1981) have reported that neurofilaments can be prepared in a manner that leaves them with little endogenous protein kinase activity.

This finding raises the question of whether, on the one hand, the neurofilament-associated protein kinase is associated with the filaments *in vivo*, or whether, on the other hand, the conditions of preparation somehow wash off the filament-bound enzyme. This question, which pertains to all organelle-bound enzymes, will be resolved only with difficulty.

Intermediate filaments composed of desmin and of vimentin become extensively phosphorylated *in vivo* (Gard *et al.*, 1979), and it thus seems likely that phorphorylation of intermediate filament subunit proteins (including the neurofilament triplet) is a generalized phenomenon. At this time the role of phosphorylation in assembly or function or both is not known.

Neurofilament preparations from bovine brain also contain substantial cyclic nucleotide phosphodiesterase activity (Runge *et al.*, 1979a). Treatment with 0.5 M buffer, and other solvents, does not remove the enzyme from the neurofilaments. The phosphodiesterase activity is stimulated when Ca^{2+} and calmodulin are added together. Under normal conditions, neurofilaments are prepared in the presence of the chelator EGTA and, as would be expected, little calmodulin is present in these preparations (its binding to phosphodiesterase and hence to neurofilaments depends upon the presence of Ca^{2+}). However, when neurofilaments are prepared in the presence of Ca^{2+} they retain the calmodulin activity that, because of its thermostability, can be easily isolated from the Ca^{2+}-prepared filaments. It seems likely that neurofilaments have a cGMP phosphodiesterase and that Ca^{2+} regulates it via calmodulin.

Finally, studies of neurofilament degradation have indicated the presence of a specific Ca^{2+}-activated protease which attacks the proteins of the neurofilament triplet selectivity (Schlaepfer and Micko, 1978; Schlaepfer, 1979; Pant and Gainer, 1980). This enzyme is probably not attached to the neurofilaments, since it is readily removed in an active state during purification. Based on morphological and biochemical studies of slow axonal transport, it appears that neurofilaments are assembled only once, in the region of the neuronal cell body, and are disassembled only at axon terminals. This scheme is apparently in contrast to the dynamic assembly and disassembly of microtubules. Present evidence suggests that the neurofilaments are disassembled in part by proteolysis (Schlaepfer, 1979) and it seems likely that this Ca^{2+}-activated protease is important in this process.

From the limited studies described in this section, it is clear that research on the enzymic activities that may be important in neurofilament structure and function is in its early stages. The increased present emphasis on this subject promises further progress.

6. Microtubule–Neurofilament Interactions

6.1. In Vivo Studies of Microtubule–Neurofilament Interactions

The existence of an "axoplasmic network" of microtubules and neurofilaments has long been suspected (Wuerker and Palay, 1969; Yamada *et al.*,

1971; Bertolini *et al.*, 1970). In thin sections of mammalian axons, neurofilaments often show wispy filamentous projections (Wuerker, 1970; Yamada *et al.*, 1971; Wuerker and Palay, 1969; Smith *et al.*, 1975). The projections appear to connect neurofilaments with microtubules and with other neurofilaments. They may act to maintain uniform interfilament distances, respectively 30 ±8 nm and 43 ±8 nm in different cell types (Wuerker and Palay, 1969; Weiss and Meyer, 1971). In recent stereo electron microscopic studies of thin (Rice *et al.*, 1980) and thick (Hodge and Adelman, 1980) sections of neurons of invertebrates, a distinct network has been clearly seen. Filamentous projections, with a longitudinal periodicity of about 35–45 nm, connect microtubules and neurofilaments into a continuous array.

Independent evidence for the existence of such a network is provided by studies of axonal transport. Initial studies of the slow component of anterograde axonal transport demonstrated that tubulin and the triplet of neurofilament proteins are transported together down the axon (Lasek and Hoffman, 1976; Hoffman and Lasek, 1975). Subsequent studies have demonstrated that the slow component of axonal transport can be subdivided into two subcomponents, a and b (Black and Lasek, 1980). One consists of tubulin and the neurofilament triplet, presumably associated together. A second component containing tubulin, actin, and possibly actin-binding proteins may also be transported together (Black and Lasek, 1980). [This finding is interesting in light of the studies indicating that microtubules and actin will interact *in vitro* to form highly viscous solutions (Griffith and Pollard, 1978). Microtubule–microfilament complexes may also be an important constituent of mammalian axoplasm.]

That the cotransport of microtubules and neurofilaments, and their probable association into a network, is of importance in the axon is supported by the studies of Griffin *et al.* (1978) of the effect of β,β′-iminodipropionitrile (IDPN) on axonal transport. As described above, IDPN administration produces large neurofilament-filled swellings in the most proximal portion of the axon (Chou and Hartmann, 1965; Chou and Klein, 1972). The transport of the neurofilament triplet, and to some extent that of tubulin, is impaired by IDPN administration, although synthesis of these proteins is apparently not inhibited. Since the neurofilament proteins are normally transported down the axon, impairment of this process by IDPN results in the neurofilament accumulations or swellings observed. These consist of morphologically normal neurofilaments that are circularly oriented, and normal, longitudinally oriented microtubules that extend through the swellings. One possible explanation for this finding is that IDPN may produce a postsynthetic alteration in components of the axonal microtubule–neurofilament network, rendering it incapable of normal transport. This is supported by observations of early events following IDPN injection that demonstrate that within 6 h of injection there is a striking change in the axonal cytoskeleton, usually within the initial segment. The normal structure, consisting of interspersed microtubules and neurofilaments, changes to one in which there is a paracentral channel containing the vast majority of microtubules, surrounded by a ring of randomly oriented neurofilaments extending to the axolemma. This alteration is recognizable even at the light microscopic level because of the tendency of mitochondria to accumulate within the microtubule channel. The neurofilaments present in these axons are of normal diameter and exhibit neurofilament–neurofilament projections, suggesting that only the neurofilament–microtubule projections had been lost. Strikingly, the orientation of the neurofilaments at early times in IDPN intoxication is random, both with respect to each other and with respect to the long axis of the axon. This phenomenon is reversible at low IDPN levels, but at higher levels of IDPN the characteristic neurofilament-filled axonal swellings develop within 24 h. It has been found that in large unmyelinated axons (such as some of those found in the vagus nerve) that contain a much higher ratio of microtubules to neurofilaments than do those that are myelinated, the effect of IDPN is minimal, suggesting that its effect is primarily upon neurofilaments (Yokoyama *et al.*, 1980).

In addition to providing an *in vivo* model for studying microtubule–neurofilament interactions, these experiments are of possible clinical importance. Accumulations of neurofilaments are the pathological hallmark of a number of disorders of the central and peripheral nervous systems. Alteration of normal neurofilament structure has been seen in Alzheimer's disease (Kidd, 1964), Pick's disease (Rewcastle and Ball, 1968), and in sporadic motor neuron disease (Schochet *et al.*, 1969). Neurofilamentous swellings in the proximal portions of axons almost identical to those seen in IDPN intoxication have been seen in patients with amyotrophic lateral sclerosis (Wohlfart, 1959; Carpenter, 1968; Chou *et al.*, 1970) and in Brittany spaniels with hereditary canine spinal muscular atrophy (Cork *et al.*, 1979). It has been suggested that the primary axonal defect in the latter two situations is a disintegration of the normal microtubule–neurofilament lattice present in the axon.

6.2. In Vitro Studies of Microtubule–Neurofilament Interactions

The results of three studies provide tentative evidence for interaction of microtubules and neurofilaments *in vitro*. It was noted by Berkowitz *et al.* (1977) that neurofilaments are tenacious contaminants of microtubule preparations. Based on the sedimentation properties of intact neurofilaments and intact microtubules, it is clear that without some sort of interaction between the two structures the extensive copurification and cosedimentation that occurs during purification of microtubule protein would not occur (although this result does not rule out nonspecific interactions). It has also been demonstrated that neurofilaments prepared from a number of different sources by different techniques contain variable amounts of copurifying tubulin (Runge *et al.*, 1979a; Schlaepfer, 1978; Shecket and Lasek, 1981). In fact it has been shown that if neurofilaments isolated from intradural nerve roots (which contain very little tubulin) are present during one cycle of microtubule assembly/disassembly, they form a complex with tubulin of fairly consistent stoichiometry (Shelanski *et al.*, 1981). This association is accompanied by a binding of microtubule associated proteins to the filament–tubulin complex (Shelanski *et al.*, 1981). The association of tubulin and neurofilaments under these conditions is stable and not easily dissociated, and the ratio of tubulin to neurofilament protein is similar to that observed for neurofilaments isolated from cycled microtubule preparations (Runge *et al.*, 1979a).

Evidence supporting *in vitro* interactions between microtubules and neurofilaments is the result of application of physical techniques to the study of this system. A recent study (Runge *et al.*, 1981c) has demonstrated that a highly viscous complex forms when microtubules and neurofilaments are incubated together in the presence of ATP, but not in its absence. Falling ball viscometry revealed that the (non-Newtonian) apparent viscosity of such an ATP-containing mixture of purified tubulin and neurofilaments is at least several 100-fold greater than that of the mixtures prepared without ATP. The magnitude of the increase in viscosity depends on the concentration of both neurofilaments and tubulin. The viscosity was reduced by stirring the mixture or by cooling to 0°C, conditions that would disrupt a fragile association or disassemble microtubules. Sedimentation velocity experiments conducted at 35°C revealed the presence of a fraction of very rapidly sedimenting material (greater than 1000 S) in the ATP-containing solution but not in those without ATP. Electron microscopy of ATP-containing solutions of microtubules and neurofilaments show areas of crossover between microtubules and neurofilaments while the appearance of the solutions without ATP was dramatically different. This study concludes that a microtubule– neurofilament complex forms *in vitro* in the presence of ATP.

In light of these findings it is relevant to note that neurofilaments purified by the method of Runge *et al.* (1981a) contain a protein kinase activity. It is possible that the ATP-mediated interaction between purified tubulin (containing no protein kinase activity) and neurofilaments may be mediated by this kinase.

ACKNOWLEDGMENTS.   We thank Drs. R. K. H. Liem, G. Shecket, and M. L. Shelanski for providing us with manuscripts in advance of publication.

References

Bennett, G. S., Tapscott, S. J., Kleinbart, F. A., Antin, P. B., and Holtzer, H., 1981, Different proteins associated with 10-nanometer filaments in cultured chick neurons and nonneuronal cells, *Science* **212:**567.

Berkowitz, S. A., Katagiri, J., Binder, H. K., and Williams, R. C., Jr., 1977, Separation and characterization of microtubule proteins from calf brain. *Biochemistry* **16:**5610.

Bertolini, B., Monaco, G., and Rossi, A., 1970, Ultrastructure of a regular arrangement of microtubules and neurofilaments, *J. Ultrastruct. Res.* **33:**173.

Black, M., and Lasek, R. J., 1980, Slow component of axonal transport: Two cytoskeletal networks, *J. Cell Biol.* **86:**616.

Blose, S. H., Shelanski, M. L., and Chacko, S., 1977, Localization of bovine brain filament antibody on intermediate filaments in guinea pig vascular endothelial cells and chick cardiac muscle cells. *Proc. Natl. Acad. Sci. USA* **74**:662.

Carpenter, S., 1968, Proximal axonal enlargement in motor neuron disease, *Neurology* **18**:841.

Chiu, F.-C., Korey, B., and Norton, W. T., 1980, Intermediate filaments from bovine, rat and human CNS: Mapping analysis of the major proteins, *J. Neurochem.* **34**:1149.

Chou, S. M., and Hartman, H. A., 1964, Axonal lesions and waltzing syndrome after IDPN administration in rats, *Acta Neuropathol.* **3**:428.

Chou, S. M., and Hartmann, H. A., 1965, Electron microscopy of focal neuroaxonal lesions produced by β,β'-iminodipropionitrile (IDPN) in rats, *Acta Neuropathol.* **4**:590.

Chou, S. M., and Klein, R. A., 1972, Autoradiographic studies of protein turnover in motoneurons of IDPN-treated rats, *Acta Neuropathol.* **22**:183.

Chou, S. M., Marton, J., Gutrecht, J. A., and Thompson, H. G., 1970, Axonal balloons in subacute motor neuron disease, *J. Neuropath. Exp. Neurol.* **29**:141.

Cooke, P., 1976, A filamentous cytoskeleton in vertebrate smooth muscle fibers, *J. Cell Biol.* **68**:539.

Cork, L. C., Griffin, J. W., Munnell, J. S., Lorenz, M. D., Adams, R. J., and Price, D. L., 1979, Hereditary canine spinal muscle atrophy, *J. Neuropath. Exp. Neurol.* **38**:209.

Czosnek, H., Soifer, D., and Wisniewski, H. M., 1980, Heterogeneity of intermediate filament proteins from rabbit spinal cord, *Neurochem. Res.* **5**:777.

Czosnek, H., Soifer, D., Mack, K., and Wisniewski, H. M., 1981, Similarity of neurofilament proteins from different parts of the rabbit nervous system, *Brain Res.* **216**:387.

Davison, P. F., 1981, Intermediate filaments: Intracellular diversities and interspecies homologies, in: *International Cell Biology, 1980–81* (H. G. Schweiger, ed.), pp. 286–292, Springer Verlag, Berlin.

Davison, P. F., and Hong, B. S., 1977, Structural homologies in mammalian neurofilament proteins, *Brain Res.* **134**:287.

Davison, P. F., and Jones, R. N., 1980, Neurofilament proteins of mammals compared by peptide mapping, *Brain Res.* **182**:470.

Davison, P., and Winslow, B., 1974, The protein subunit of calf brain neurofilament, *J. Neurobiol.* **5**:119.

Delacourte, A., Filliatreau, G., Boutteau, F., Biserte, G., and Schrevel, J., 1980, Study of the 10-nm-filament fraction isolated during the standard microtubule preparation, *Biochem. J.* **191**:543.

Gard, D. L., Bell, P. B., and Lazarides, E., 1979, Coexistence of desmin and the fibroblastic intermediate subunit in muscle and nonmuscle cells, *Proc. Natl. Acad. Sci. USA* **76**:3894.

Goldman, J. E., Schaumburg, H. H., and Norton, W. T., 1978, Isolation and characterization of glial filaments from human brain, *J. Cell Biol.* **78**:426.

Goldman, R. D., Milsted, A., Schloss, J. A., Starger, J., and Yerna, M.-J., 1979, Cytoplasmic fibers in mammalian cells: cytoskeletal and contractile elements, *Ann. Rev. Physiol.* **41**:703.

Griffin, J. W., Hoffman, P. N., Clark, A. W., Carroll, P. T., and Price, D. L., 1978, Slow axonal transport of neurofilament proteins: Impairment by β,β'-iminodipropionitrile administration, *Science* **202**:633.

Griffith, L. M., and Pollard, T. D., 1978, Evidence for actin filament-microtubule interaction mediated by microtubule-associated proteins, *J. Cell Biol.* **78**:958.

Hodge, A. J., and Adelman, W. J., Jr., 1980, The neuroplasmic network in *Loligo* and *Hermissenda* neurons, *J. Ultrastruct. Res.* **70**:220.

Hoffman, P. N., and Lasek, R. J., 1975, The slow component of axonal transport. Identification of major structural polypeptides of the axon and their generality among mammalian neurons, *J. Cell Biol.* **66**:351.

Huiatt, T. W., Robson, R. M., Arakawa, N., and Stromer, M. H., 1980, Desmin from avian smooth muscle, *J. Biol. Chem.* **255**:6981.

Iqbal, L., Grundke-Iqbal, I., Wisniewski, H. M., and Terry, R. D., 1977, On neurofilament and neurotubule proteins from human autopsy tissue, *J. Neurochem.* **29**:417.

Iqbal, K., Grundke-Iqbal, I., Wisniewski, H. M., and Terry, R. D., 1978, Chemical relationship of the paired helical filaments of Alzheimer's dementia to normal human neurofilaments and neurotubules, *Brain Res.* **142**:321.

Kidd, M., 1964, Alzheimer's disease: An electron microscopical study. *Brain* **87**:307.

Kim, H., Binder, L. I., and Rosenbaum, J. L., 1979, The periodic association of MAP-2 with brain microtubules *in vitro, J. Cell Biol.* **80**:266.

Lasek, R. J., and Hoffman, P. N., 1976, The neuronal cytoskeleton, axonal transport and axonal growth, in: *Cell Motility* (R. Goldman, T. Pollard, and J. Rosenbaum, eds.), p. 1021. Cold Spring Harbor Press, Cold Spring Harbor, N.Y.

Lazarides, E., 1980, Intermediate filaments as mechanical integrators of cellular space, *Nature (London)* **283**:249.

Liem, R. K. H., and Turey, M., 1980, Purification of soluble neurofilament triplet polypeptides from brain (abstract), *J. Cell Biol.* **87**:180a.

Liem, R. K. H., Yen, S.-H., Loria, C. J., and Shelanski, M. L., 1977, Immunological and biochemical comparison of tubulin and intermediate brain filament protein, *Brain Res.* **132**:167.

Liem, R. K. H., Yen, S.-H., Salomon, G. D., and Shelanski, M. L., 1978, Intermediate filaments in nervous tissues, *J. Cell Biol.* **79**:637.

Micko, S., and Schlaepfer, W. W., 1978, Protein composition of axons and myelin from rat and human peripheral nerves, *J. Neurochem.* **30**:1041.

Milstone, L. M., 1981, Isolation and characterization of two polypeptides that form intermediate filaments in bovine esophageal epithelium, *J. Cell Biol.* **88**:317.

Moon, H. M., Wisniewski, T., Merz, P., DeMartini, J., and Wisniewski, H. M., 1981, Partial purification of neurofilament subunits from bovine brains and studies on neurofilament assembly, *J. Cell Biol.* **89**:560.

Mori, H., Komiya, Y., and Kurokawa, M., 1979, Slowly migrating axonal polypeptides. Inequalities in their rate and amount of transport between two branches of bifurcating axons, *J. Cell Biol.* **82**:174.

Pant, H. C., and Gainer, H., 1980, Properties of a calcium-activated protease in squid axoplasm which selectively degrades neurofilament proteins, *J. Neurobiol.* **11**:1.

Pant, H. C., Shecket, G., Gainer, H., and Lasek, R. J., 1978, Neurofilament protein is phosphorylated in the squid giant axon, *J. Cell Biol.* **78**:R23.

Pant, H. C., Yoshioka, T., Tasaki, I., and Gainer, H., 1979, Divalent cation dependent phosphorylation of proteins in squid giant axon, *Brain Res.* **162**:303.

Renner, W., Franke, W. W., Schmid, E., Geisler, N., Weber, K., and Mandelkow, E., 1981, Reconstitution of intermediate-sized filaments from denatured monomeric vimentin, *J. Mol. Biol.* **149**:285.

Rewcastle, N. B., and Ball, M. J., 1968, Electron microscopic structure of the inclusion bodies in Pick's disease, *Neurology* **18**:1205.

Rice, R. V., Roslansky, P. F., Pascoe, N., and Houghton, S. M., 1980, Bridges between microtubules and neurofilaments visualized by stereoelectron microscopy, *J. Ultrastruct. Res.* **71**:303.

Runge, M. S., Hewgley, P. B., Puett, D., and Williams, R. C., Jr., 1979a, Cyclic nucleotide phosphodiesterase activity in 10-nm filaments and microtubule preparations from bovine brain, *Proc. Natl. Acad. Sci. USA* **76**:2561.

Runge, M. S., El-Maghrabi, M. R., Claus, T. H., Pilkis, S. J., and Williams, R. C., Jr., 1979b, Phosphorylation of MAP-2 by a neurofilament-associated protein kinase, *J. Cell Biol.* **83**:352a.

Runge, M. S., Schlaepfer, W. W., and Williams, R. C., Jr., 1981a, Isolation and characterization of neurofilaments from mammalian brain, *Biochemistry* **20**:170.

Runge, M. S., El-Maghrabi, M. R., Claus, T. H., Pilkis, S. J., and Williams, R. C., Jr., 1981b, A MAP-2-stimulated protein kinase activity associated with neurofilaments, *Biochemistry* **20**:175.

Runge, M. S., Laue, T. M., Yphantis, D. A., Lifsics, M. R., Saito, A., Altin, M., Reinke, K., and Williams, R. C., Jr., 1981c, ATP-indiced formation of an associated complex between microtubules and neurofilaments, *Proc. Natl. Acad. Sci. USA* **78**:1431.

Schlaepfer, W. W., 1971, Stabilization of neurofilaments by vincristine sulfate in low ionic strength media, *J. Ultrastruct. Res.* **36**:367.

Schlaepfer, W. W., 1977a, Studies on the isolation and substructure ᴜᴏ mammalian neurofilaments, *J. Ultrastruct. Res.* **61**:149.

Schlaepfer, W. W., 1977b, Immunological and ultrastructural studies of neurofilaments isolated from rat peripheral nerve. *J. Cell Biol.* **74**:226.

Schlaepfer, W. W., 1978, Observation on the disassembly of isolated mammalian neurofilaments, *J. Cell Biol.* **76**:50.

Schlaepfer, W. W., 1979, Nature of mammalian neurofilaments and their breakdown by calcium. *Prog. Neuropathol.* **4**:101.

Schlaepfer, W. W., and Freeman, L. A., 1978, Neurofilament protein of rat peripheral nerve and spinal cord, *J. Cell Biol.* **78**:653.

Schlaepfer, W. W., and Lynch, R. G., 1977, Immunofluorescence studies of neurofilaments in the rat and human peripheral and central nervous system, *J. Cell Biol.* **74**:241.

Schlaepfer, W. W., and Micko, S., 1978, Chemical and structural changes of neurofilaments in transected rat sciatic nerves, *J. Cell Biol.* **78**:369.

Schlaepfer, W. W., and Micko, S., 1979, Calcium-dependent alterations of neurofilament proteins of rat peripheral nerve, *J. Neurochem.* **32**:211.

Shecket, G., and Lasek, R. J., 1979, Phosphorylation of neurofilament protein. *J. Cell Biol.* **83**:143a.

Shecket, G., and Lasek, R. J., 1980, Preparation of neurofilament protein from guinea pig peripheral nerve and spinal cord, *J. Neurochem.* **35**:1335.

Shecket, G., and Lasek, R. J., 1981, Neurofilament protein phosphorylation: species generality and reaction characteristics, *J. Biol. Chem.* **257**:4788.

Shelanski, M. L., Albert, S., De Vries, G. H., and Norton, W. T., 1971, Isolation of filaments from brain, *Science* **174**:1242.

Shelanski, M. L., Gaskin, F., and Cantor, C. R., 1973, Microtubule assembly in the absence of added nucleotides, *Proc. Natl. Acad. Sci. USA* **70**:765.

Shelanski, M. L., Leterrier, J.-F., and Liem, R. K. H., 1981, Evidence for interactions between neurofilaments and microtubules, *Neurosci. Res. Program Bull.* **19**:32.

Schocket, S. S., Hardman, J. N., Ladewig, T. P., and Earle, K. M., 1969, Intraneural conglomerates in sporadic motor neuron disease, *Arch. Neurol.* **20**:548.

Small, J. V., and Sobieszek, A., 1977, Studies on the function and composition of the 10-nm (100Å) filaments of vertebrate smooth muscle, *J. Cell Sci.* **23**:243.

Smith, D. S., Jalfors, V., and Cameron, B. F., 1975, Morphological evidence for participation of microtubules in axonal transport, *Ann. N.Y. Acad. Sci.* **253**:472.

Steinert, P., Idler, W. W., and Zimmerman, S. B., 1976, Self-assembly of bovine epidermal keratin filaments *in vitro*, *J. Mol. Biol.* **108**:547.

Walsh, D. A., Ashby, C. D., Gonzalez, C., Calkins, D., Fisher, E. H., and Krebs, E. G., 1971, Purification and characterization of a protein inhibitor of adenosine 3′-5′-monophosphate-dependent protein kinases, *J. Biol. Chem.* **246**:1977.

Weingarten, M. D., Lockwood, A. H., Hwo, S.-Y., and Kirschner, M. W., 1975, A protein factor essential for microtubule assembly, *Proc. Natl. Acad. Sci. USA* **72**:1858.

Weiss, P. A., and Mayr, R., 1971, Organelles in neuroplasmic ("axonal") flow: neurofilaments, *Proc. Natl. Acad. Sci. USA* **68**:846.

Willard, M. B., 1976, A genetically-determined protein polymorphism in the rabbit nervous system, *Proc. Natl. Acad. Sci. USA* **73**:3641.

Willard, M. B., and Hulebak, K. L., 1977, The intra-axonal transport of polypeptide H: Evidence for a fifth (very slow) group of transported proteins in the retinal ganglion cells of the rabbit, *Brain Res.* **136**:289.

Willard, M., and Simon, C., 1981, Antibody decoration of neurofilaments, *J. Cell Biol.* **89**:198.

Wohlfart, G., 1959, Degenerative and regenerative axonal changes in the ventral horns, brainstem and cerebral cortex in amyotrophic lateral sclerosis. *Acta Univ. Lundiensis (New Series)* **2**:3.

Wuerker, R. B., 1970, Neurofilaments and glial filaments, *Tissue and Cell* **2**:1.

Wuerker, R., and Palay, S., 1969, Neurofilaments and microtubules in anterior horn cells of the rat, *Tissue and Cell* **1**:387.

Yamada, K. M., Spooner, B. S., and Wessells, N. K., 1971, Ultrastructure and function of growth cones and axons of cultured nerve cells, *J. Cell Biol.* **49**:614.

Yen, S., and Fields, K. L., 1981, Antibodies to neurofilament, glial filament, and fibroblast intermediate filament proteins bind to different cell types of the nervous system, *J. Cell Biol.* **88**:115.

Yen, S., Dahl, D., Schachner, M., and Shelanski, M., 1976, Biochemistry of the filaments of brain, *Proc. Natl. Acad. Sci. USA* **73**:529.

Yokoyama, K., Tsukita, S., Ishikawa, H., and Kurokawa, M., 1980, Early changes in the neuronal cytoskeleton caused by IDPN: Selective impairment of neurofilament polypeptides, *Biomed. Res.* **1**:537.

3

Organization of Contractile Fibers in Smooth Muscle

Peter Cooke

1. Introduction

The smooth muscle of vertebrates is chiefly a mural tissue of the large and small hollow organ systems. The "smoothness" of this musculature presumably describes the impression of a uniformity in the optical properties of the cytoplasm that contrasts with the regular optical anisotropy that is characteristic of the striated muscles. The quality of optical uniformity is obtained from a labyrinth of three distinct systems of cytoplasmic filaments that comprise the bulk of the contractile apparatus, and a major objective in contemporary morphological analysis of smooth muscle is to elucidate the molecular organization of the constituent filaments and define the symmetry of the functional elements. Structural studies have been undertaken primarily with the view of establishing correlations between the composition and organization of the filaments in smooth muscle fibers and the structural composition of the sarcomere in striated muscle (Shoenberg and Needham, 1976; Chamley-Campbell *et al.*, 1979; Small and Sobieszek, 1980; Somlyo, 1980). The basis for this approach is provided by some quantitative and especially qualitative similarities in the traditionally defined functional characteristics shared by smooth and striated muscle, and by the remarkable correlations between structural changes and functional variation that are formulated or otherwise implicit in the sliding filament concept and the cross-bridge model of contraction for the striated muscles.

Peter Cooke • Department of Physiology, University of Connecticut Health Center, Farmington, Connecticut 06032.

2. General Morphology of Smooth Muscle Fibers

The cellular form of smooth muscle fibers was determined in classical observations on macerated and manually teased tissues. The fibers are fusiform cells that are sometimes branched along the apices or helicoidal, especially in the very small hollow organs (Rhodin, 1967). The nucleus is centrally located and ellipsoidal, and the juxtanuclear regions are conical and predominantly contain membranous organelles (Fig. 1). Fibers range from 1–20 μm in diameter around the nuclear region and gradually taper along the fiber length, which varies from 25 to several 100 μm (Burnstock, 1970). The axial ratio of relaxed fibers calculated from fibers in various sources of tissue ranges from 25–100. These values indicate that the degree of asymmetry in cell form approaches the axial ratio that is expressed by many skeletal muscle fibers. Nearly all of the cytoplasm in the fibers contains a multitude of filaments that are axially aligned and very densely packed, but there is no evidence of axial or transverse registration within the three populations of filaments (Fig. 2).

The surface of the fibers is a mosaic of two axially elongated regions of structural specialization: nearly half of the surface is elaborate in the form of patches composed of regular small involutions or vesicles of the plasma membrane, termed "micropinocytotic vesicles" or "caveolae intracellulares." They are often apposed to flattened cisternae of the endoplasmic reticulum (Devine *et al.,* 1972; Gabella, 1973). This relationship gives these structures the outline of a rudimentary system of transverse tubules. The alternate areas of the mosaic are discrete stretches of unit membrane subtended by amorphous patches of electron-dense material described as attachment sites for the myofilaments (Pease and Molinari, 1960). These regions are continuous with bundles of myofilaments observed in longitudinal sections (Ashton *et al.,* 1975) and profiles of thin filaments and intermediate filaments are embedded within the amorphous density. The proteins α-actinin (Schollmeyer *et al.,* 1976) and vinculin (Geiger, 1979) are localized by immunochemical methods in these regions, but the mechanism of integration between the filaments and the membrane is not known. There are no obvious intramembraneous particles localized in fracture planes of the membrane that correspond to the areas occupied by the amorphous density (Fig. 3).

Comparison of the two areas composing the cell surface in isotonically contracted and relaxed isolated muscle fibers indicates that these regions are differentially displaced during shortening. The vesicular regions bulge out from the cell surface, whereas the regions with subtending density are displaced inward, suggesting that the shortening force is differentially borne by these areas that are randomly arranged along the entire length of the muscle fiber (Fay and Delise, 1973; Gabella, 1976). In addition to a likely role in transmitting the force of contraction to the cell surface, the areas of the membrane subtended by this density can also be correlated with regions of intercellular contact that are mediated by focal condensations of the inter-

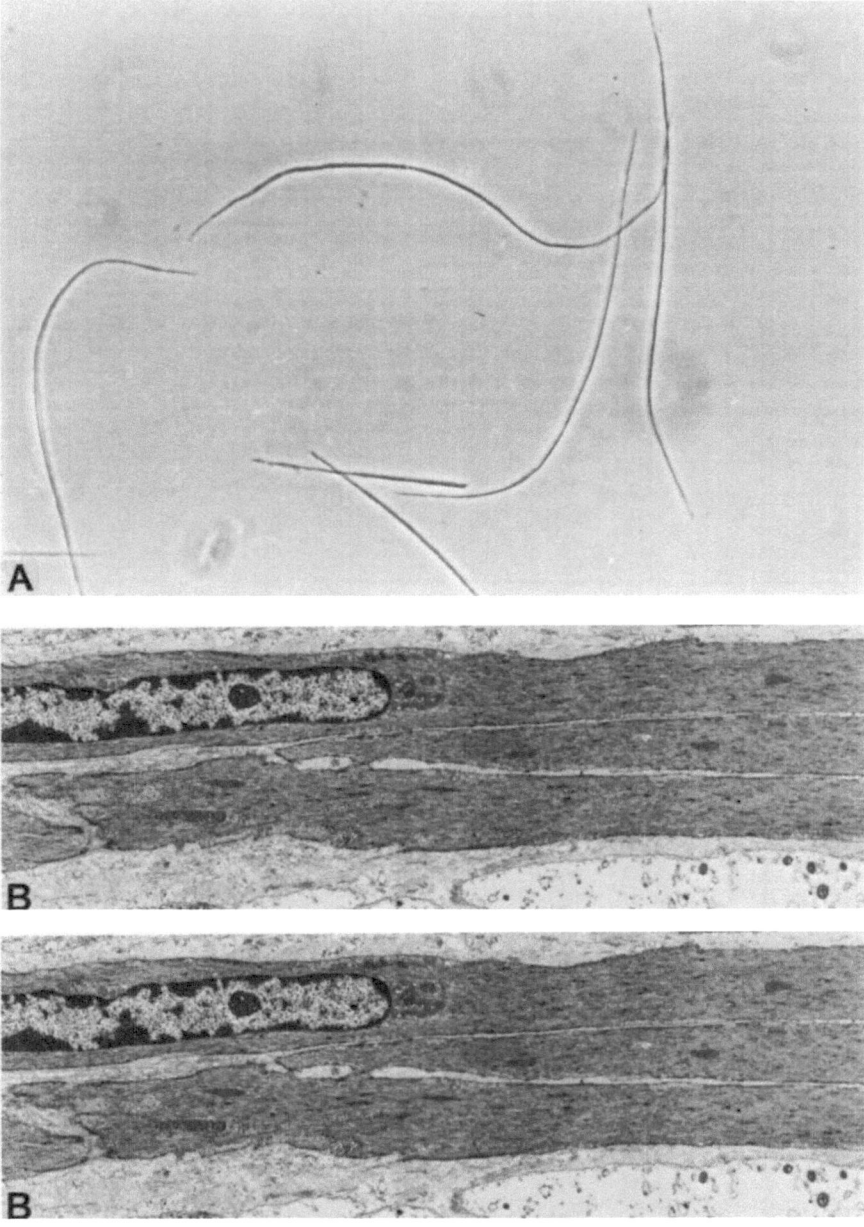

Figure 1. (A) Phase-contrasted optical micrograph of isolated smooth muscle fibers obtained from the taenia coli of a guinea pig illustrating the highly asymmetric cell shape and optical uniformity of the cytoplasm. ×300. (B) Longitudinal section of taenia coli smooth muscle fibers that contain an ellipsoidal nucleus and a cytoplasm packed with axially aligned filaments. ×2500. (C) Cross section of smooth muscle fibers illustrating the range diameters that reflect the gradual taper along the length of the cells and the uniform and marked density distribution of the cytoplasmic filaments. ×3000.

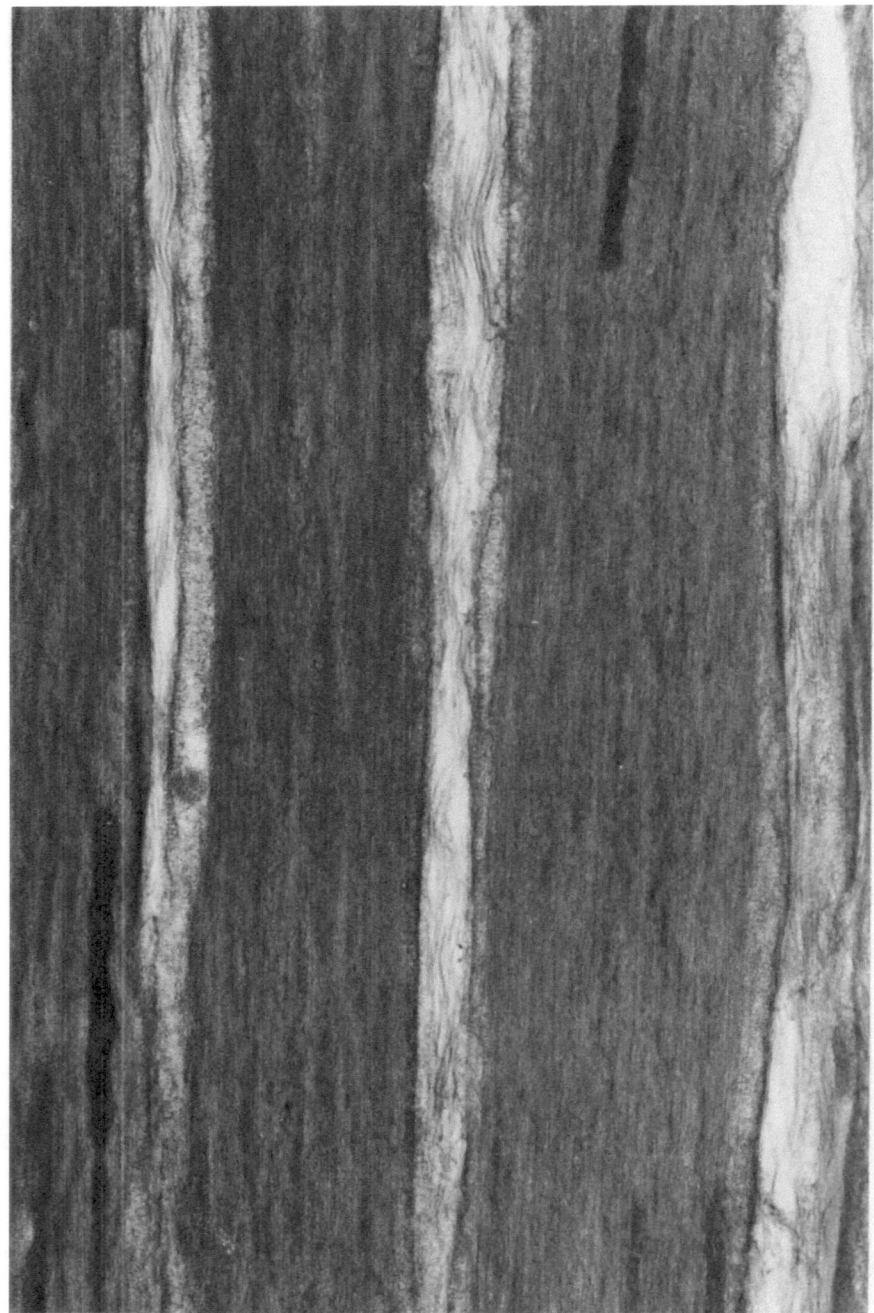

Figure 2. Smooth muscle fibers contain a uniform and compact complement of myofilaments that lack clear indications of registration in a 0.5-μm-thick section recorded with a 1 MEV electron microscope. ×10,200.

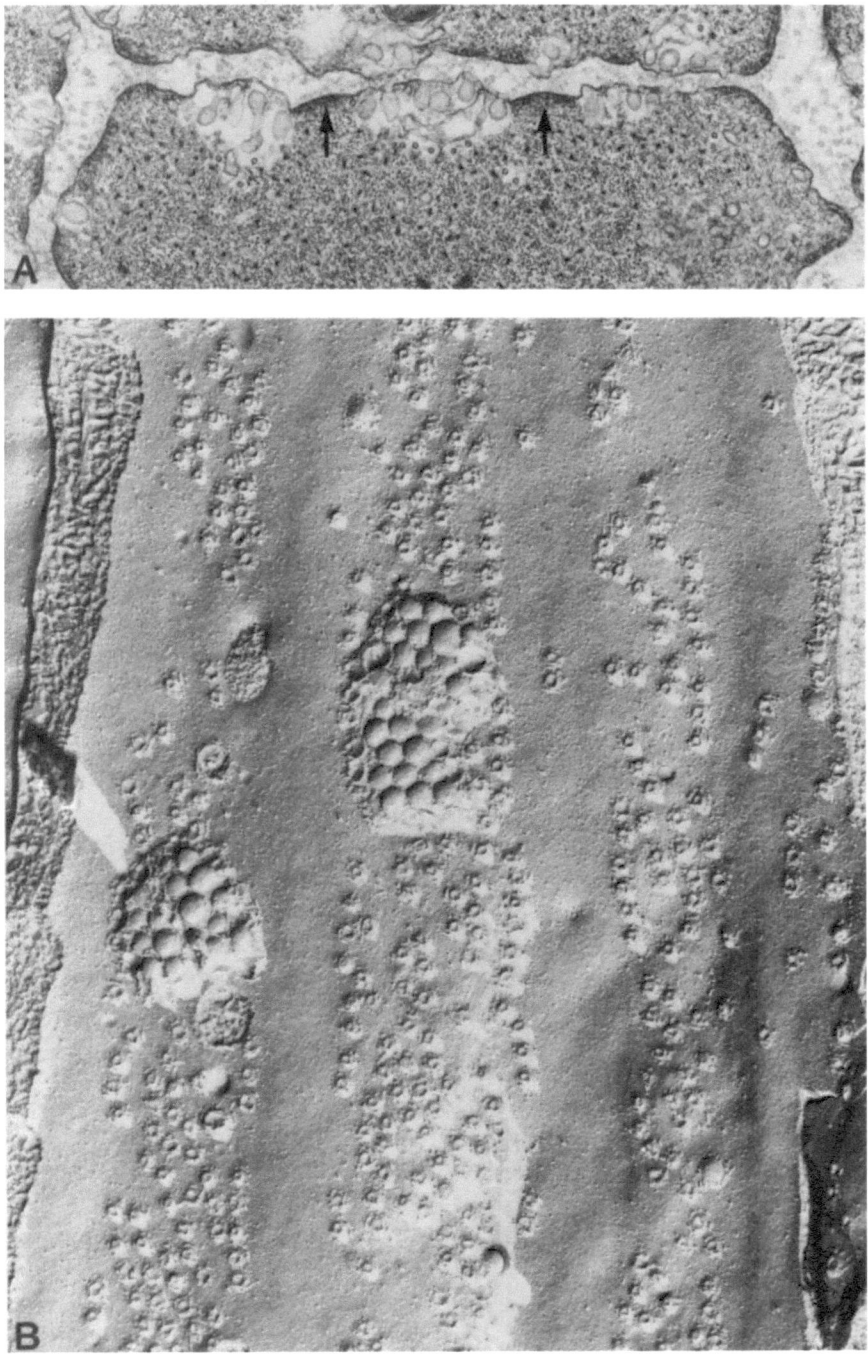

Figure 3. (A) Cross section of a fiber showing the two areas of the cell surface comprising a mosaic: areas with regular invaginations of surface vesicles, and discrete stretches of membrane subtended by amorphous dense regions that represent sites of filament attachment (arrows). ×23,000. (B) Corresponding structures of the cell surface in a fracture plane of the membrane that contains regions of surface vesicles, and areas of the membrane with filament attachments. ×25,000.

cellular matrix. These regions (that resemble hemidesmosomes) are often coordinated with corresponding structures in adjacent fibers and with condensations of the intercellular matrix (Fig. 3A). Morphometric correlation of the detailed and varied intercellular junctions in comparative studies to analyse the relationship of these connections to length changes in the fibers is complicated by the possibility of nonuniformity of strain in the tissue (i.e., mechanical forces may not be uniformly transmitted through the fibers) (Gabella, 1976). Large-scale shifts occur in the relative positions of muscle fibers over the working range of some hollow organs, and it is likely that smooth muscles contain a pattern of microscopic mechanical connections that is best modeled as an admixture of series and parallel arrangements. This greatly complicates the results of comparative structural studies on cells in tissues.

3. The System of Intermediate Filaments

The demonstration of filamentous components of the contractile apparatus has followed a trend in which the determinants are based upon the relative abundance of distinct forms and the intrinsic stability of the components in enduring the methods of preparation. The least abundant and most recently recognized class of filaments in smooth muscle is the cytoskeletal or intermediate filament (Ishikawa *et al.*, 1968), so described because the regular cross-sectional diameter of 10 nm is between the diameters usually found for the well-defined thin actin-containing filaments (5–8 nm diameter) and thick myosin-containing filaments (15–20 nm diameter).

A distinctive structural feature of native intermediate filaments in smooth muscle is the very regular diameter and great (indeterminant) length (Rice *et al.*, 1970; Cooke and Chase, 1971; Uehara *et al.*, 1971). The center of these long cylindrical filaments is often described as electron lucent, but the filaments contain no regular fine structure when examined under conditions that resolve subunit detail in other types of cytoplasmic filaments (Figure 4). The intermediate filaments are further characterized generally by a consistent structural association with the amorphous dense bodies both at the plasma membrane and scattered throughout the cytoplasm. These structures are homologous with Z-bands and intercalated discs that collimate thin filaments in skeletal and cardiac muscles (Rosenbluth, 1965; Popescu and Ionescu, 1970).

Another distinction between the intermediate filaments and the actin- and myosin-containing filaments is based upon differences in solubility in aqueous salt solutions. Intermediate filaments are insoluble in salt solutions of high ionic strength (Cooke and Chase, 1971), and this property provides a general method for the isolation of a cell fraction that is enriched with intermediate filaments in bulk quantities. The major protein component in these cell-fractions, named *desmin* by Lazarides and Hubbard (1976) and *skeletin* by Small and Sobieszek (1977), has a molecular weight around 55,000 (Cooke,

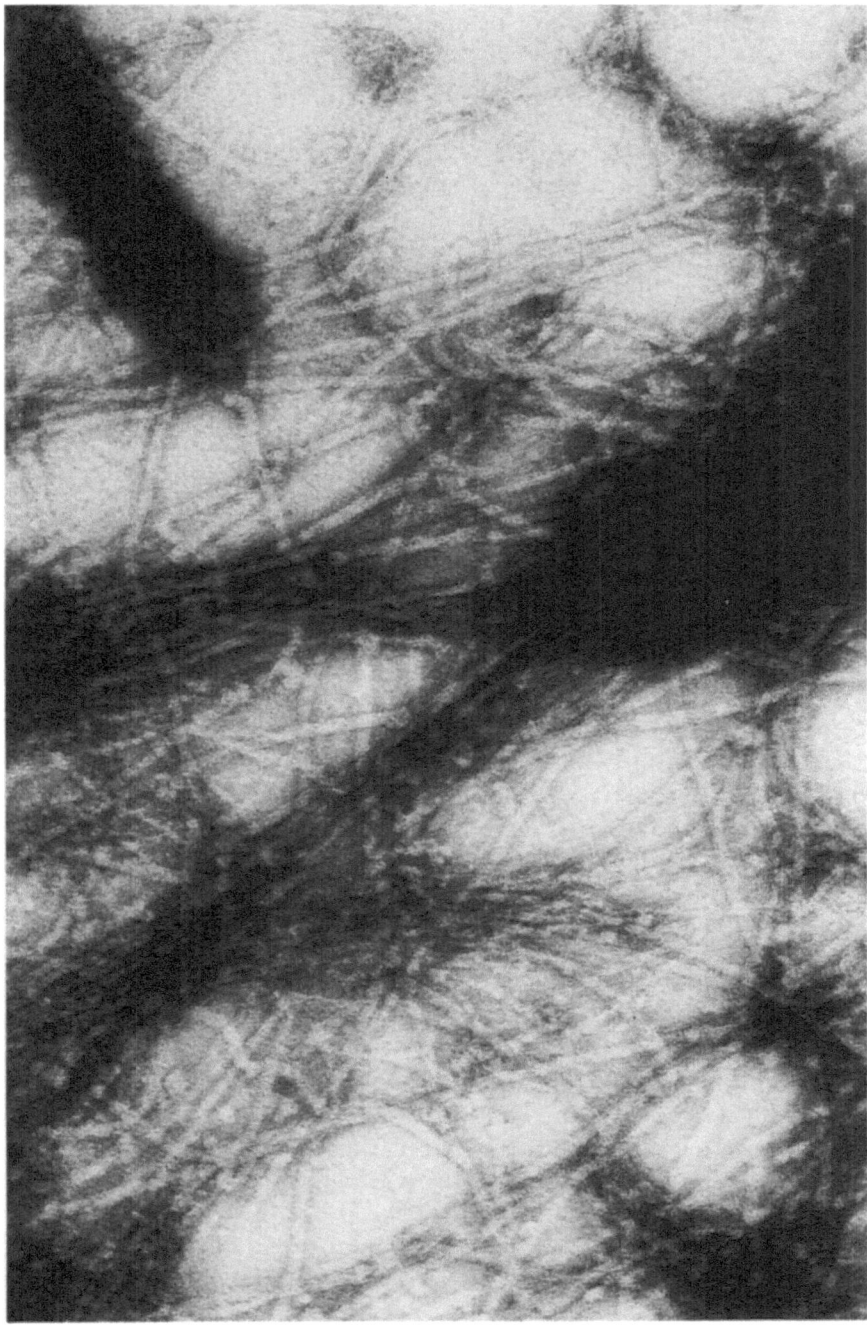

Figure 4. Isolated, negatively stained intermediate filaments from avian gizzard smooth muscle. The diameter is regularly 100 Å and the filaments are highly convoluted and of indeterminant length. ×116,000.

1976). This protein is qualitatively distinct from the major muscle proteins and the microtubular protein, tubulin (Cooke, 1976; Huiatt *et al.*, 1980; Geisler *et al.*, 1980). A similar, but antigenically distinct protein, *vimentin*, is a major component of intermediate filaments in some single-unit smooth muscles (Gabbiani *et al.*, 1981).

Systematic morphological sampling of homogenates and sectioned preparations of selected smooth muscles suggests that intermediate filaments are integrated into a network (Fig. 5) that extends throughout the muscle fibers (Cooke and Chase, 1971; Cooke and Fay, 1972; Cooke, 1976; Small and Sobieszek, 1977). The conformation of the cytoplasmic network is dependent on the length of the muscle fibers over a range of fiber lengths that correlates with the force–length diagram (see Cooke and Fay, 1972), and the density distribution of intermediate filaments increases during chronic conditions that stimulate hypertrophy and tissue remodeling (Gabella, 1979; Berner *et al.*, 1981). There is, however, no substantial evidence to indicate that intermediate filaments play only a passive, cytoskeletal role serving to selectively damp the distribution of mechanical strain in the cytoplasm, although cellular muscles in a tissue would clearly require some structural mechanism for storing or transmitting localized or graded contractures.

The dense bodies are localized at the nodes in the network of intermediate filaments. These structures are considered as being homologous to the Z-discs in striated muscles, and a consistent relationship between dense bodies and thin filaments is recorded in a variety of early studies. Isolated examples of systematic probes to determine the relationship of the dense bodies to intermediate and thin filaments are disparate. In taenia coli smooth muscle fibers, there is no apparent continuity between bundles of thin filaments and the dense bodies (Cooke, 1976), but in vascular smooth muscles, some filament profiles in the dense bodies are represented as thin filaments (Ashton *et al.*, 1975). Isolated dense bodies from avian gizzard smooth muscle display a variety of forms (Fig. 6): (1) discrete cores lacking cytoplasmic filaments, (2) cores with thin filaments (Panner and Honig, 1967), (3) cores with intermediate filaments (Cooke and Chase, 1971), and (4) cores with intermediate and thin filaments (Nonomura and Ebashi, 1975). These various forms could represent derived structures from a single type of dense body that is complexed with both intermediate and thin filaments, or they could represent functionally distinct types in a mixed population of dense bodies that separately regulate their association with the cytoplasmic filaments in some spatial or temporal pattern. Some measure of differential expression would seem to be necessary in order to explain the diversity in the form of these structures, although a broad view of the ultrastructure of the dense bodies would suggest that the variety of forms are related. The apparent pleomorphism of the fusiform dense bodies is further manifest in developing smooth muscle fibers of embryonic chicken gizzard. Smooth muscle myoblasts contain precursors of the dense bodies scattered along the developing myofibrils. These structures are amorphous, electron-dense patches that compare with the precursor Z-band material in developing cardiac and skeletal muscles. Structures of

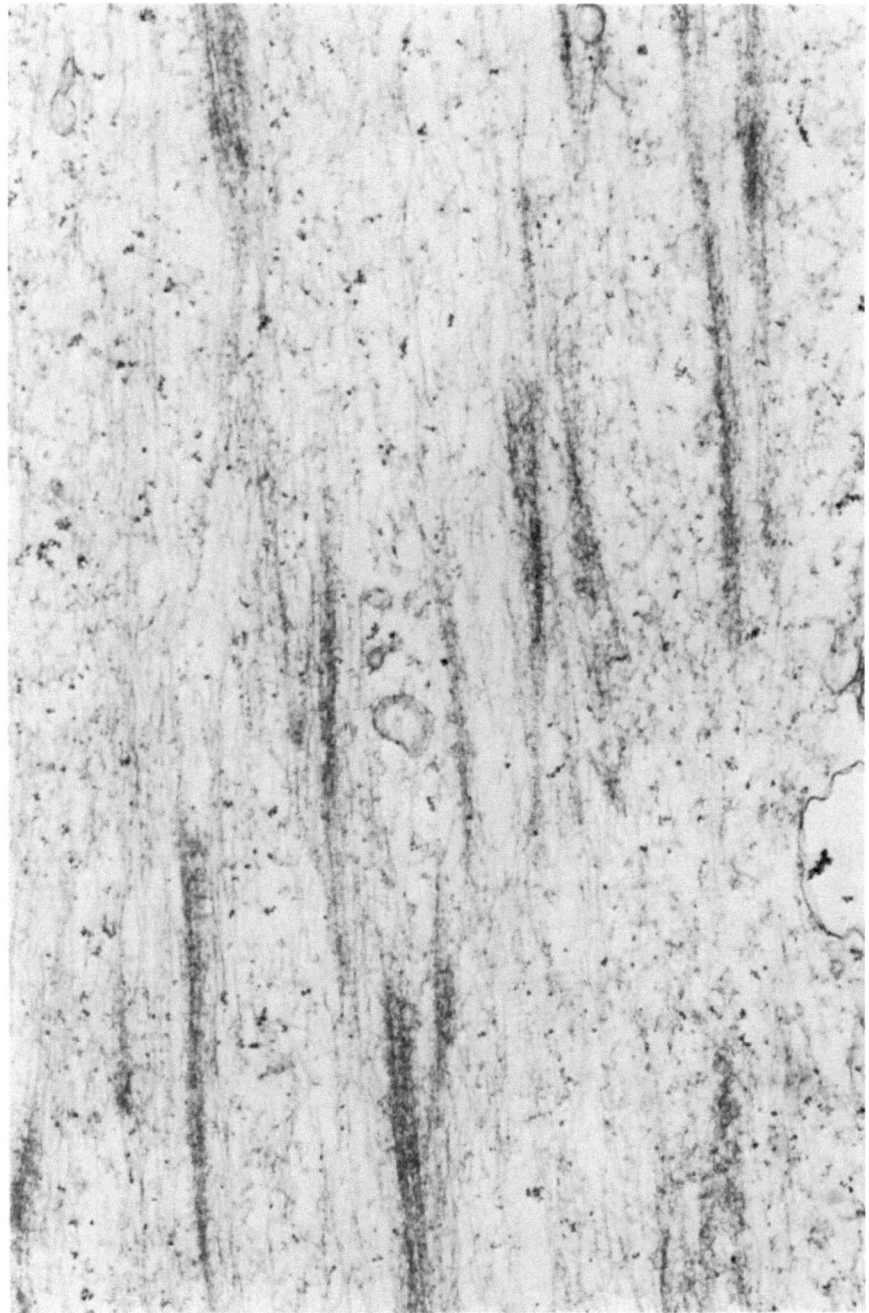

Figure 5. Cytoskeletal network of intermediate filaments and fusiform dense bodies in a portion of a taenia coli smooth muscle fiber following the selective removal of thick and thin filaments. ×29,000

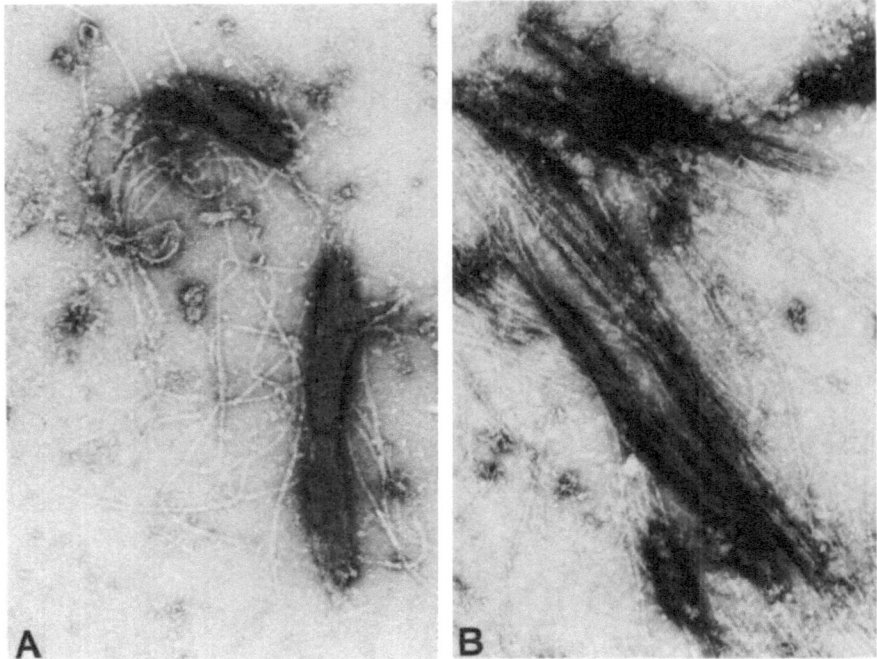

Figure 6. Isolated, negatively stained fusiform dense bodies from avian gizzard smooth muscle with (A) cores and attending intermediate filaments, and (B) cores associated with segments of thin filaments. ×46,000.

similar morphology are adjacent to developing myofibrils or free in the cytoplasm and unattended by recognizable cytoplasmic filaments (Fig. 7). Structures of comparable size and density are condensed around the cyto-centrum as pericentriolar satellites in the developing myoblasts. The relative abundance of pericentriolar satellites and dense body precursors throughout the differentiation process would suggest that these structures are related to the development of the contractile apparatus in smooth muscle, and it further emphasizes the potential for selective expression in the association of cytoplasmic filaments with the dense bodies and their precursors.

4. Thin Filaments

The first class of filaments to be identified with a muscle protein in smooth muscle were the thin filaments observed in homogenized and thin-sectioned preparations. They do not have a fixed length unless it is very long (> 5 μm), but they are very similar to the thin filaments in striated muscle and were conclusively demonstrated to contain F-actin by Hanson and Lowy (1963) on the basis of distinctive subunit structure. X-ray diffraction patterns of smooth muscles are composed predominantly of a set of reflections re-

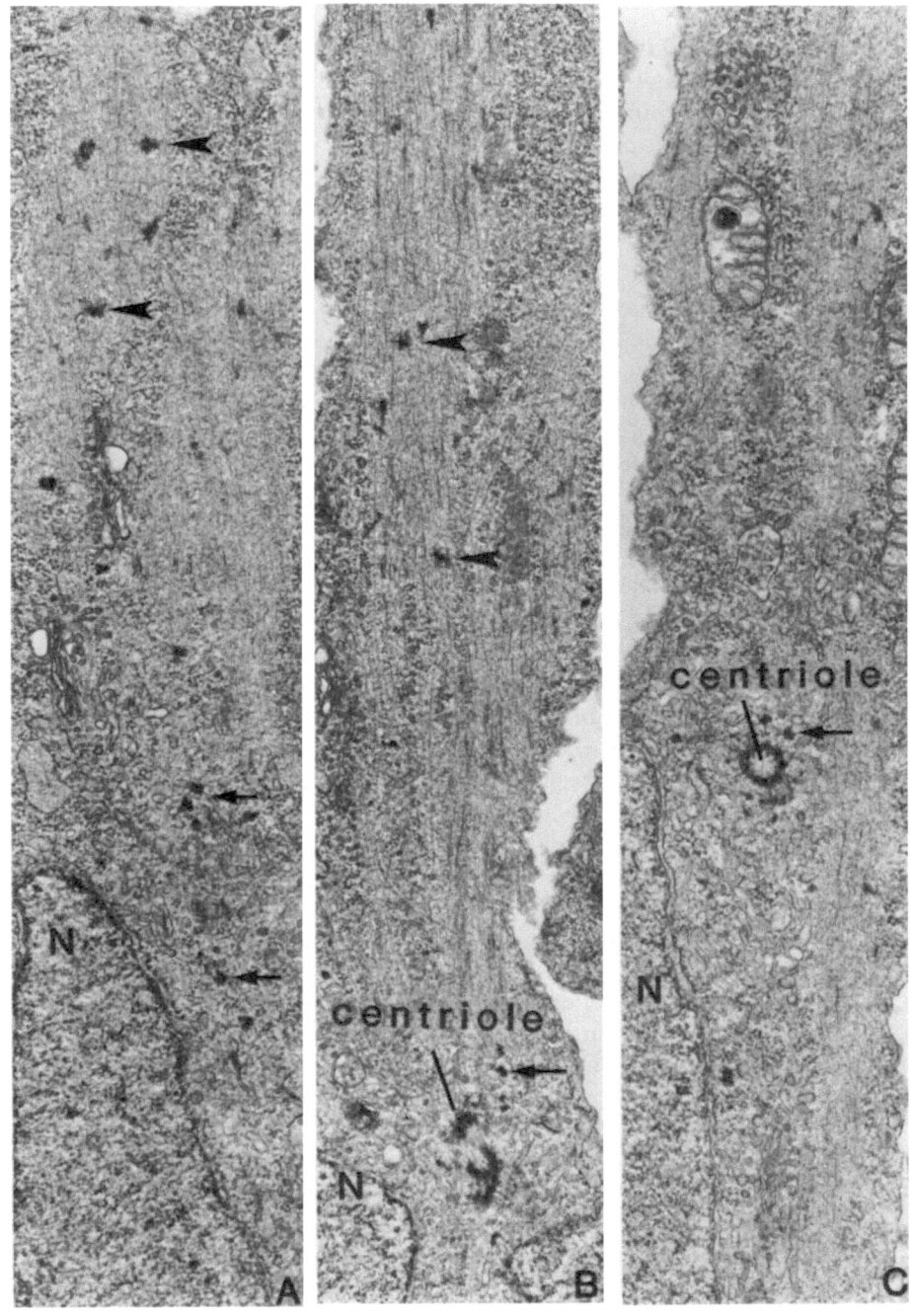

Figure 7. Developing avian gizzard smooth muscle myoblasts (representing a portion of the population of developing fibers at 6–8 days of incubation) in longitudinal section showing the nucleus (N) and developing myofibrils containing precursor structures of the dense bodies that are (A) associated with filaments (darts) and free of the cytoplasm. (B,C) The nuclear region of the developing fibers contains a diplosome with one centriole serving as the basal body for a rudimentary cilium, and dense bodies free in the cytoplasm and condensed around the diplosome as pericentriolar satellites (arrows) and microtubule-organizing centers. ×15,000.

solved to 0.9 nm that is characteristic of axially aligned F-actin (G. Elliott, 1964). A dominant feature of the organization within the population of thin filaments in smooth muscle is related to the constituent packing arrangement. The low angle x-ray diffraction spectrum from smooth muscles chiefly reflects the actin component of the thin filaments, and the equatorial region contains a relatively prominent diffraction maximum that corresponds to a Bragg spacing of 11.5–12.0 nm that is independent of muscle length but sensitive to changes in osmotic conditions (Elliott and Lowy, 1968). Correlative observations utilizing electron microscopy reveal that thin filament profiles are loosely condensed into numerous uncoordinated hexagonal arrays where the interfilament spacings roughly match the 1,0 lattice reflection in the x-ray diffraction patterns (Fig. 8). There is some indication from the spacing calculated from the patterns that this form of packing does not substantially change over a range of muscle length that includes the physiological region of the length–tension relationship. The ultrastructural consequences of this behavior would mean that the thin filaments remain axially oriented even with a 50% change in muscle length and that the density of the thin filaments are separated by a parenchymal layer of thick myosin-containing filaments forming a more or less uniform pattern throughout the cytoplasm. Under conditions that show this type of packing of the thin filaments, the regular organization of "rosettes" of thin filaments ensheathing thick filaments is frequently absent or very localized in extent. The localization of thin filaments in rosettes surrounding individual thick filaments is, however, the predominant form of organization in certain preparations of smooth muscle, especially those in which the muscle has been glycerinated or treated with nonionic detergents to render the membrane more permeable to ions in solution. The factors involved in this shift in the distribution of thin filaments from hexagonal arrays to rosettes are unknown, and they appear to be unrelated to the contractile state (Small, 1977). The dramatic difference in the packing arrangements of these two forms must reflect major differences in the type of filament interactions. Heuman (1971) has integrated these two forms into a contractile model by suggesting that the hexagonal arrays are analogous to I-bands and that the rosettes represent an irregular but overlapping arrangement of thin and thick filaments comparable to the A-bands of striated muscle. Accordingly, shortened smooth muscle fibers should contain thin filaments in the form of hexagonal arrays. The analysis of smooth muscle fibers in terms of such a model is further complicated by the lack of a reliable index of filament orientation along the axis of the muscle fibers. The bundles or rosettes of thin filaments do not contain obvious indications of their polarity. In homogenized preparations, native thin filaments compare closely with F-actin and thin filaments derived from striated muscle (Fig. 9). They differ in at least one respect: the prominent 40-nm stripe that is characteristic of the tropomyosin–troponin complex in the isotropic band of striated muscle (Hanson *et al.*, 1973) is absent along the hexagonal arrays of thin filaments in sectioned preparations and from Mg^{2+}-induced paracrystals of isolated, native thin filaments from smooth muscle (Figs. 9 and 10). There is evidence

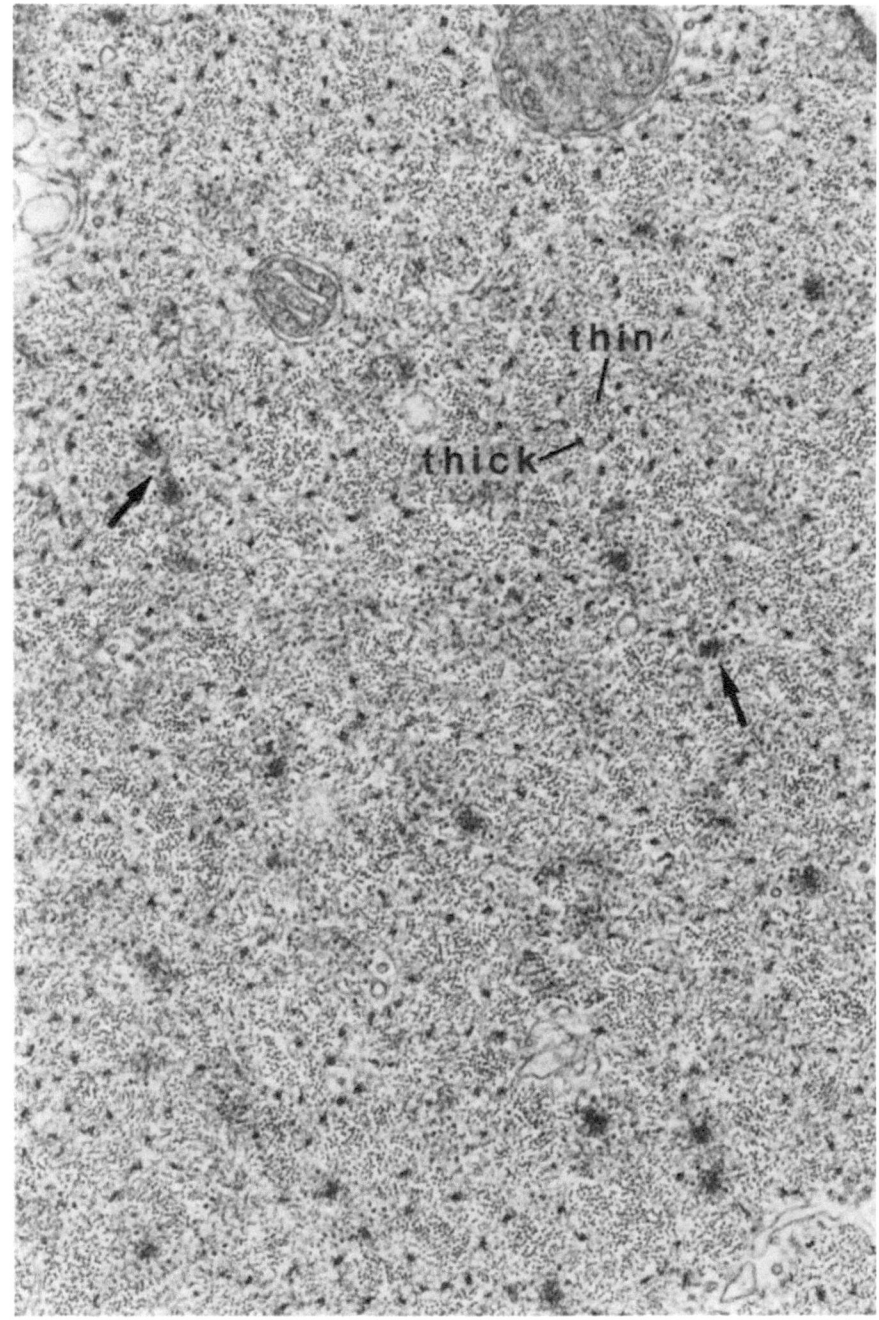

Figure 8. Cross-sectional view of a taenia coli smooth muscle fiber with abundant profiles of thin filaments condensed into latticelike bundles surrounded by a parenchymal layer of thick filaments and a uniform distribution of dense bodies with attending profiles of intermediate filaments (arrows). ×42,000.

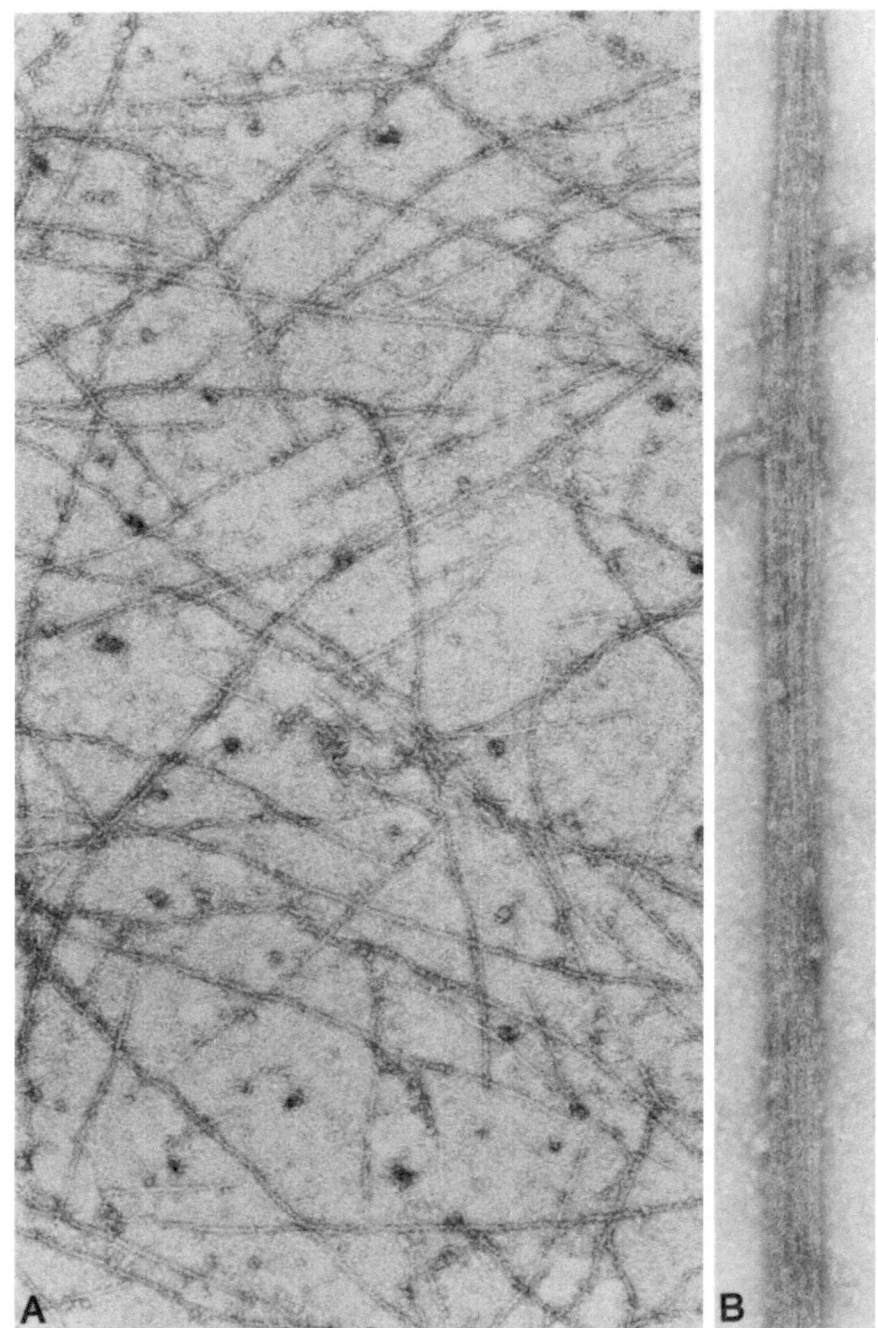

Figure 9. (A) Negatively stained homogenates of avian gizzard smooth muscle contain profuse thin filaments that correspond to F-actin. (B) Mg^{2+}-induced paracrystal of native, isolated thin filaments lacking the 40-nm stripe that is characteristic of the tropomyosin–troponin complex associated with the native thin filaments in striated muscles. ×62,000.

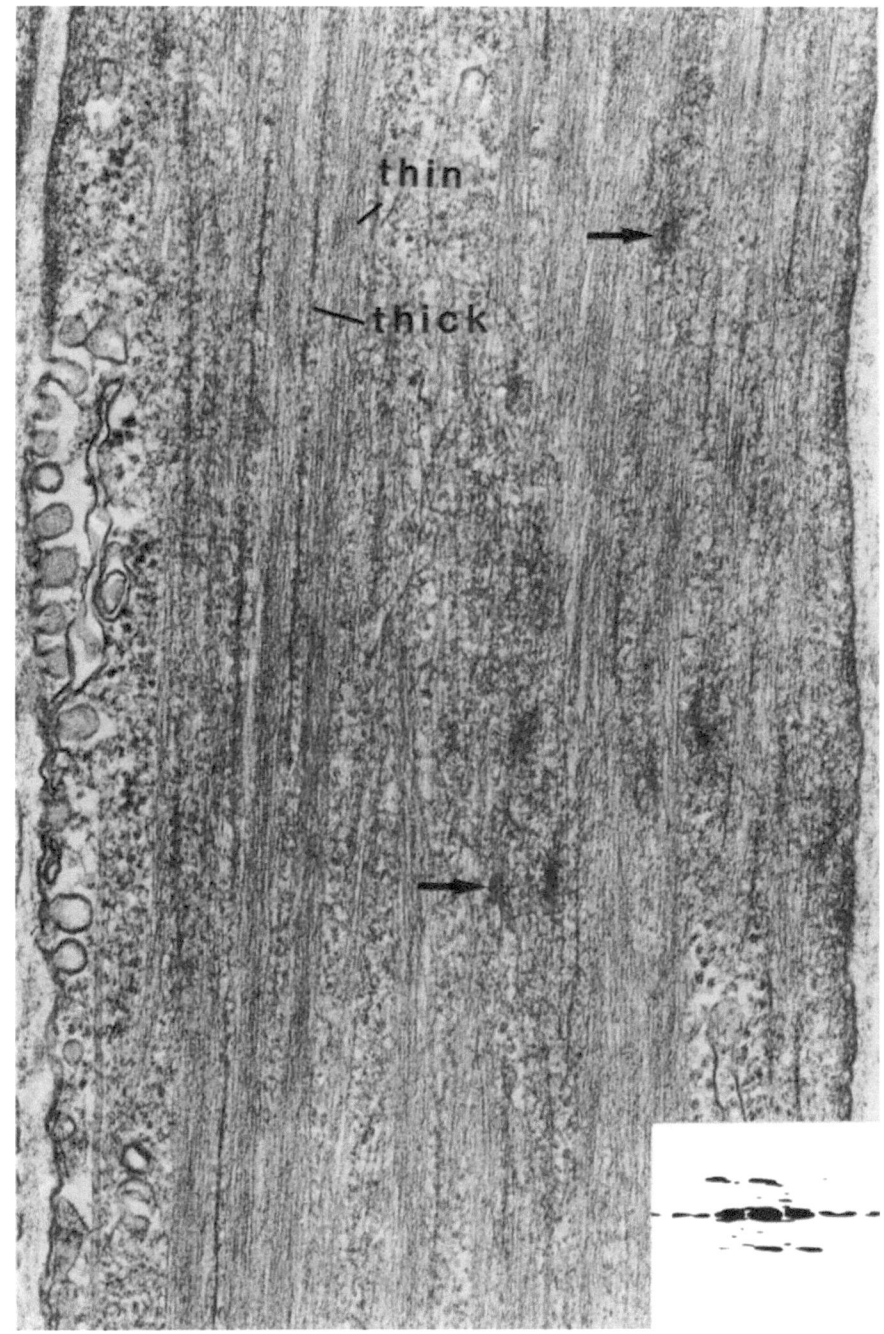

Figure 10. Longitudinal section of taenia coli smooth muscle fiber illustrating the bundles of thin filaments and intervening thick filaments bearing cross-bridge projections with a period of ~14 nm that extend along the entire length of the filaments (inset), and dense bodies with attendant intermediate filaments. ×42,000.

that the tropomyosin in smooth muscle (Cummins and Perry, 1974; Driska and Hartshorne, 1975; Ebashi *et al.*, 1975) is an integral part of isolated thin filaments (Sobieszek and Small, 1976), and low angle x-ray diffraction diagrams of smooth muscle show the same functionally related changes in the system of layer lines that indicate a shift in the radial distribution of mass in the thin filaments that are ascribed to tropomyosin (Vilbert *et al.*, 1972). The constancy in molar ratios (7 : 1) of extracted actin and tropomyosin also suggests a linkage between these two major proteins in the thin filaments of smooth muscles (Cohen and Murphy, 1978). The structural analog of proteins that functionally resemble troponin (Ebashi *et al.*, 1975; Carsten, 1971) has not been localized, but native thin filaments might selectively bind exogenous troponin (Bremel, 1974). Overall, the distribution of the elements in the contractile apparatus that regulate the sensitivity of the contractile mechanism to Ca^{2+} is not clear at this time, and the contribution of components in the thin filaments is uncertain. Correlations may be found between functional variation and the three isoelectric forms of smooth muscle actin (Elce *et al.*, 1981; Gabbiani *et al.*, 1981).

5. Thick Filaments

The organization of myosin as thick filaments in smooth muscle has been the most difficult aspect of the contractile apparatus to reliably demonstrate and analyze; so questions surrounding this component have preoccupied investigators several years (Shoenberg and Needham, 1976). Most of the attention has been fixed upon the demonstration of an organized form of myosin that compares with the thick filaments in striated muscles. Myosin isolated from smooth muscle is quite similar to other muscle myosins in terms of physiochemical properties (Kendrick-Jones, 1973; Elliott *et al.*, 1976; Frederikson, 1980) and it aggregates into the traditional forms of bipolar filaments (Hanson and Lowy, 1964; Kaminer, 1969; Wachsberger and Pepe, 1974; Cooke, 1975). Kelly and Rice (1968) first demonstrated the localization of thick myosin-containing filaments under controlled conditions, and, subsequently, preparations derived from physiologically incubated smooth muscles from a variety of tissue sources were found by many investigators to contain abundant thick filaments. The size, shape, and structural composition of these filaments together suggested that they correspond to aggregates of myosin. The filaments are roughly cylindrical in cross-sectional contour and are 15–20 nm in diameter. Filament profiles are composed of a central dense core and an irregular or asymmetric cortex (Craig and Megerman, 1978) identified as the myosin cross-bridges that in axial projection are periodically distributed at intervals of about 14 nm along the entire length (2–8 μm) of the filaments (Figs. 8 and 10). No central "bare-zone" corresponding to the region of antiparallel packing of the constituent myosin molecules or a structure analogous to the M-line has been clearly identified (Small and Squire, 1972). Filaments prepared from isolated myosin contain many structural features related to the cross-bridge projections on the filaments *in situ* (Hinssen *et al.*, 1978), but the detailed molecular plan is not uniform: there is evidence for side-polarity and mixed polarity of cross-bridges on the surface of the filaments. This issue is complicated from an analytical point of view by the irregular axial distribution of the thick filaments and the lack of registration between the filaments; so these structures offer very little regularity to the methods that provide for high spatial resolution. The only feature related to the thick filaments obtained in x-ray diffraction diagrams is a single, axial reflection measuring 14.3 nm that represents the principal (cross-bridge) periodicity in the thick filaments (Lowy *et al.*, 1970; Shoenberg and Haselgrove, 1974). The evidence that myosin exists in filamentous form in contracting and relaxed smooth muscle establishes that the relatively low and variable content of myosin (Cohen and Murphy, 1978) is arranged into extremely long filaments bearing cross-bridge projections (Ashton *et al.*, 1975; Small, 1977).

6. Models of the Contractile Apparatus

In classical histological studies, the contractile material of smooth muscle fibers was characterized as either optically homogeneous with undifferentiated staining methods or composed of longitudinally oriented myofibrils. The myofibrils were continuous structures, markedly basophilic in localized contraction nodes, and convoluted only passively in isotonic contractures when differentiated staining was employed (McGill, 1909).

Recent structural models of the contractile mechanism have been based on a unique structural feature of smooth muscle. The oblique orientation of filaments is a characteristic that is repeatedly raised in connection with models of contraction. Rosenbluth (1965) combined the organizational scheme of obliquely striated muscle found in certain groups of invertebrates with a single class of obliquely oriented (thin) filaments in amphibian smooth muscle fibers into the "parallel orientation" model. This model would suitably fit the schematic models developed from contemporary optical studies of isolated single fibers (Bagby *et al.*, 1971; Fay and Delise, 1973; Small, 1974) given that closely spaced attachments of filaments occur along the entire length of muscle fibers and are concentrated at the termini (Gabella, 1976). Obliquely oriented filaments at some point during active shortening might be organized into a long-range helical conformation as suggested by the observations of Fisher and Bagby (1977). A helical pattern of organization for the myofilaments could explain why some smooth muscle fibers, especially in small hollow organs, are helicoidal, but a helical arrangement of filaments is not particularly obvious in the (usually) spindle-shaped muscle fibers.

The homology between the thin filaments in smooth muscle and actin-containing filaments in striated muscles strongly suggests that the fusiform dense bodies are homologous to Z-bands (Shoenberg and Needham, 1976); so the implication is that the bulk of the contractile apparatus in smooth muscle is composed of structures that are roughly equivalent to the isotropic bands in striated muscles. By introducing myosin as bipolar dimers, Panner and Honig (1967) further developed Rosenbluth's model to contain the essential elements of a sarcomere as defined in striated muscle. A general appreciation for the lability of myosin in filamentous form led to the development of preparative methods for electron microscopy that facilitated the consistent localization of myosin as discrete, long (2–8 μm) filaments. The constituent molecules are packed "antiparallel" in these filaments and are aligned normal to the usual orientation in bipolar filaments where packing is both antiparallel and parallel (Small and Sobieszek, 1980), and they contain an asymmetric or mixed polarity of cross-bridge projections. Thick (myosin-containing) filaments are illustrated in Heuman's (1971) model of the contractile apparatus (see Figure 11) as irregular or skewed arrays of overlapping thick and thin filaments that provides a structural basis for the broad range of lengths at which a maximal active tension is generated by smooth muscle fibers (Fay, 1976). The basis for all these models depends on a functional homology between the dense bodies in smooth muscle and the Z-disc in striated muscle, and the recent introduction of a distinctive system of intermediate filaments to our understanding of the structural organization of smooth muscle fibers that has a counterpart in skeletal and cardiac muscle further supports the idea that the contractile mechanisms have a similar structural basis. Immediate progress in defining a sarcomere-equivalent in smooth muscle may require a clear definition of the respective molecular symmetry within the populations of thick and thin filaments and additional information on the functional roles of the intermediate filaments and associated dense bodies.

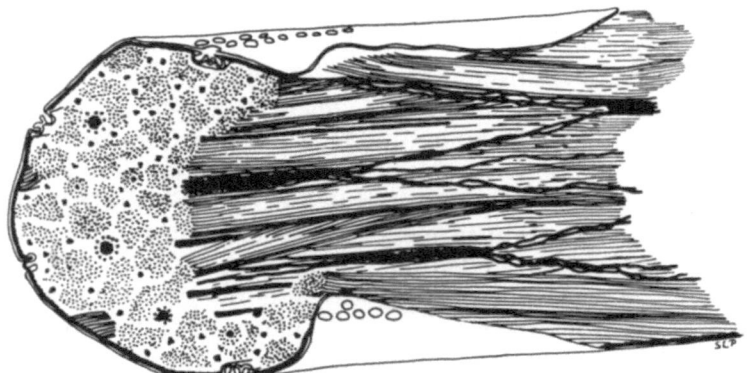

Figure 11.   Illustration of the three-dimensional relationships between the three classes of filaments based upon serial reconstructions of taenia coli smooth muscle fibers. The thin filaments are grouped into bundles that are integrated into a spatially complex network through branches and anastomoses between adjacent bundles. The thick filaments are parenchymal to the bundles of thin filaments and interdigitate with some bundles of the network that assume the form of rosettes. The intermediate filaments are complexed with the dense bodies to form a cytoskeletal network extending throughout the muscle fiber.

References

Ashton, F., Somlyo, A., and Somlyo, A., 1975, The contractile apparatus of vascular smooth muscle. Intermediate high voltage stereo electron microscopy, *J. Mol. Biol.* **98**:17–29.

Bagby, R., Young, A., Dotson, R., Fisher, B., 1971, Contraction of single smooth muscle cells from Bufo marinus, *Nature* **234**:351–353.

Berner, P., Somlyo, A., and Somlyo, A., 1981, Hypertrophy induced increase of intermediate filaments in vascular smooth muscle, *J. Cell Biol.* **88**:96–101.

Bremel, R., 1974, Myosin linked calcium regulation in vertebrate smooth muscle, *Nature* **252**:405–407.

Burnstock, G., 1970, Structure of smooth muscle and its innervation, in: *Smooth Muscle* (E. Bulbring, A. Brading, A. Jones, and T. Thomas, eds.), pp. 1–69, E. Arnold, London.

Carsten, M., 1971, Uterine smooth muscle: troponin, *Arch. Biochem. Biophys.* **147**:353–357.

Chamley-Campbell, J., Campbell, G., and Ross, R., 1979, The smooth muscle cell in culture, *Physiol. Rev.* **59**:1–61.

Cohen, D., and Murphy, R., 1978, Differences in contractile protein contents among porcine smooth muscles: Evidence for variation in the contractile system, *J. Gen. Physiol.* **72**:369–380.

Cooke, P., 1975, Filamentous aggregates of purified myosin from smooth muscle, *Cytobiologie* **11**:346–357.

Cooke, P., 1976, A filamentous cytoskeleton in vertebrate smooth muscle fibers, *J. Cell Biol.* **68**:539–556.

Cooke, P., and Chase, R., 1971, KCl-insoluble myofilaments in vertebrate smooth muscle cells, *Exp. Cell Res.* **66**:417–425.

Cooke, P., and Fay, F., 1972, Correlation between fiber-length, ultrastructure and the length-tension relationship of mammalian smooth muscle, *J. Cell Biol.* **52**:105–116.

Craig, R., and Megerman, J., 1978, Assembly of smooth muscle myosin into side-polar filaments, *J. Cell Biol.* **75**:990–996.

Cummins, P., and Perry, S., 1974, Chemical and immunochemical characteristics of tropomyosins from striated and smooth muscle, *Biochem. J.* **141**:43–49.

Devine, C., Somlyo, A., and Somlyo, A., 1972, Sarcoplasmic reticulum and excitation–contraction coupling in mammalian smooth muscle, *J. Cell Biol.* **52**:690–718.

Driska, S., and Hartshorne, D., 1975, The contractile proteins of smooth muscle. Properties and components of a Ca^{2+}-sensitive actomyosin from chicken gizzard, *Arch. Biochem. Biophys.* **167**:203–212.

Ebashi, S., Toyo-Oka, T., and Nonomura, Y., 1975, Gizzard troponin, *J. Biochem. Tokyo* **78**:859–861.

Elce, J., Elbrecht, A., Middlestadt, M., McIntyre, E., and Anderson, P., 1981, Actin from pig and rat uterus, *Biochem. J.* **193**:891–898.

Elliott, A., Offer, G., and Burridge, K., 1976, Electron microscopy of myosin molecules from muscle and nonmuscle sources, *Phil. Trans. R. Soc. London B* **193**:45–53.

Elliott, G., 1964, X-ray diffraction studies on striated and smooth muscles, *Proc. Roy. Soc. London B* **160**:467–472.

Elliott, G., and Lowy, J., 1968, Organization of actin in a mammalian smoth muscle, *Nature* **219**:156–157.

Fay, F., 1976, Structural and functional features of isolated smoth muscle cells, in: *Cell Motility* (R. Goldman, T. Dollard, and J. Rosenbaum, eds.), pp. 185–201, Cold Spring Harbor Laboratory.

Fay, F., and Delise, C., 1973, Contraction of isolated smooth muscle cells-structural changes, *Proc. Natl. Acad. Sci.* **70**:641–645.

Fisher, B., and Bagby, R., 1977, Reorientation of myofilaments during contraction of a vertebrate smooth muscle, *Am. J. Physiol.* **232**:C5–C13.

Frederikson, D., 1980, Physical properties of myosin from aortic smooth muscle, *Biochemistry* **18**:1651–1656.

Gabbiani, G., Schmid, E., Winter, S., Chaponnier, C., Chastonary, C., Vanderkerckhove, J., Weber, K., and Franke, W., 1981, Vascular smooth muscle cells differ from other smooth muscle cells: Predominance of vimentin filaments and a specific α-type actin, *Proc. Natl. Acad. Sci.* **78**:298–302.

Gabella, G., 1973, Cellular structures and electrophysiological behavior. Fine structure of smooth muscle, *Phil. Trans. Roy. Soc. London B* **265**:7–16.

Gabella, G., 1976, Structural changes in smooth muscle cells during isotonic contraction, *Cell Tiss. Res.* **170**:187–201.

Gabella, G., 1979, Hypertrophic smooth muscle. IV. Myofilaments, intermediate filaments and some mechanical properties, *Cell Tiss. Res.* **201**:277–288.

Geiger, B., 1979, A 130 K protein from chicken gizzard. Its localization at the termini of microfilament bundles in cultured chicken cells, *Cell* **18**:193–205.

Geisler, N., and Weber, K., 1980, Purification of smooth muscle desmin and a protein-chemical comparison of desmins from chicken gizzard and hog stomach, *Eur. J. Biochem,* **111**:425–433.

Hanson, J., and Lowy, J., 1963, The structure of F-actin and actin filaments isolated from muscle, *J. Mol. Biol.* **6**:46–60.

Hanson, J., and Lowy, J., 1964, The structure and origin of the axial periodicity in the I substance of vertebrate striated muscles, *Proc. Roy. Soc. London B* **160**:449–460.

Hanson, J., Ledner, V., O'Brien, E., and Bennett, P., 1973, Structure of actin-containing filaments in vertebrate skeletal muscle, *Cold Spring Harbor Symp. Quant. Biol.* **37**:311–318.

Heuman, H.-G., 1971, Mechanism of muscle contraction. An electron microscope study of the mouse large intestine, *Cytobiologie* **3**:259–281.

Hinssen, H., D'Haese, J., Small, J., and Sobieszek, A., 1978, Mode of filament assembly of myosins from muscle and non-muscle cells, *J. Ultrastruct. Res.* **64**:282–302.

Huiatt, T., Robson, R., Arakawa, N., and Stromer, M., 1980, Desmin from avian smooth muscle, *J. Biol. Chem.* **255**:6981–6989.

Ishikawa, A., Bischoff, R., and Holtzer, H., 1968, Formation of arrowhead complexes with HMM in a variety of cell types, *J. Cell Biol.* **43**:312–328.

Kaminer, B., 1969, Synthetic myosin filaments from vertebrate smooth muscle, *J. Mol. Biol.* **39**:257–264.

Kelly, R., and Rice, R., 1968, Localization of myosin filaments in smooth muscle, *J. Cell Biol.* **37**:105–116.

Kendrick-Jones, J., 1973, The subunit structure of gizzard myosin, *Phil. Trans. Roy. Soc. London B* **265**:183–189.

Lazarides, E., and Hubbard, B., 1976, Immunological characterization of the subunit of the 100 Å filaments from muscle cells, *Proc. Natl. Acad. Sci.* **73**:4344–4348.

Lowy, J., Poulsen, F., and Vibert, P., 1970, Myosin filaments in vertebrate smooth muscle, *Nature* **225**:1053–1054.

McGill, C., 1909, The structure of smooth muscle in the resting and in the contracted condition, *Am. J. Anat.* **9**:494–545.

Nonomura, Y., and Ebashi, S., 1975, Isolation and identification of smooth muscle contractile proteins, in: *Methods in Pharmacology* (E. Daniel and D. Paton, eds.), pp. 141–162, Plenum Press, New York.

Panner, B., and Honig, C., 1967, Filament ultrastructure and organization in vertebrate smooth muscle. Contraction hypothesis based on localization of actin and myosin, *J. Cell Biol.* **35**:303–321.

Pease, D., and Molinari, S., 1960, Electron microscopy of muscular arteries: pial vessels of the cat and monkey, *J. Ultrastruct. Res.* **3**:447.

Popescu, L., and Ionescu, N., 1970, On the equivalent between dense bodies and Z bands, *Experientia* **26**:624–643.

Rice, R., Moses, J., McManus, G., Brady, A., and Blasik, L., 1970, The organization of contractile filaments in a mammalian smooth muscle, *J. Cell. Biol.* **47**:183–196.

Rhodin, J., 1967, The ultrastructure of mammalian arterioles and precapillary sphincters, *J. Ultrastruct. Res.* **18**:181–223.

Rosenbluth, J., 1965, Smooth muscle, an ultrastructural basis for the dynamics of its contraction, *Science* **148:**1337–1339.

Schollmeyer, J., Furcht, L., Goll, D., Robson, R., and Stromer, M., 1976, Localization of contractile proteins in smooth muscle cells and in normal and transformed fibroblasts, in: *Cell Motility* (R. Goldman, T. Pollard, and J. Rosenbaum, eds.), pp. 361–388, Cold Spring Harbor Laboratory.

Shoenberg, C., and Haselgrove, J., 1974, Filaments and ribbons in vertebrate smooth muscle, *Nature* **249:**152–154.

Shoenberg, C., and Needham, D., 1976, A study of the mechanism of contraction in vertebrate smooth muscle, *Biol. Rev.* **51:**53–104.

Small, J., 1974, Contractile units in vertebrate smooth muscle, *Nature* **249:**324–327.

Small, J., 1977, Studies on isolated smooth muscle cells: the contractile apparatus, *J. Cell Sci.* **24:**327–350.

Small, J., and Squire, J., 1972, Structural basis of contraction in vertebrate smooth muscle, *J. Mol. Biol.* **67:**117–149.

Small, J., and Sobrieszek, A., 1977, Studies on the function and composition of the 10-nm (100 Å) filaments of vertebrate smooth muscle, *J. Cell Sci.* **23:**243–268.

Small, J., and Sobieszek, A., 1980, The contractile apparatus of smooth muscle, *Intl. Rev. Cytol.* **64:**241–306.

Sobieszek, A., and Small, J., 1976, Myosin linked calcium regulation in vertebrate smooth muscle, *J. Mol. Biol.* **102:**75–92.

Somlyo, A., 1980, Ultrastructure of vascular smooth muscle, in: *Handbook of Physiology* (D. Bohr, A. Somlyo, and H. Sparks, eds.), Section 2, Vol. 2, pp. 33–67, American Physiological Society, Bethesda.

Uehara, Y., Campbell, G., and Burnstock, G., 1971, Cytoplasmic filaments in developing and adult vertebrate smooth muscle, *J. Cell Biol.* **50:**484–497.

Vibert, P., Haselgrove, J., Lowy, J., and Poulsen, F., 1972, Structural changes in actin-containing filaments of muscle. *J. Mol. Biol.* **71:**757–767.

Wachsberger, P., and Pepe, F., 1974, Purification of uterine myosin and synthetic filament formation, *J. Mol. Biol.* **88:**385–391.

4

Regulation of Muscle Contraction

Setsuro Ebashi

1. Introduction

The discovery of troponin (Ebashi and Kodama, 1965; cf. Ebashi *et al.*, 1969) established the concept of regulation in muscle contraction and the term *regulatory proteins* (Maruyama and Ebashi, 1970). The absence of troponin or other specific regulatory proteins in molluscan striated muscle (Kendrick-Jones *et al.*, 1970) suggested that the mode of regulation by Ca^{2+} is not as simple as it was expected to be. The complexity of Ca^{2+} regulation was further confirmed by research on regulatory mechanisms in smooth muscle (cf. Ebashi, 1979).

Indeed, Ca^{2+} is amazingly versatile, exerting its effects on numerous aspects of biological function, and the mode of its action is quite variable. Modulator protein, or calmodulin, discovered independently by Kakiuchi (Kakiuchi *et al.*, 1970) and Cheung (Cheung, 1970) is certainly utilized by many Ca^{2+}-dependent enzyme reactions as a common mediator. On the other hand, an increasing number of reactions that are independent of calmodulin have been discovered. One of the aims of this chapter is to suggest the diversity of regulatory mechanisms.

2. The Troponin Mechanism and Troponin T

There is no doubt that the troponin mechanism has the most important role in the body movement of vertebrates. Even in protostomias, highly developed species use the troponin system for regulation. As has recently been

Setsuro Ebashi • Department of Pharmacology, Faculty of Medicine, University of Tokyo, Bunkyo-ku, Tokyo, Japan.

emphasized (Ebashi, 1980), skeletal muscle is not representative of all kinds of muscle but is uniquely developed for its function, e.g., precise and rapid contraction. In this respect, troponin composed of three different kinds of subunits [tropinin I (TnI), inhibitory factor; troponin C (TnC), Ca^{2+}-binding factor; and troponin T (TnT), tropomyosin-binding factor] represents the most advanced mechanism of regulation in biological systems from the evolutionary point of view.

We now have abundant knowledge about troponin. The role of TnT has not yet been understood, but it has become clearer with the discovery of calmodulin. If this common mediator is used, the contraction–relaxation cycle can be completed by TnI alone (Fig. 1) (Amphlett *et al.*, 1977); TnT is not required at all in this system.

A question then naturally arises: Why has muscle developed TnT, instead of using a much simpler system, i.e., the TnI–calmodulin system? To answer this question, we should note the differences between troponin C and calmodulin. Calmodulin loses its affinity for TnI in the absence of Ca^{2+} and is liberated from the actin filament. In contrast, TnC is not detached from TnI in any circumstances, even in the presence of 6 M urea (Perry *et al.*, 1972). To meet the requirements of the rapid contractile processes in muscle, the Ca^{2+}-binding protein must be continuously fixed at an appropriate position. In this respect TnC is satisfactory, but it also prevents TnI from exhibiting its inhibitory action in the absence of Ca^{2+} (Fig. 1). Consequently, TnT should mediate functional detachment of TnC from TnI in the absence of Ca^{2+}. The next problem to pursue is how TnT plays this role.

Herein we introduce the long-term work of I. Ohtsuki (cf. Ohtsuki, 1980) on the structural arrangement of the thin filament, placing emphasis on the fine localization of TnT in relation to tropomyosin. Although his work has not yet fully revealed the secret of TnT, it is undoubtedly a promising approach to the final goal.

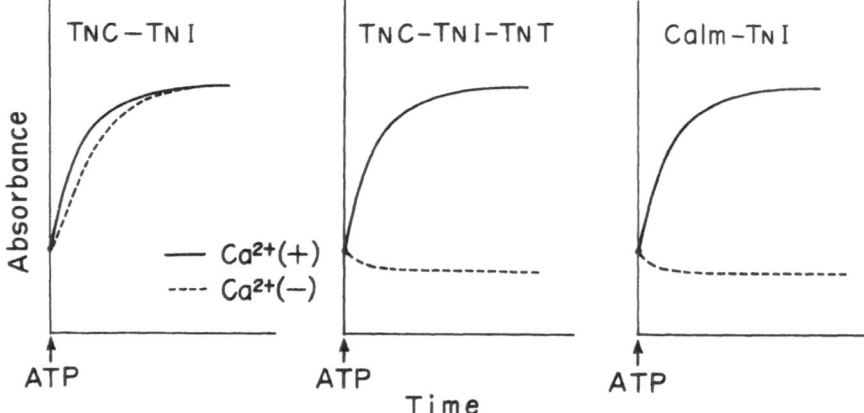

Figure 1. Schematic illustration of the difference between the TnC–TnI–TnT system and the calmodulin–TnI system as expressed in superprecipitation.

Table 1. Chymotryptic Fragments of TnT

	N-terminal sequence	Electrolyte	Binds to	Physiological activity[a]
TnT	1–259	Basic	Tropomyosin and TnI	+++
T_1	1–158	Basic	Tropomyosin	−
$T_2\alpha$	159–259	Neutral	TnI and tropomyosin	+
$T_2\beta$	159–242	Basic	TnI	−

[a] +++, High. ++, Moderate. +, Low, but significant. −, None.

Ohtsuki has shown that TnT is split by tropomyosin into two fragments, TnT_1 and TnT_2, N- and C-terminal fragments, respectively (Table 1). It is interesting that TnT_2 can bind to TnI (Katayama, 1979), though not as strongly as TnT_1, and still retain its function to a certain degree, about one-third that of the original TnT (Nakamura *et al.*, 1981).

Slightly increased digestion of TnT_2 completely abolishes its function and is accompanied by the loss of its affinity for tropomyosin (Ohtsuki *et al.*, 1981). All these observations indicate the physiological importance of the binding between TnT and TnI. It was first thought that the fragment released from TnT_2 by further chymotryptic digestion should be the chain continuous to TnT_1. The idea that the C-terminal fragment, which could bind to TnI, would be floating, seemed very attractive though somewhat puzzling. However, the small fragment digested by further treatment has been found to be located at the C-terminal region. Thus TnT binds to tropomyosin at two sites (Table I and Fig. 2) (Tanokura *et al.*, 1981).

In summary, the main role of TnT is binding both to tropomyosin and to TnI. TnT_1 is mainly responsible for the former and TnT_2 for the latter. The C-terminal region of TnT_2 can bind to tropomyosin, too; this binding also seems to be crucial for the function of Tn.

Ohtsuki, in collaboration with Nagano, has been continuing the effort to

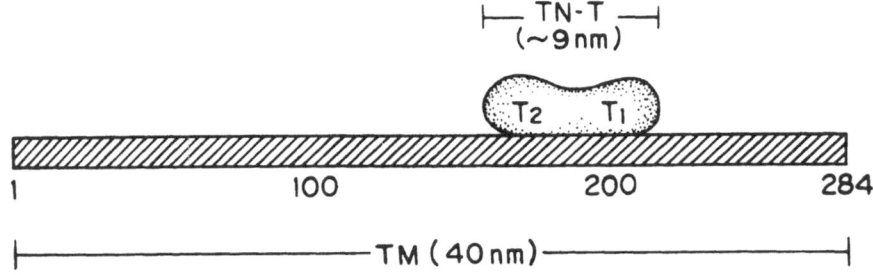

Figure 2. Model for the tropomyosin–TnT complex proposed by Ohtsuki. Revised from Nagano *et al.* (1980).

reconstruct the TnT–tropomyosin complex by supersecondary prediction technique. One of their conclusions thus far is that a large part of TnT_1 makes a triple helix together with the two chains of the tropomyosin molecule (Nagano *et al.*, 1980).

At present, the concept of regulation is commonly understood, and we have fairly extensive knowledge about the primary steps of regulation. In the case of function based on structural or architectural arrangement, such as muscle contraction or membrane-related phenomena, the mechanism of regulation must be more complicated. It is time to inquire into such advanced types of regulation. TnT may be a model subject for such investigations.

3. The Leiotonin System

One of the important results of the studies on smooth muscle regulation was to reveal the unique nature of the smooth muscle contractile system. We had been too much accustomed to the idea that skeletal muscle would be the most typical muscle and that it would typify in an elegant manner the contractile mechanism common to all kinds of muscles. This idea, however, was misleading. We have already noted the singular nature of smooth muscle contractile system in previous publications (cf. Ebashi *et al.*, 1976; Ebashi, 1980). The most important singularity is that its actomyosin system does not actively respond to ATP; to induce contraction, a regulatory system is necessary. Ca^{2+} is a kind of derepressor in the troponin system, but it is a real activator in smooth muscle, in collaboration with the regulatory system (Fig. 3).

Mikawa *et al.* (1977a) have revealed that in smooth muscle, at least in gizzard, the regulation of contraction is actin-linked. The regulatory protein leiotonin can activate the thin filament in quantities only one hundredth that

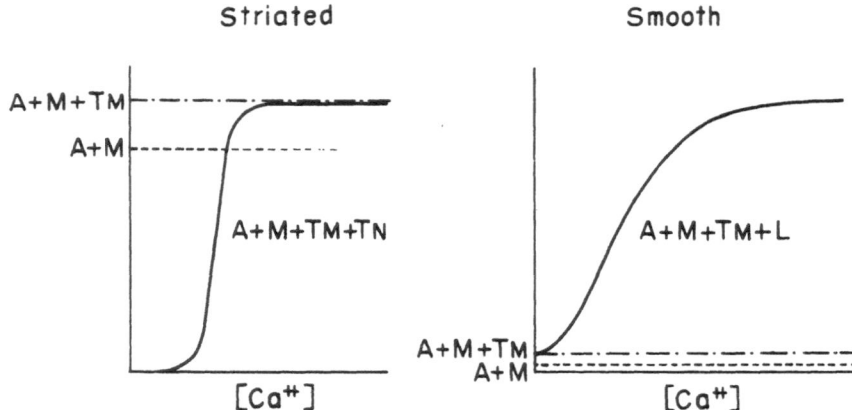

Figure 3. Schematic illustration of the fundamental mechanism of regulation in striated and smooth muscles. A: actin. M: myosin. TM: tropomyosin. TN: troponin. L: regulatory factor in smooth muscle, i.e., leiotonin, according to Ebashi's group, or light chain kinase, according to the majority of workers.

Table 2. Cross-Linked Thin Filaments

	ATPase activities (relative values)	
	1×10^{-7} M Ca^{2+}	1.3×10^{-5} M Ca^{2+}
Original thin filament	0.13	(1)
Cross-linked with Ca^{2+} (a)	0.94	0.92
Cross-linked without Ca^{2+} (b)	0.21	0.19
(a) plus leiotonin and tropomyosin	0.92	0.96
(b) plus leiotonin and tropomyosin	0.17	0.21
(a) plus native tropomyosin	0.96	1.11
(b) plus native tropomyosin	0.17	0.28

of actin by weight. Like troponin, leiotonin is a complex, composed of leiotonin A, with a molecular weight of 80,000 and leiotonin C, the Ca receptor with a molecular weight of 17,000. The role of leiotonin C can be replaced almost completely by calmodulin, whereas leiotonin C, like troponin C, cannot be a substitute for calmodulin in enzymic reactions. Almost all the findings in gizzard (and bovine stomach) have been confirmed by Hirata *et al.* (1980) in bovine aorta.

It has been shown that if the actin filament, together with leiotonin and tropomyosin, is fixed or frozen by glutaraldehyde in the absence or presence of Ca^{2+}, the repressed and activated states of the actin–leiotonin–tropomyosin complex are stabilized, respectively; i.e., the former is repressed even in the presence of Ca^{2+} and the latter is activated even in the absence of Ca^{2+} (Mikawa, 1979). Essentially the same results have been demonstrated with thin filament isolated directly from fresh gizzard muscle (Table 2; Mikawa, 1980). On the other hand, it has also been shown that the contraction of gizzard actomyosin system can be activated without the light chain kinase (Mikawa *et al.*, 1977b).

The regulatory mechanism in smooth muscle is the focus of serious controversy. The majority of biochemists studying smooth muscle are of the opinion that the phosphorylation of myosin light chain is the key reaction in activating its contractile system (cf. the paper of Dr. Hartshorne in Vol. 2 of this series). Chacko *et al.* (1977)* have shown that the phosphorylated myosin isolated from activated gizzard myosin B interacts vigorously with gizzard and rabbit skeletal actin in the absence of Ca^{2+}. Using pure myosin, we have confirmed this result in our laboratory. On the other hand, we have also reconfirmed the findings of Mikawa *et al.* (1977b) that the contractile response of gizzard actomyosin does not occur in spite of full phosphorylation of myosin, unless provided with the leiotonin system. We have to reconcile these two apparently inconsistent results.

The argument against the role of leiotonin asks how leiotonin can regu-

---

*We had the pleasure of having Dr. Chacko in our laboratory in the summer of 1980; he beautifully demonstrated this experiment to us.

Table 3. Preparation of Leiotonin

1. Gizzard mince homogenized with 0.05M KCl
2. Residue (0.08)[a] washed with 1 mM NaHCO$_3$
3. Residue (0.02) washed with 0.02 M KCl twice
4. Residue (0.02) extracted with 1 M KCl
5. Extract (0.6) subjected to inverse ammonium sulfate fractionation

[a] Numbers within parentheses denote ionic strength.

late the actin in such a small quantity as one hundredth that of actin (Mikawa *et al.*, 1978; Ebashi, 1979). We cannot answer this question directly, but we must emphasize that gelsolin, one of the Ca^{2+}-dependent regulatory proteins in nonmuscle cells can affect actin in a quantity only one hundredth that of actin (Yin and Stossel, 1979). Furthermore, β-actinin can exert its full activity in a quantity one hundredth that of actin (Maruyama *et al.*, 1977). Though the situation is somewhat different, fragmin (Hasegawa *et al.*, 1980) and α-actinin (Ebashi and Ebashi, 1965) can be effective even in a quantity one fiftieth that of actin. Therefore, it is not surprising that leiotonin could exhibit its full effect in such a minute quantity. On the contrary, it is a common phenomenon of actin-related regulation. Perhaps the event in the troponin system is an unusual case.

Another criticism is the difficulty in preparing leiotonin. We must accept this criticism. The yield of leiotonin by the method of Mikawa *et al.* (1977a, 1978) is very small. Furthermore, their work was done at a time when we were ignorant about the role of calmodulin in light chain phosphorylation and therefore did not add calmodulin to the testing system. Besides, we were using a phosphate buffer system in carrying out the sodium dodecyl sulfate–polyacrylamide gel, which is now shown to have much less resolution than the Tris-glycine system. Therefore, the preparation believed to be fairly homogeneous at that time is no longer regarded as a pure sample.

We have again started to search for a simple preparation method that could be performed by everyone and that would bring about a high yield.

Table 4. Inverse Ammonium Sulfate Fractionation of Extract (0.6) in Table 3[a]

Concentration of ammonium sulfate (M)	Leiotonin activity	Kinase activity	Leiotonin activity/ kinase activity
Original	(100)	(100)	(1)
3.1–2.2	1	11	0.1
2.2–1.8	20	2.8	7
1.8–1.4	12	0.5	24
1.4	6	0.4	15
	39	14.7	

[a] All values are expressed in a relative manner. Total activity of the original extract is considered as being 100 and the activity of each fraction is expressed by its ratio to 100.

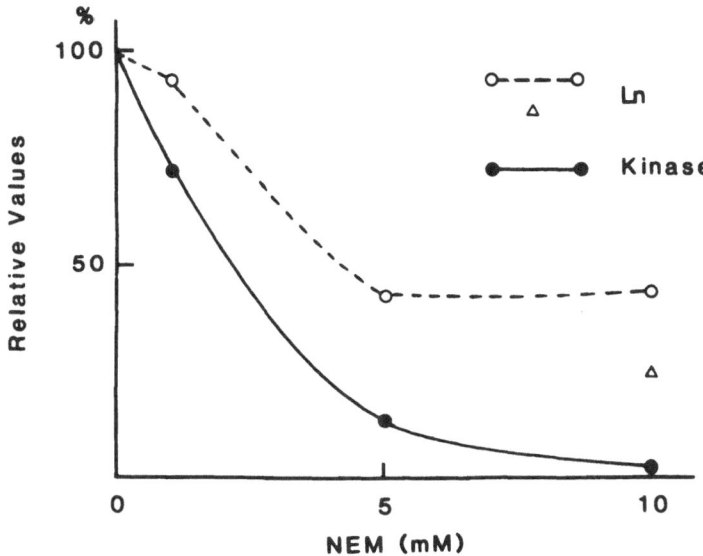

Figure 4.   Effect of *N*-ethylmaleimide (NEM) on leiotonin activity and light chain kinase activity of native tropomyosin. Native tropomyosin in 2 M urea was incubated with 10 mM NEM in the presence of 2 mM EGTA (○, ●) or 1.3×10^{-5} M Ca^{2+} (△).

Although our preparation has not proved completely satisfactory, particularly because of its nonhomogeneity, the method described in Tables 3 and 4 can give rise to a preparation with a relatively high yield of leoiotonin.

On the other hand, our effort to prepare kinase free of leiotonin has not yet been successful (in the case of our previous attempt [Mikawa *et al.*, 1977b], the kinase used was 80,000 molecular weight, perhaps being a proteolytic fragment; although this fact does not affect our conclusion, the experiment should be repeated with intact enzyme). The difficulty is that the kinase is more labile in general than leiotonin, especially to oxidation, as indicated in Fig. 4.

4. Conclusions

In contrast with the very advanced state of research on the regulatory mechanism in striated muscle, smooth muscle studies lag very much behind. There has been still no agreement on the identity of the regulatory factor in smooth muscle. This is partly due to the reason stated above: most muscle biochemists were not enthusiastic about the study of smooth muscle, tacitly believing that research on skeletal muscle would eventually solve every secret of muscle contraction. However, this gap is also due to the fact that protein preparation from smooth muscle is far more difficult than that from skeletal muscle; one of the reasons for this difficulty is the abundance of membrane fraction that comprises various kinds of enzyme activities, including different kinds of strong ATPase.

Progress in smooth muscle research is important not only for its own sake but also for research on contractile and regulatory proteins in nonmuscle tissues, where the preparation of proteins is far more difficult than in smooth muscle. Despite such difficulties, we must isolate such proteins in a pure and native state to establish the actual regulatory mechanisms, i.e., the mechanisms wherein the organism utilizes Ca^{2+}.

ACKNOWLEDGMENTS.   This work was supported in part by research grants from the Muscular Dystrophy Association of America; the Ministry of Education, Science and Culture, Japan; the Ministry of Health and Welfare, Japan; and the Iatrochemical Foundation.

References

Amphlett, G. W., Vanaman, T. C., and Perry, S. V., 1977, Effect of the troponin C-like protein from bovine brain (brain modulator protein) on the Mg^{2+}-stimulated ATPase of skeletal muscle actomyosin, *FEBS Lett.* **72**:163–168.

Chacko, S., Conti, M. A., and Adelstein, R. S., 1977, Effect of phosphorylation of smooth muscle myosin on actin activation and Ca^{2+} regulation, *Proc. Natl. Acad. Sci. USA* **74**:129–133.

Cheung, W. Y., 1970, Cyclic 3′,5′-nucleotide phosphodiesterase, Demonstration of an activator, *Biochem. Biophys. Res. Commun.* **38**:533–538.

Ebashi, S., 1979, Ca ion and muscle contraction, in: *Advances in Pharmacology and Therapeutics*, Vol. 3 (J. C. Stoclet, ed.) pp. 81–98, Pergamon Press, Oxford.

Ebashi, S., 1980, Regulation of muscle contraction (The Croonian Lecture 1979). *Proc. Roy. Soc. London B* **207**:259–286.

Ebashi, S., and Ebashi, F., 1965, α-Actinin, a new structural protein from striated muscle, *J. Biochem.* **58**:7–12.

Ebashi, S., and Kodama, A., 1965, A new protein factor promoting aggregation of tropomyosin, *J. Biochem.* **58**:107–108.

Ebashi, S., Endo, M., and Ohtsuki, I., 1969, Control of muscle contraction, *Quart. Rev. Biophys.* **2**:351–384.

Ebashi, S., Nonomura, Y., Toyo-oka, T., and Katayama, E., 1976, Regulation of muscle contraction by the calcium-troponin-tropomyosin system, in: *Calcium in Biological Systems*. Symp. Soc. Exp. Biol., No. 30, pp. 349–360.

Hasegawa, T., Takahashi, S., Hayashi, H., and Hatano, S., 1980, Fragmin: A calcium ion sensitive regulatory factor on the formation of actin filaments, *Biochemistry* **19**:2677–2683.

Hirata, M., Mikawa, T., Nonomura, Y., and Ebashi, S., 1980, Ca^{2+} regulation in vascular smooth muscle, *J. Biochem.* **78**:369–378.

Kakiuchi, S., Yamazaki, R., and Nakajima, H., 1970, Properties of a heat stable phosphodiesterase activating factor isolated from brain extract, *Proc. Japan Acad.* **46**:587–592.

Katayama, E., 1979, Interaction of troponin-I with troponin-T and its fragment, *J. Biochem.* **85**:1379–1381.

Kendrick-Jones, J., Lehman, W., Szent-Györgyi, A. G., 1970, Regulation in molluscan muscles, *J. Mol. Biol.* **54**:313–326.

Maruyama, K., and Ebashi, S., 1970, Regulatory proteins of muscle, in: *The Physiology and Biochemistry of Muscle as a Food*, Vol. 2, E. J. Briskey, R. G. Cassens, and B. B. March, eds., The Univ. of Wisconsin Press, Madison, pp. 373–382.

Maruyama, K., Kimura, S., Ishii, T., Kuroda, M., Ohashi, K., and Muramatsu, S., 1977, β-Actinin, a regulatory protein of muscle. Purification, characterization and function, *J. Biochem.* **81**:215–232.

Mikawa, T., 1979, 'Freezing' of the calcium-regulated structures of gizzard thin filaments by glutaraldehyde, *J. Biochem.* **85:**879–881.

Mikawa, T., 1980, "Freezing" of activated and depressed thin filaments of skeletal and smooth muscle, in: *Muscle Contraction: Its Regulatory Mechanisms,* (S. Ebashi, K. Maruyama and M. Endo, eds.) Japan Sci. Soc. Press, Tokyo pp. 347–357.

Mikawa, T., Toyo-oka, T., Nonomura, Y., and Ebashi, S., 1977a, Essential factor of gizzard 'Troponin' fracation, *J. Biochem.* **81:**273–275.

Mikawa, T., Nonomura, Y., and Ebashi, S., 1977b, Does phosphorylation of myosin light chain have direct relation to regulation in smooth muscle, *J. Biochem.* **82:**1789–1791.

Mikawa, T., Nonomura, Y., Hirata, M., Ebashi, S., and Kakiuchi, S., 1978, Involvement of an acidic protein in regulation of smooth muscle contraction by the tropomyosin-leiotonin system, *J. Biochem.* **84:**1633–1636.

Nagano, K., Miyamoto, S., Matsumura, M., and Ohtsuki, I., 1980, Possible formation of a triple-stranded coiled-coil region in tropomyosin-troponin T binding complex, *J. Mol. Biol.* **141:**217–222.

Nakamura, S., Hashimoto, K., and Ohtsuki, I., 1981, Effect of chymotryptic troponin T subfragments on the Ca^{2+} sensitivity of superprecipitation, *J. Biochem.* **89:**1639–1641.

Ohtsuki, I., 1980, Functional organization of the troponin-tropomyosin system, in: *Muscle Contraction: Its Regulatory Mechanisms* (S. Ebashi, K. Maruyama and M. Endo, eds.) Japan Sci. Soc. Press, Tokyo, pp. 237–249.

Ohtsuki, I., Yamamoto, K., and Hashimoto, K., 1981, Effect of two C-terminal side chymotryptic troponin T subfragments on the Ca^{2+} sensitivity of superprecipitation and ATPase activities of actomyosin, *J. Biochem.* **90:**259–261.

Perry, S. V., Cole, H. A., Head, J. F., and Wilson, F. J., 1972, Localization and mode of action of the inhibitory protein component of the troponin complex, Cold Spring Harbor Symposia, **32:**251–262.

Tanokura, M., Tawada, Y., Onoyama, Y., Nakamura, S., and Ohtsuki, I., 1981, Primary structure of chymotryptic subfragments from rabbit skeletal troponin T, *J. Biochem.* **90:**263–265.

Yin, H. L., Stossel, T. P., 1979, Control of cytoplasmic actin gel-sol transformation by gelsolin, a calcium-dependent regulatory protein, *Nature* **281:**583–586.

5

The Membrane Systems and Cytoskeletal Elements of Mammalian Myocardial Cells

M. S. Forbes and N. Sperelakis

1. Introduction

This chapter reviews various findings, made primarily by means of electron microscopy, on the "tubular" (membrane) systems and structural fibrils of mammalian cardiac muscle cells. The term "membrane systems," as used herein, does not necessarily imply possession of cylindric structure by all parts of the so-called "transverse" (T)-tubules or the sarcoplasmic reticulum (SR), but instead implies solely that these structures are composed of membranes and possess some form of organization within each cell.

Numerous descriptions of myocardial membrane systems have appeared since the advent of transmission electron microscopy, either as a feature of reviews devoted to the myocardial cell in general (e.g., Stenger and Spiro, 1961; Slautterback, 1963; Simpson *et al.*, 1973, 1974; McNutt and Fawcett, 1974; Sommer and Johnson, 1979) or as a subject in its own right (e.g., Sperelakis *et al.*, 1974; Sommer and Waugh, 1976). The body of information relating to the T-system and the SR continues to expand, aided by the development of new techniques and investigations of different species, and this chapter is intended to provide an update of this important subject.

Cytoskeletal elements of the myocardium, such as microtubules and intermediate filaments, have not thus far been as well studied as membrane systems. In this chapter, we consider both our own and other recent findings that deal with the intra- and extracellular supporting structures in heart.

M. S. Forbes and N. Sperelakis • Department of Physiology, University of Virginia School of Medicine, Charlottesville, Virginia 22908.

2. Membrane Systems

2.1. Transverse-Axial Tubular System (TATS)

2.1.1. Terminology

Reviews that are concerned with striated muscle cells often divide the membrane system components into "extracellular" and "intracellular" elements. Traditionally the extracellular membrane system of the mammalian myocardial cell is described as elongated, regularly distributed, approximately cylindrical invaginations of the surface membrane, which project inward in an orientation that is primarily perpendicular to the long axis of each cell. The term "transverse tubules" is therefore widely used to describe this system. Electron micrographs that depict the surfaces of myocardial cells as revealed by freeze–fracture replication (Fig. 1) indeed indicate a regular array of openings. In thin section, these openings usually are found positioned adjacent to the levels of Z-lines of the nearest myofibrils, and lead into transverse (T)-tubules, which generally maintain their location in the "Z-level myoplasm" (Forbes and Sperelakis, 1980a) (Figs. 2 and 3). With the advent of electron-opaque "tracer" techniques (see Section 5.1), it was found that the so-called "T-system" could be delineated at all depths of the myocardial cell. Tracing of the T-system has to date been accomplished by the use of ferritin (Forssmann and Girardier, 1966), thorium dioxide (Rubio and Sperelakis, 1971), colloidal lanthanum hydroxide (Simpson et al., 1973; Sommer and Waugh, 1976), horseradish peroxidase–diaminobenzidine (HRP–DAB) reaction product (Forssman and Girardier, 1970; Sperelakis and Rubio, 1971), ferrocyanide-reduced osmium tetroxide ("osmium ferrocyanide" [OsFeCN]: Sybers and Gann, 1975; Forbes et al., 1977; Forbes and Sperelakis, 1977, 1980a), and tannic acid (Leeson, 1978, 1980). Such studies have confirmed the existence of a great number of truly *transverse* tubules in "working" ventricular myocardial cells of mammalian hearts, and in addition have demonstrated the existence of *longitudinally*—or "axially"—arranged tubules that are confluent with transverse tubules. For example, in guinea pig ventricle there is a latticework of interconnected channels that extends throughout each cardiac muscle cell (Rubio and Sperelakis, 1971; Sperelakis and Rubio, 1971; Sperelakis et al., 1974). This latticework has been given the name "transverse-axial tubular system" (Sperelakis and Rubio, 1971); in view of similar findings in other

——————————————————————————————→

Figure 1. Right ventricular wall of squirrel monkey. The longitudinal axis of the cell replicated by freeze–fracture runs horizontally across this micrograph. Several Z-line levels are indicated (Z), and it is frequently in this region of the sarcomeres that the apertures of transverse tubules (TT) appear, here projecting inward from the E face of the sarcolemma. Scale bar represents 1.0 μm.

Figure 2. Rhesus monkey right papillary muscle. Two longitudinal rows of T-tubules (TT) are visible, each example located at the Z-line level of the adjacent sarcomeres, thus comparing favorably with the sarcolemmal replica shown in Fig. 1. Scale bar represents 1.0 μm.

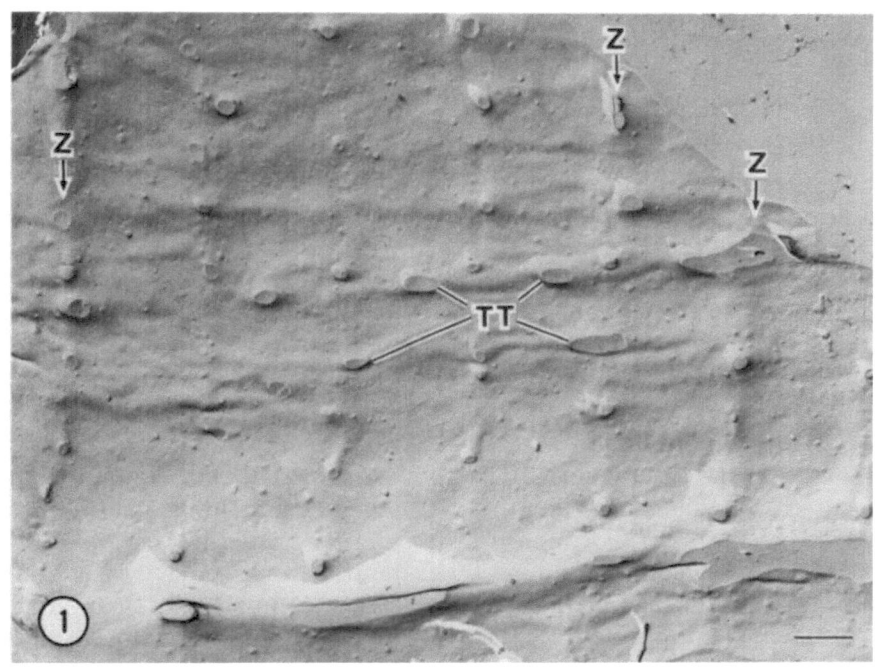

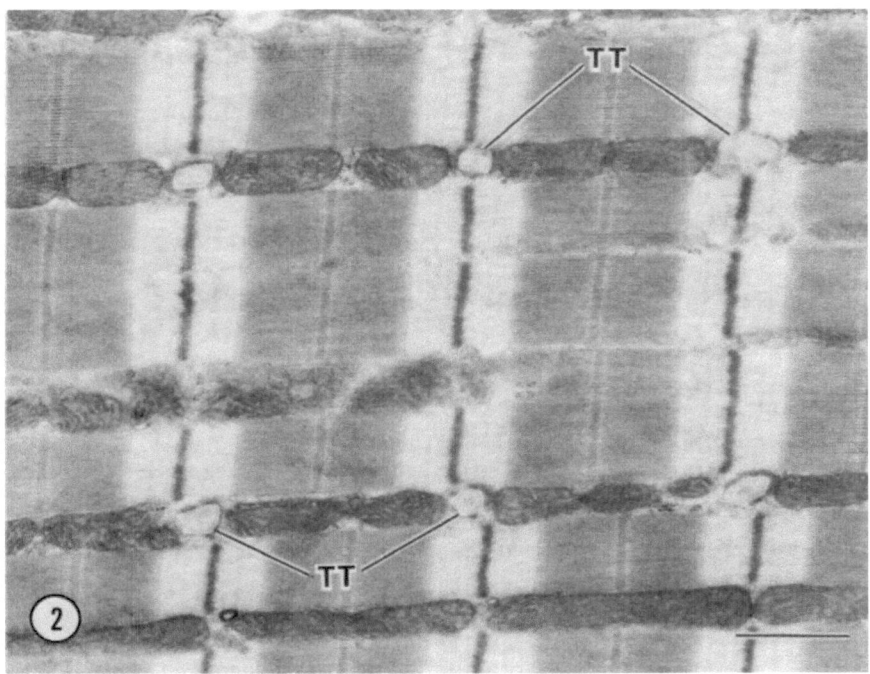

animals (Simpson, 1965; Simpson and Rayns, 1968; Forssmann and Girardier, 1970; Forbes and Sperelakis, 1977), we will use the acronym "TATS" to denote all homologous elements in various regions of the mammalian hearts considered here.

2.1.2. Distribution and Structure of TATS Elements in Adult Mammalian Heart

A substantial portion of the TATS is composed of transverse elements that lie at or near the Z-lines of myofibrils. The distribution implied by 70- to 90-nm thin sections (Fig. 3) can be deceptive, however, as can readily be seen by inspection of stereoscopic pairs of micrographs taken of "semithin" to "thick" (0.3–2.0 μm) plastic sections of myocardium impregnated with an extracellular tracer substance (Fig. 4). The three-dimensional views obtained by this combination of techniques (see Sections 5.1 and 5.2) indicate that in mouse heart the TATS is a complex set of interconnected channels oriented transversely, longitudinally, or obliquely with respect to the long axis of the cell (Fig. 4). This observation differs somewhat from that of Sommer and his colleagues (Sommer and Waugh, 1976; Sommer and Johnson, 1979), who suggest that the TATS of mouse heart consists primarily of transverse tubules. The architecture of the TATS in the transverse cell axis is also better appreciated in thick slices than in thin sections (compare Figs. 5 and 6).

Much of the mouse TATS is composed of small-diameter (up to ca. 180 nm) tubules that are frequently anastomosed and possess irregular dilatations (Figs. 4, 6, and 16). In contrast, the TATS of guinea pig ventricle is composed almost entirely of large-bore (ca. 310-nm diameter) tubules, most of which are either round or slightly ovoid in cross section (Figs. 7, 8, 13, and 14). This propensity to a "finished" TATS seems to prevail in the majority of mammals, and tubule diameters ranging up to 500 nm have been reported for myocardial cells of seal (Ayettey and Navaratnam, 1980) and golden hamster (Ayettey and Navaratnam, 1981) (Table 1).

In the guinea pig the ventricular TATS is known to form a latticework that assures its own presence at all cell levels. Because of this, no point in the myoplasm is further than ca. 0.7 μm from an extracellular fluid space (Rubio and Sperelakis, 1971; Sperelakis *et al.*, 1974). An appreciation of this phe-

$\longrightarrow$

Figure 3.   Mouse right ventricular wall. Tissue prepared with ferrocyanide-reduced osmium tetroxide (OsFeCN) postfixation so as to fill the system of extracellular fluid spaces, including the lumina of the various elements of the T-axial tubular system (TATS) of the myocardial cells, which in this thin section appear in profile as numerous opaque profiles, many of which are located near the Z-lines of the myofibrils, indicating that there is a preponderance of transverse tubules in such cells. Scale bar represents 5 μm.

Figure 4.   Stereoscopic pair (12° tilt angle) of micrographs of "thick" (ca. 2 μm) section from the same block of tissue shown in Fig. 3. In contrast to the seeming predominance of T-tubules indicated in thin sections, a significant population of longitudinal (axial) tubules is revealed, these being anastomosed with the transverse elements. The irregular contours of the mouse TATS are clearly demonstrated, and a number of obliquely directed tubules can be discerned. Scale bar represents 2 μm.

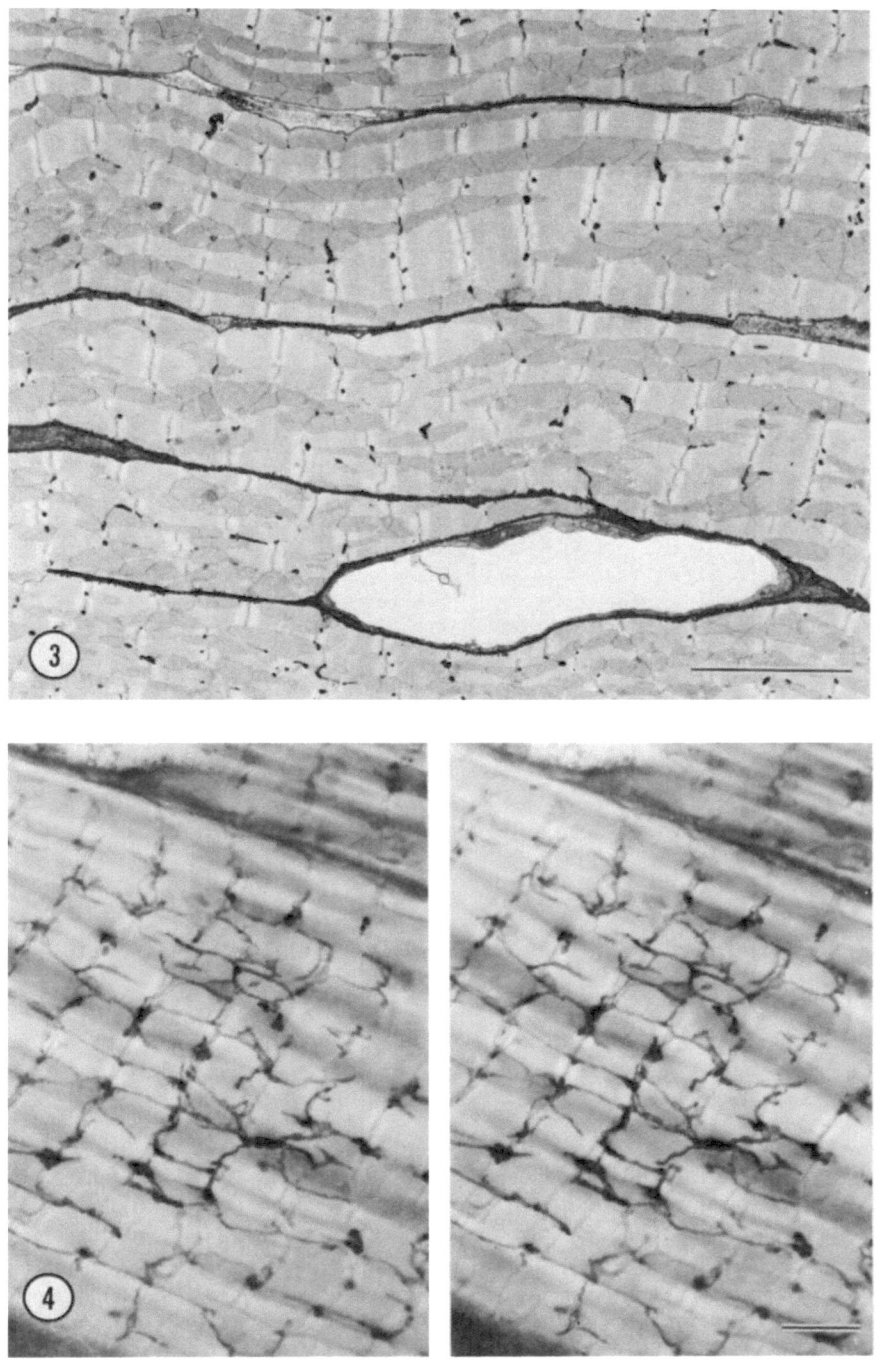

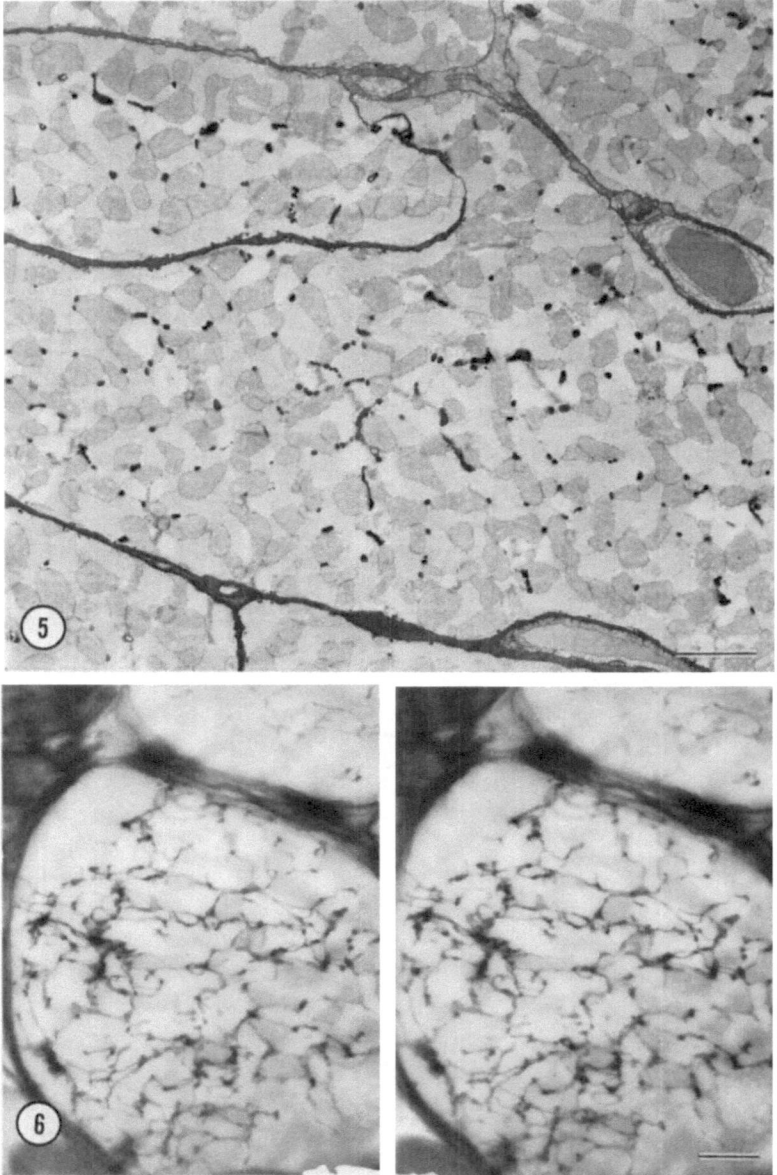

Figure 5. Transverse thin section through mouse ventricle, treated as described in Fig. 3. Though the TATS is traced throughout the depth of these cells, its elements appear only as isolated profiles, many of which appear to represent axial tubules (thus confirming the TATS structure demonstrated in Fig. 4). Scale bar represents 2 μm.

Figure 6. Stereo pair of micrographs (tilt angle 12°) of transverse section (ca. 2 μm in thickness) of TATS-traced mouse ventricular myocardium (same block as in Fig. 5). Substantial interconnection is visible among the elements of the TATS, indicating that a complex latticework exists both in the transverse and longitudinal axes (cf. Fig. 4) of these cells. The dilatations of the mouse TATS are evident in this plane, as they are in the longitudinal plane (Fig. 4). Scale bar represents 2 μm.

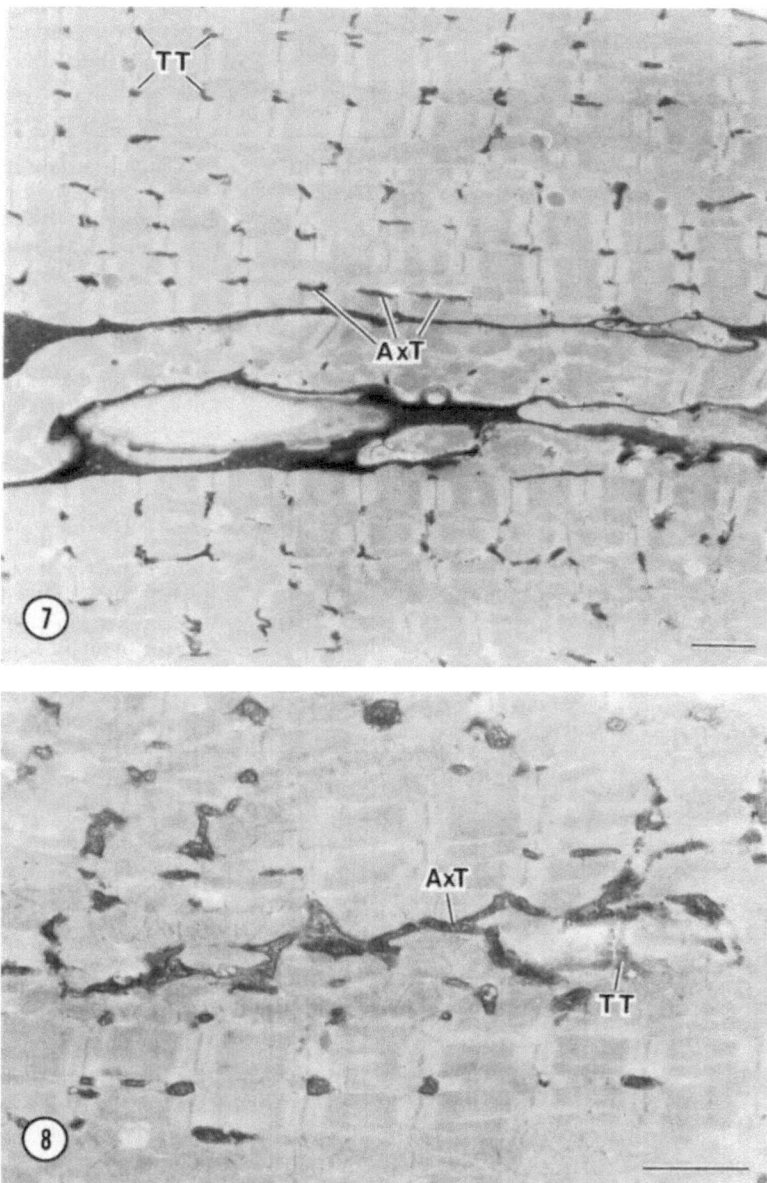

Figure 7. Guinea pig papillary muscle, infiltrated with horseradish peroxidase and reacted with diaminobenzidine and H_2O_2 to produce electron-opaque precipitate. Compare this longitudinal thin section with that of the mouse shown in Fig. 3. Numerous axial elements (AxT) are seen, and the transversely oriented profiles (TT) are noticeably larger and more regular in their placement than are their murine counterparts. Scale bar represents 2 μm.

Figure 8. Extensive anastomosis of transverse (TT) and axial tubules (AxT), substantiating reports that organized TATS latticeworks exist in guinea pig ventricular myocardial cells. The interconnected TATS profile extends for approximately six sarcomere lengths. Scale bar represents 2 μm.

Table 1. Diameters of Transverse-Axial Tubular System (TATS) Elements

Species	Region	Diameter (nm)	Source
Cat	(1) Ventricle (papillary)	150–200	Fawcett and McNutt (1969)
	(2) Atrium	150–200	McNutt and Fawcett (1969)
Cat	Ventricle	104.8	Rubio and Sperelakis (1971)
Rat	(1) Ventricle	>100[a]	Forssmann and Girardier (1970)
		100[a]	
		50–80[a]	
		<50[a]	
	(2) Atrium	80	
Ferret	Ventricle (papillary)	50–80	Simpson and Rayns (1968)
Guinea pig	Ventricle	310	Rubio and Sperelakis (1971)
Guinea pig	Ventricle (papillary)	340	Rayns et al. (1975)
Golden hamster	(1) Ventricle	450–500[a]	Ayettey and Navaratnam (1981)
		200–250[a]	
	(2) Atrium	250	
	(3) AVCS[b]	Absent	
Grey seal	Ventricle	450	Ayettey and Navaratnam (1980)
Sand rat (gerbil)	Ventricle (papillary)	300	Myklebust et al. (1978)
Mouse	Ventricle (papillary)	200	Myklebust et al. (1978)
Mouse	Ventricle	65–180	Forbes (unpublished)
Mouse	Ventricle ("labyrinth")	50–80[a]	Forbes and Sperelakis (1973a)
		80–120[a]	
Opossum	Ventricle	100–200	Hirakow and Krause (1980)

[a] Populations of tubules that were categorized according to diameters.
[b] AVCS, atrioventricular conducting system

nomenon can easily be gained, even from inspection of thin sections of TATS-traced guinea pig ventricle, in which frequent axial tubules and extensively anastomosed TATS profiles can be found (Figs. 7 and 8).

Scanning electron microscopy (SEM) offers potential advantage in the study of myocardial membrane systems (though certain disadvantages of the necessary preparative techniques should be kept in mind: see Section 5.3). Various methods of fracturing, breaking, or tearing cardiac tissue (e.g., Sybers and Ashraf, 1973; Myklebust et al., 1975) have been tested, many of which produce SEM images of regularly spaced, transversely oriented vermiform bodies (Figs. 9 and 10). Some of these structures very probably represent elements of the TATS. Filigrees appear on the surfaces of myofibrils (Fig. 10); it is likely that these result from exposure of the perimyofibrillar sleeves of sarcoplasmic reticulum (SR) (cf. Figs. 24–28).

———————————————————————————————→

Figure 9. Scanning electron micrograph of guinea pig ventricular papillary muscle, prepared by longitudinal tearing of the tissue after alcohol dehydration and critical-point drying from CO_2 (see Section 4.3). The surfaces of individual myofibrils are exposed by this procedure; regularly spaced tubular bodies (arrows) appear, the periodicity of which suggests alignment with the Z-lines and therefore their identification as transverse tubules. Scale bar represents 2 μm.

Figure 10. Detail of Fig. 9, showing structures which correspond, seemingly, to a transverse tubule (TT), an axial branch (AxT), and networks of sarcoplasmic reticulum (N-SR). Scale bar represents 1.0 μm.

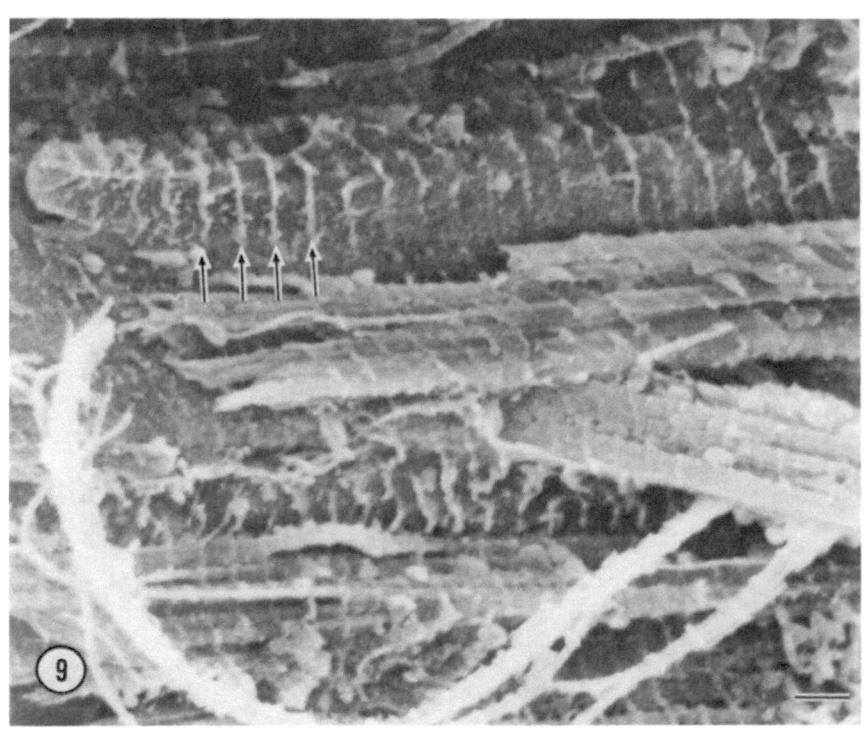

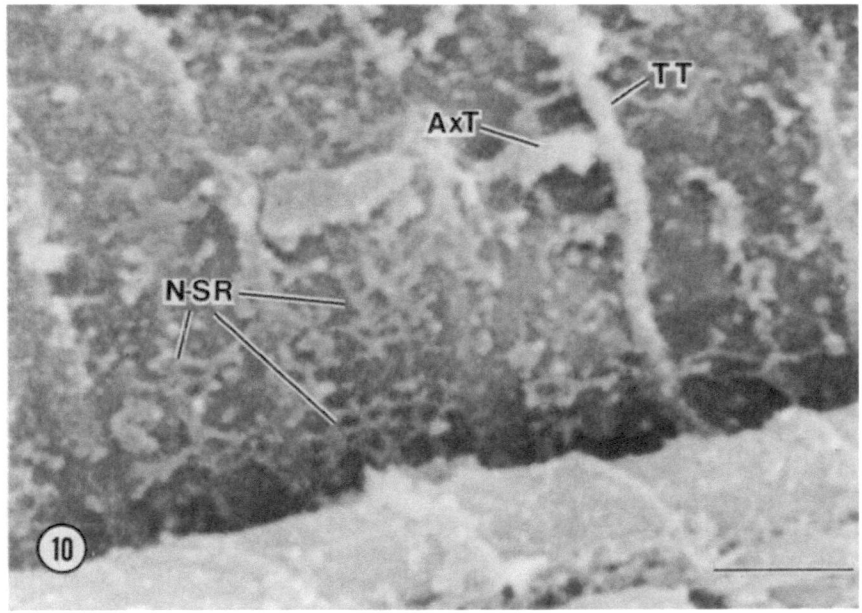

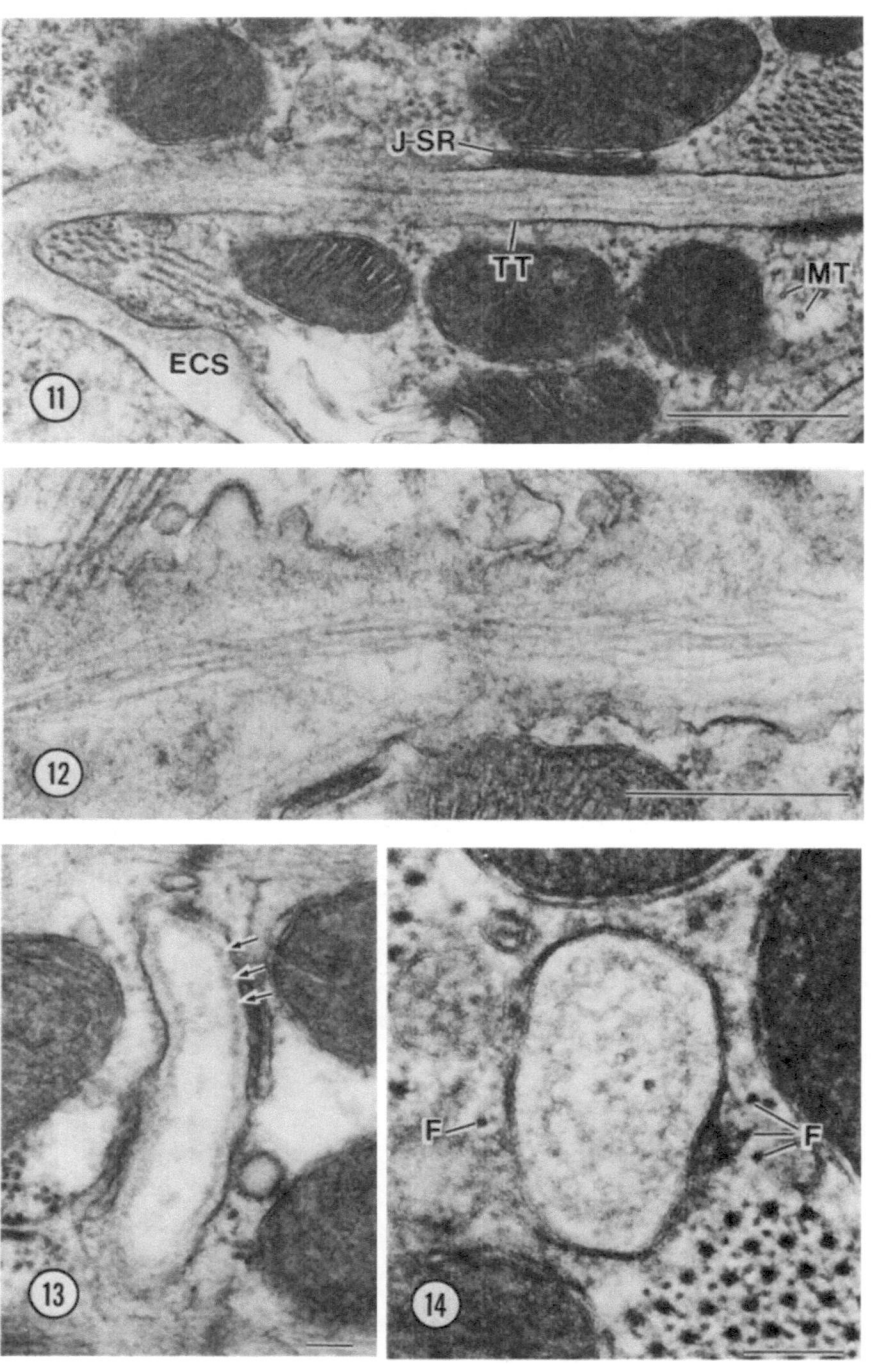

The lumina of those transverse and axial tubules over 100 nm in diameter are usually invested with an extracellular lining of glycoprotein (surface coat, cell coat, glycocalyx, "basal lamina," "basement membrane"). It is as yet unclear whether the smallest TATS constituents, such as the small-bore axial tubules of rat heart (Forssmann and Girardier, 1970) and the "beaded tubules" of mouse myocardium (Figs. 15–17, 19, and 21) are so endowed. The cell coat of the T-tubule may appear in the electron microscope as a linear entity, separated by a space of 20 nm from the tubule's extracellular surface. Extracellular contents of the TATS vary widely, however, forming in many cases a delicate reticulum that is distributed throughout the luminal space (Figs. 14 and 36). The cell coat attains a particularly high degree of organization in many elements of the guinea pig TATS, appearing as an internal veneer composed of discrete fibrils (Figs. 11–14) that course along the longitudinal axis of each tubule, and that are revealed in transverse section as 18- to 25-nm tubular bodies, poised 20–25 nm from the surface of the outer unit-membrane leaflet of the T-axial tubule (Fig. 13).

Although the classic concept of the TATS has included the dismissal of the system as primarily comprising a collection of transverse tubules, it has become quite clear to us that this is not necessarily—or even typically—the case. TATS elements in rat atrium appear to be found only in certain cells, and there are represented primarily by longitudinal (axial) tubules (Forssmann and Girardier, 1970; cf. Leeson, 1980). The absence of a TATS was reported both in guinea pig atrium (Sperelakis and Rubio, 1971) and in cells of the atrioventricular conducting system (AVCS) (Sommer and Johnson, 1968; Thornell and Erikson, 1981). In the latter this remains a moot point, however, since distinct invaginations can be seen in some conducting cells (Osculati *et al.,* 1978).

2.1.3. Caveolae

The development of reliable techniques (see Section 5.1.) for the tracing out of the extracellular fluid space (ECS) and all its contiguous cavities and

---

← ───

Figures 11–14. Features of the TATS in guinea pig ventricle.

Figure 11. Transverse section. A typical large-bore transverse tubule (TT) projects into the myocardial cell; its lumen displays a longitudinal collection of fibrillar structures, unlike the relatively clear extracellular space (ECS) abutting the surface sarcolemma. J-SR, saccule of junctional SR sandwiched between the T tubule and a mitochondrion; MT, microtubules that run longitudinally within the cell. Scale bar represents 0.5 μm.

Figure 12. T-tubule lumen in which fibrillar bodies, ca. 20 nm in diameter, are well defined. Scale bar represents 0.5 μm.

Figure 13. Longitudinally sectioned cell, with a T-tubule caught in cross-section. At the inner rim of the T-tubule there appears a lining of opaque material that at certain points (arrows) can be resolved into tubular profiles, 18–25 nm in diameter. Scale bar represents 0.1 μm.

Figure 14. Cross section of axial tubule in which the extracellular material is organized into a delicate reticulum throughout most of the lumen. F, intermediate filaments oriented longitudinally within the cell and associated with the axial tubule. Scale bar represents 0.1 μm.

recesses—such as the surface-connected caveolae and definitive TATS elements—has provided a "quantum jump" in the potential comprehension of the myocardial cell. The collection of seemingly incidental invaginations of the surface sarcolemma, which are by now known by a variety of terms, such as caveolae, *caveolae intracellulares* (or *intracellularis*), inpocketings, surface vesicles, surface pits, pinocytotic vesicles, among others (Gabella, 1978; Forbes *et al.*, 1979; Forbes, 1982), are now known to be important and conspicuous components of the adult mammalian myocardial cell (yet they are not present in great numbers during its early developmental stages: Figs. 18–20). The occurrence and function of caveolae are considered below.

2.1.3a. Contribution of Caveolae to Formation of the TATS. The hearts of mouse and guinea pig represent the opposite ends of a spectum of TATS development. The guinea pig ventricle begins to develop transverse and axial tubules before birth (Forbes and Sperelakis, 1976; Hirakow and Gotoh, 1980). In the newborn animal these tubules are generally fully formed and of large diameter, and thus equivalent to those of the adult TATS latticework (see Sperelakis and Rubio, 1971). In contrast, the TATS elements of mouse ventricle retain irregular, often "beaded" profiles (Figs. 15, 16, and 21), including the extensive tubulovesicular, three-dimensional arrays termed "labyrinths" (Fig. 17) (see also Forbes and Sperelakis, 1973a, 1977; Forbes *et al.*, 1977; Ishikawa and Yamada, 1976). This "roughed-in" look of the mouse TATS patently betrays its origin from the proliferation of caveolae, and indicates that it has not progressed substantially in architecture since the time of its initial formation.

In the ventricle of the newborn mouse heart, the sarcolemmata of many muscle cells are nearly devoid of infoldings or invaginations. In certain cells, however, there occur extensive ramifications of the sarcolemma—presumably derived from caveolae—that are open to the extracellular space (ECS) (Fig. 18). Distinct "beaded tubules," similar to those of adult heart (Fig. 19; cf. Figs. 15 and 16), are soon developed, and complex entities, reminiscent of labyrinths, can be detected as well (Fig. 20). The caveolar nature of the TATS persists in the adult mouse heart, at times manifesting itself in bizarre silhouettes that form interior couplings with junctional SR (Figs. 15, 16, and 21).

---→

Figure 15.   Mouse heart. In this longitudinal section of conventionally fixed and stained ventricle, a transverse tubule (TT) originates at the surface sarcolemma (aperture indicated by arrow) adjacent to an A-I junction of the nearest myofibril and forms a sinuous, "beaded" profile which leads into a coupling (C) at the Z-line level. Scale bar represents 0.2 μm.

Figure 16.   Field equivalent to that in Fig. 15, but with the TATS traced by means of the OsFeCN technique. Clearly demonstrated are the distinct beaded contour of the T-tubule (TT) as it leads into a coupling, as well as its pronounced flattening between junctional SR saccules. Note the close five-membered association ("pentad": Prasad and Singal, 1977) between the single T-tubule, the pair of J-SR elements, and two mitochondria. Scale bar represents 0.2 μm.

Figure 17.   TATS-traced mouse ventricular cell. A complex ramification of the TATS, known as a "labyrinth," is shown in longitudinal section. Its origin from the TATS is demonstrated both by its filling of opaque material and its association with junctional SR (J-SR). Z, Z-line. Scale bar represents 0.2 μm.

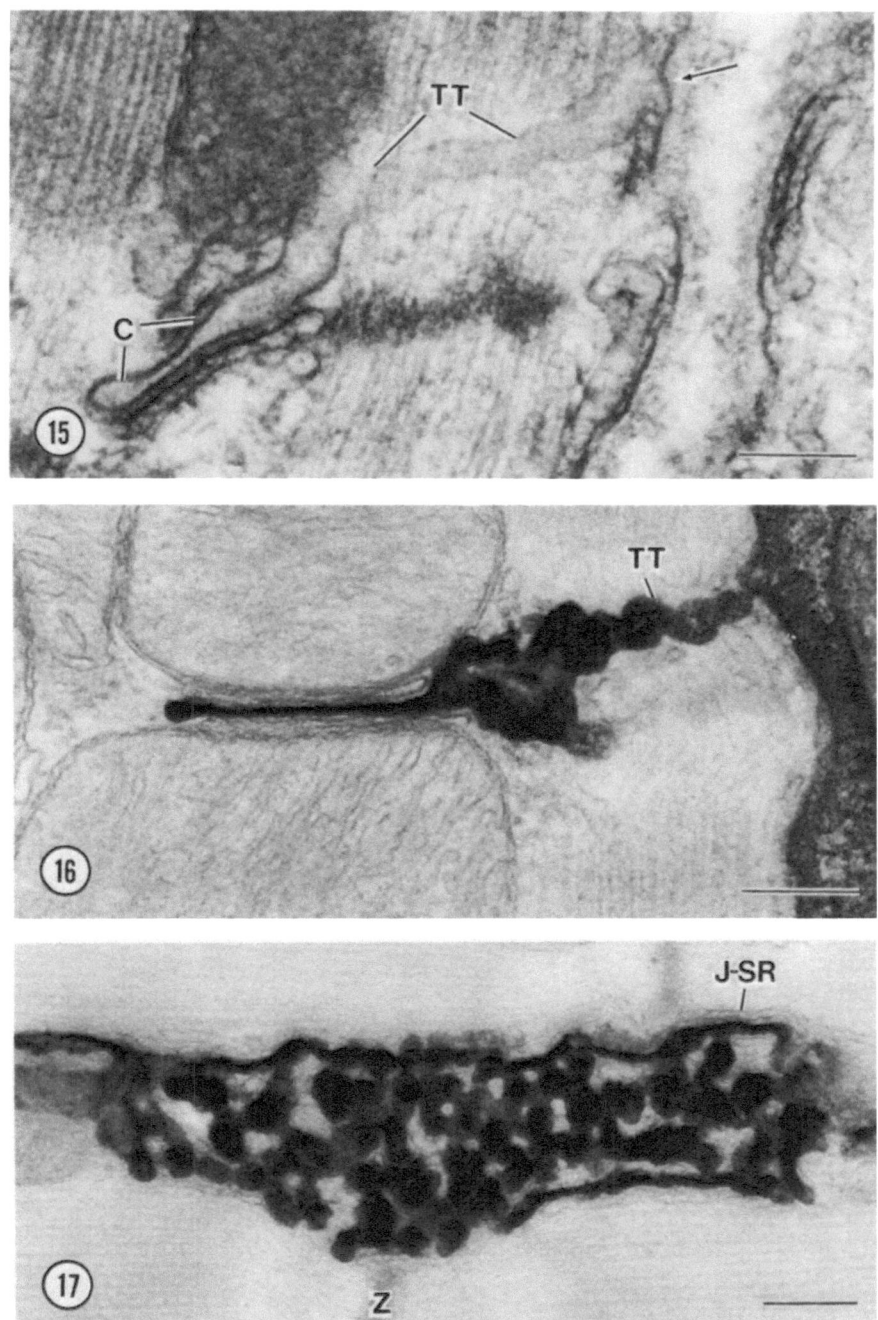

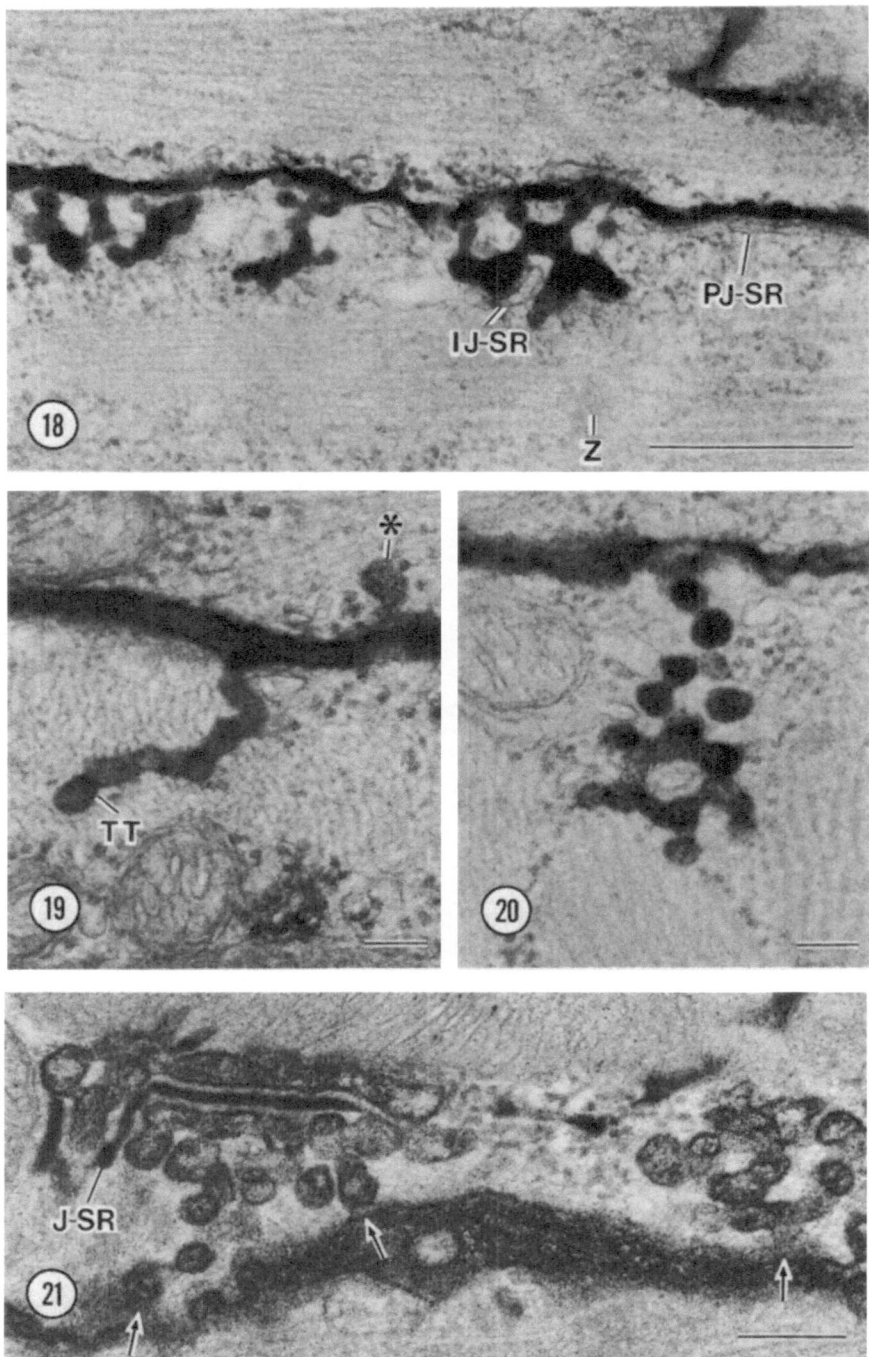

2.1.3b. Caveolae in Adult Heart. In the mouse, caveolae appear in thin sections, taken vertically through the sarcolemma, as individual amphoralike profiles (Figs. 19 and 29). In sections, taken *en face*, that encompass the subsarcolemmal myoplasm, the caveolae are frequently revealed as alveolar bodies (Fig. 22), composed of 2–5 fused caveolar units that interdigitate with the surface arrays of network and peripheral junctional SR (Fig. 22) forming couplings with the latter (Forbes and Sperelakis, 1982b). It is likely that some of these alveolar complexes lead, at deeper levels, into T-tubules, labyrinths, and "in-between" systems such as those illustrated in Fig. 21. It is obvious from inspection of freeze–fracture replicas that the caveolar openings in mouse ventricle tend to collect into groups of up to 40–50 profiles lying in arrays having 100- to 150-nm center-to-center spacing. This clustering of surface invaginations has previously been demonstrated by Ishikawa and Yamada (1976). These authors suggested that there is a preferential location of caveolae at the I-band, but we have not been able to confirm this in our studies (see, for example, Fig. 23). We agree with Ishikawa and Yamada (1976), however, that the unequivocal identification of caveolar depressions vs. T-tubule openings is quite difficult, especially in light of the fact that vectorial caveolar proliferation is responsible for the formation of the TATS.

"Caveolar" openings are distributed primarily on the lateral surfaces of the ventricular myocardial cells. The sarcolemma of the intercalated disc contains relatively few caveolae. Another fact that emerges from such examina-

←

Figure 18. Newborn mouse right ventricular wall; tissue treated with OsFeCN to trace out the system of extracellular spaces, together with any forming elements of the TATS. A profusion of opaque, tubulovesicular profiles is seen connected with the border of the lower, longitudinally sectioned myocardial cell. Untraced, membrane-bound bodies are present, including a subsarcolemmal saccule of peripheral junctional SR (PJ-SR) and a forming saccule of interior junctional SR (IJ-SR), which already has developed junctional processes in the space between it and the opacified tubule, and in addition contains a small amount of intraluminal opaque material. Z, Z-line of myofibril. Scale bar represents 0.5 μm.

Figure 19. Right ventricular wall from 3-day-old mouse. A beaded tubule (TT) has developed as an unbranched invagination, apparently composed of approximately ten caveolae fused end-to-end. A single, surface-connected caveola (*) is present in the adjacent cell. Scale bar represents 0.1 μm.

Figure 20. Right ventricular wall of newborn mouse. Though the components of this opacified TATS element seem at several points to be discontinuous in thin-section profile, their filling with osmium precipitate indicates that continuity among them exists at other levels. The three-dimensional configuration of this body would seem to be that of a beaded tubule emanating from the surface sarcolemma, thence forming a ring-like segment composed of the caveolar units on which the TATS is based. Note the relatively electron-lucent membrane profile in the center of the ring, indicating interdigitation of the SR with the TATS at this early stage of development, perhaps preparatory to forming a labyrinth (cf. Fig. 17). Scale bar represents 0.1 μm.

Figure 21. Adult mouse ventricular myocardium. Persistence of caveolar contribution to the TATS is implied by the presence of beaded silhouettes, many of which can be followed to their connection with the surface sarcolemma (arrows). The intensely opacified structure interleaved with the doublet of TATS components is a saccule of junctional SR (J-SR), which in this section forms a "reversed triad," a combination frequently found in mouse heart (Forbes and Sperelakis, 1977). Scale bar represents 0.2 μm.

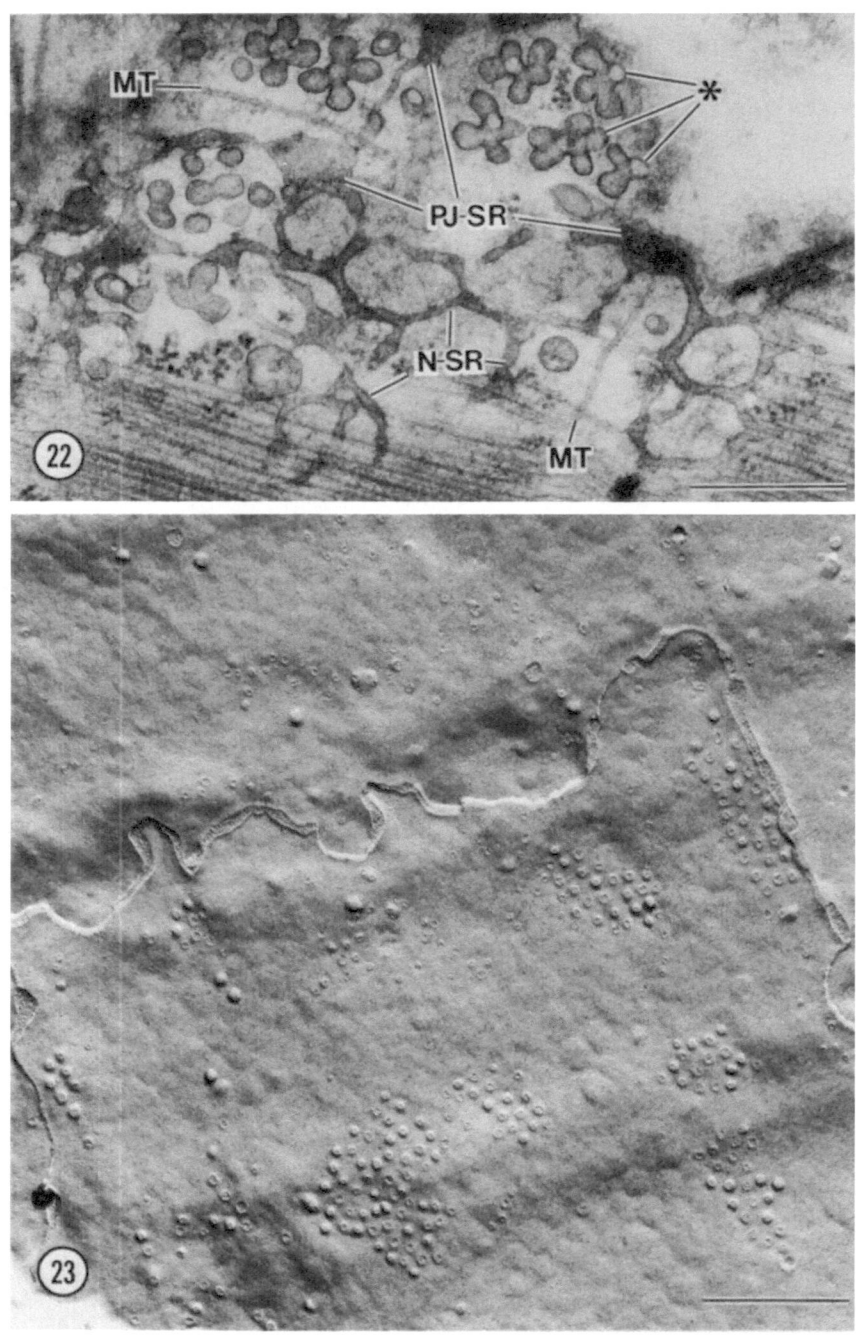

tion is that the surfaces of some cells are remarkably free of surface pits of any category, though other features (such as subsarcolemmal network SR) show them to be otherwise equivalent to adjacent caveola-rich cells (Fig. 23).

There remains a need for definitive comparative work to be carried out, addressed specifically to the occurrence and detailed morphology of caveolae in a spectrum of mammalian hearts. Gabella (1978) can readily distinguish between the T-system and caveolae in the rat by means of freeze–fracture. Likewise, the openings of T-tubules are obvious in the guinea pig ventricular myocardium (Rayns *et al.*, 1967, 1968, 1975; Simpson *et al.*, 1973).

Caveolae are particularly numerous in cells of the atrioventricular conducting system (AVCS) (Sommer and Johnson, 1979; Masson-Pévet *et al.*, 1980) and constitute a significant fraction of the sarcolemmal surface area. It is now generally agreed that caveolae are relatively static entities (as are TATS elements themselves), in contrast to the pinocytotic vesicles of actively transporting cells, such as those of the vascular endothelium (Forbes *et al.*, 1979). One function proposed for the caveolae is the maintenance of a high surface-to-volume ratio in AVCS cells. Caveolae are especially prominent in cells that lack a TATS, such as those of reptilian myocardium (Forbes and Sperelakis, 1971) and certain myocardial cells in rat atrium (Forssmann and Girardier, 1970).

Caveolar proliferation seems largely responsible for production of labyrinths in mouse heart (Fig. 17). Preliminary examination of another fast-beating mammalian heart, that of the least shrew (*Cryptotis parva*), has revealed large labyrinths that, like those of mouse, are open to the ECS, form couplings with junctional SR saccules, and are composed of anastomosed tubules and caveola-like vesicles (Forbes and Mock, unpublished observations). Evidence suggests that there is hyperproduction of myocardial TATS membrane under pathological conditions, such as administration of norepinephrine (Ferrans *et al.*, 1972).

2.1.4. Physiological Role of the TATS (Including Caveolae)

The complex ramifications of the TATS, in combination with caveolar systems, provide a supply of surface membrane to all depths of the myocar-

---

Figure 22.   Mouse left ventricle. Grazing longitudinal section of myocardial cell exposes subsarcolemmal, alveolar groups of fused caveolae (*), platelike expansions of peripheral junctional SR (PJ-SR)—viewed *en face* (cf. Fig. 40)—arrays of network SR (N-SR), and longitudinally and transversely oriented microtubules (MT). Scale bar represents 0.5 μm.

Figure 23.   Freeze–fracture replica of right ventricular wall of mouse heart, the plane of fracture passing along lateral walls of two adjoining myocardial cells. Though presumably these cells are equivalent in internal structure, their sarcolemmata are noticeably dissimilar in that the lower cell bears prominent groups of invaginations (the E face of both cells is replicated here) which in all probability are derived from surface-connected caveolae (though, because of the primitive nature of many mouse myocardial TATS elements, T-tubule openings may be interspersed with caveolar mouths), whereas the upper cell is by comparison bereft of such surface decorations. Note the meshwork pattern in both cells, which corresponds to the subsarcolemmal array of network SR (cf. Figs. 22, 27, and 28). Scale bar represents 1.0 μm.

dial cell. Whether these systems of invaginated membranes are physiologically identical to the surface sarcolemma is less clear, however. The depth to which the cell coat penetrates into caveolae is difficult to determine, since the small diameters of individual caveolae (70–90 nm) lead—in thin sections—to the frequent superposition of their lateral walls upon their lumina. This imparts additional electron opacity, which may be mistakenly interpreted as representing cell coat material. Gabella (1978) and Masson-Pévet *et al.* (1980) consider the cell coat ("basal lamina") to pass over the openings of caveolae without penetrating into their lumina. In some of the smaller, "beaded" examples of the mouse TATS that have small dimensions similar to those of caveolae, it has been difficult to discern a cell coat. However, silver methenamine-stained sections reveal opaque tubulovesicular profiles (Forbes, unpublished observations), thus suggesting that glycoprotein exists throughout the labyrinths and, by extrapolation, within most of the mouse TATS. The question as to the presence of cell coat material in caveolae remains an especially pertinent one, in view of the developmental relationship between caveolae and definitive TATS elements. It has already been suggested that caveolae of muscle cells (Forbes *et al.,* 1979) may represent a specialized population of invaginations, in terms of the properties of their included membrane and/or luminal contents. Carrascal *et al.* (1981) speculate that myocardial caveolae are instrumental in the renewal of cell coat material. Other authors implicate caveolae of various muscle cells in the provision of additional membrane area under conditions of stretch, or suggest that they behave as actual stretch receptors (see Gabella, 1978).

It seems likely that some component—albeit rarely stainable—of the cell coat lines caveolae; frequently the opaque material in TATS elements is found a small distance away from the outer sarcolemmal surface (see Section 2.1.2), indicating the presence of a lucent, close-fitting lining. This may be the case in beaded tubules of developing and adult mouse heart as well. The importance of the cell coat may reside in its ability to capture cations, thus providing a reservoir of Ca^{2+} for the initiation of contraction during excitation (Frank and Langer, 1974; Langer, 1974, 1976; Frank *et al.,* 1977). Several lines of evidence indicate that the cell coat is not vital for excitation or excitation–contraction coupling, however. These include experiments on neuraminidase-treated guinea pig papillary muscle (Sperelakis *et al.,* unpublished observations), as well as observations made on enzymatically dispersed heart cells, either in culture (Sperelakis *et al.,* unpublished observations) or in isolated single cells (Isenberg and Klöckner, 1980). In all these cases the glycocalyx was removed from the cells, but this had no effect on the intensity of the slow inward current or on the properties of the slow action potential.

It remains to be determined what chemical differences, if any, exist between the fluid residing in the ECS proper (that in contact with the "surface"—i.e., noninvaginated—sarcolemma) and that which occupies the collective TATS/caveolar system complex. The distinct architectural disparities that we have demonstrated to exist between the TATS of mouse and guinea pig suggest that different rates of fluid exchange may occur in the two species. The mouse heart, which is endowed with a primitive-appearing TATS, possesses electrophysiological properties that allow the attainment of a contraction rate upwards of 600 beats/min (Geddes, 1970), twice that of the guinea pig. The mouse myocardium thus seems superior—in terms of maximum performance—to the ventricular myocardium of the guinea pig, the TATS latticework of which is a geometric phenomenon probably unequalled among other mammals. However, it has not yet been established what contribution is made to the myocardial surface area by structures such as labyrinths. This figure is likely to be considerable, in view of the data of Masson-Pévet *et al.* (1980), who calculate that in rat heart, caveolae alone increase the surface area of cells of the sinus node and atrium by 115 and 56%, respectively. The existence of labyrinths, dilatations, and various axial and oblique excursions of the TATS may prove to impart a surface-to-volume ratio to myocardial cells of the mouse that is superior to that of more "conventional" mammalian hearts.

2.2. Sarcoplasmic Reticulum (SR)

2.2.1. Terminology

Following the recognition that systems of membranes formed prominent and regular arrays within skeletal and cardiac muscle cells (see Porter, 1961), attention was turned to detailed description of the various membrane system components. In skeletal muscle, there arose a system of terminology that was automatically applied to each corresponding structure in cardiac muscle, such as "sarcoplasmic reticulum" (SR). Some misnomers inevitably arose, "triad" prominent among them, since the SR and transverse (or axial) tubules of cardiac muscle were not always arranged in a tripartite complex (e.g., Forbes and Sperelakis, 1977; Sommer and Johnson, 1979).

Because of certain structural differences that exist among skeletal, cardiac, and smooth muscle cells, that nomenclature which has proven most broadly applicable usually has prevailed. Terms such as *coupling*, denoting any of the specialized appositions of SR with the sarcolemma or its derivatives (Johnson and Sommer, 1967), and *junctional SR* (Sommer and Johnson, 1968), referring to the SR component of the coupling, are likely to remain in vogue. Accordingly, the substructures of junctional SR are frequently referred to as *junctional granules* (the opaque luminal contents) and *junctional processes* (the bodies that extend into the myoplasm from the J-SR). We used the term "network SR" (Forbes and Sperelakis, 1972b) to describe the nonjunctional SR of mouse ventricular cells, and the term has proven applicable in a number of vertebrate hearts and in skeletal muscle (Forbes and Sperelakis, 1974, 1977, 1980a,b; Forbes *et al.*, 1977). Further terminological considerations will be brought up in the following sections as the need arises.

2.2.2. "Network" ("Longitudinal," "Free") SR (N-SR)

The majority of the SR of mammalian heart consists of complex anastomoses of membrane-limited tubules, about 20–35 nm in diameter (Sommer and Johnson, 1979). These have the appearance of "torn sleeves," closely applied to the surface of each collection of myofilaments (myofibrils, or "myo-filamentous masses": McNutt and Fawcett, 1974). Following conventional preparation for TEM (Figs. 22, 24, 40, 49, 53, and 54), the lumina of these SR components bear little discernible particulate material, but when exposed to the appropriate OsFeCN postfixation technique (Section 5.1) they are uniformly opacified (Figs. 25–27; 30–32; and 48, 56, and 57). Semithin sections, taken from selectively stained tissues, have been found to be most efficient for the visualization of myocardial SR (Figs. 25–27; also see Sections 5.1 and 5.2). Freeze–fracture replication is not especially effective in producing images of SR surfaces, but its propensity to cleave within the cell surface membrane frequently reveals the topography of the underlying SR (Figs. 23 and 28).

In mouse and guinea pig heart, the term "network SR" (N-SR) is particularly appropriate when applied to the collection of tubules that lie immediately beneath the surface sarcolemma (Figs. 25, 27, and 28). The configuration of the N-SR can vary substantially, however; around the deeper myofibrils of the cell there frequently appear N-SR profiles that are collected into closely packed retes (Figs. 24 and 26). Such highly structured arrays are segmented according to the underlying sarcomeric pattern. A highly concentrated, single thickness of tubules surrounds A-bands; a looser meshwork of tubules appears over the I-bands (Figs. 24 and 26). In a number of mammals, perforated sheets at the M-line levels have been described (Forbes *et al.*, 1977; Van Winkle, 1977; Sommer and Waugh, 1976; Sommer and Johnson, 1979), similar in appearance to the "fenestrated collars" of skeletal muscle (e.g., Peachey, 1965; Forbes and Sperelakis, 1980b).

The marked pleiomorphism of the N-SR, an entity that collectively incorporates many oblique and transverse elements, invalidates the term "longitudinal SR" that is often applied to these tubules. Another alternative term,

---

→

Figure 24. Longitudinal thin section of mouse ventricular cell in which the surface of a myofibril is grazed, capturing a portion of the perimyofibrillar sleeve of sarcoplasmic reticulum, the segmentation of which across the A-band can be discerned, given the presence of a Z-band at the right (Z) and the beaded profile of a transverse tubule (TT) at the left (note the superposition of myofilaments and membrane systems in the left half of the micrograph). A similarly corrugated axial tubule (AxT) can be made out, which appears to merge with the transverse element. Smooth-surfaced SR tubules, anastomosed into a meshwork ("network" SR: N-SR), are arranged in a planar sheet on the myofibrillar face. Much of the N-SR is this field is tightly packed into parallel arrays of longitudinally oriented tubules. Microtubules (MT) run obliquely and transversely across the myofibril, and one bends to follow the contour of an SR tubule (at arrows). Scale bar represents 0.5 μm.

Figures 25 and 26. Fields from longitudinal "semithin" (ca. 0.3 μm thick) section of mouse right ventricular myocardium, in which large expanses of the SR sleeves (selectively stained with OsFeCN postfixation) are exposed. Two distinct configurations of network SR are evident; in Fig. 25, much of the N-SR is disposed in a rather loose meshwork, whereas in Fig. 26 it forms a close-order plexus of tubules which are particularly concentrated over the central portion of the sarcomere A-band. Other features of the SR made obvious by this technique include expanded regions ("cisternal SR": *) and spherules of corbular SR (*C-SR*) (also see Figs. 48–50 and 57). Scale bars in both micrographs represent 0.5 μm.

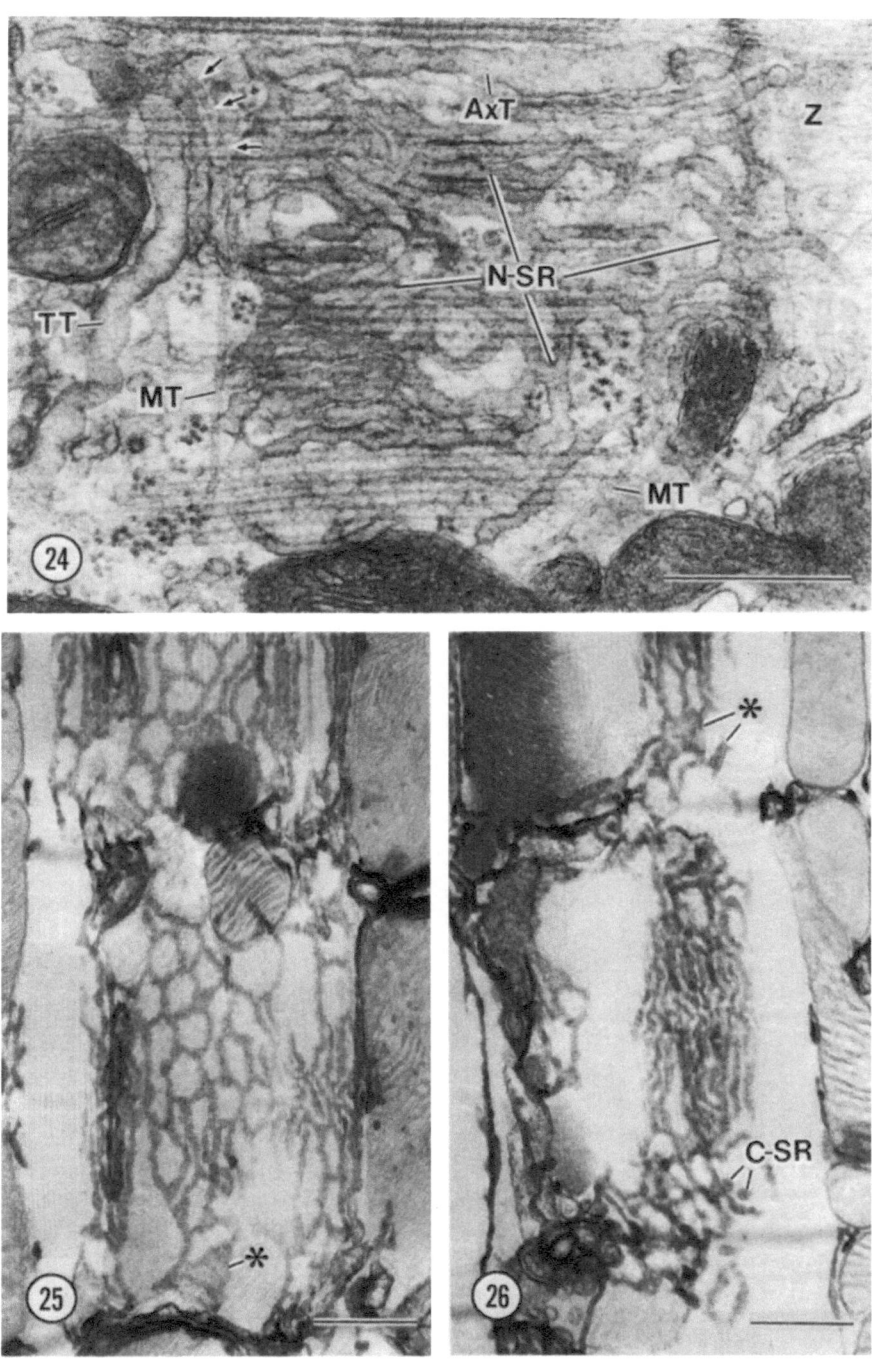

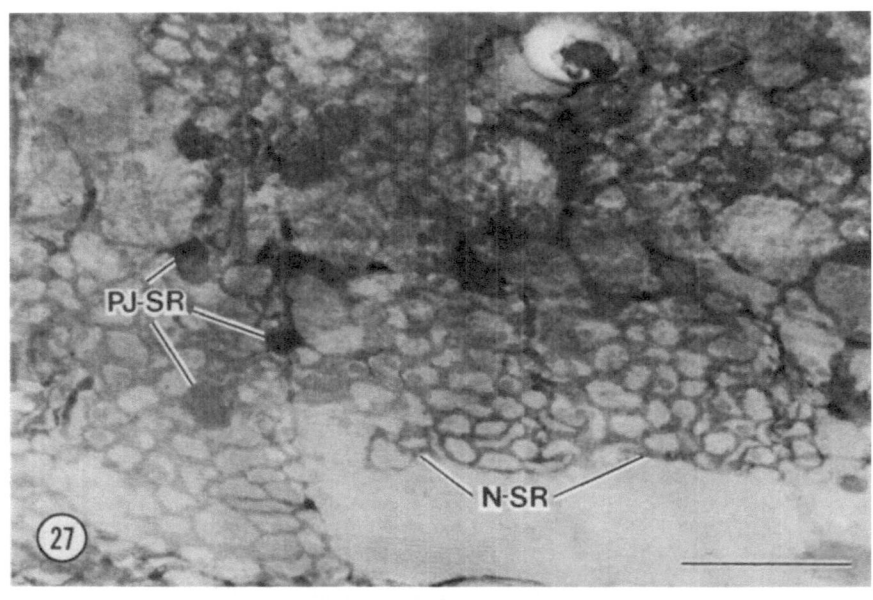

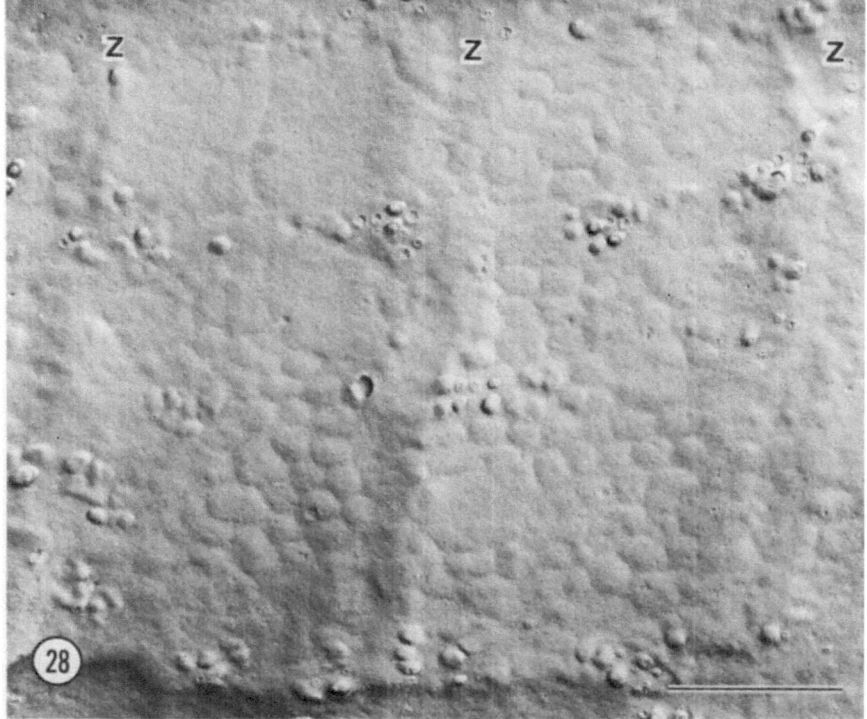

"free SR," seems inappropriate as well, since the N-SR may well be attached to the myofibrils at certain points, such as along the circumference of the Z-discs (see Section 2.2.4a).

2.2.3. *Junctional SR (J-SR) of Couplings*

The second major structural division of the SR is by far the more architecturally complex. The saccules now known as *junctional SR* (J-SR) are analogous to the terminal cisternae of skeletal muscle cells, since they form junctional complexes with the TATS or surface sarcolemma. The limiting membranes of the J-SR are continuous with those of the N-SR (Figs. 29–31). In mammalian myocardium, each typical example of J-SR takes the conformation of a flattened saccule, rather than that of a dilated cisterna as found in skeletal and lower vertebrate cardiac myocytes. J-SR saccules are characterized by opaque luminal contents (junctional granules) that form a more-or-less linear array in thin sections that pass vertically through the coupling (Figs. 29, 33, 36, 37, 42, and 44). J-SR saccules are the contribution of the internal myocardial membrane system to the complexes known as *couplings,* which in their simplest form (the subsarcolemmal, or *peripheral,* couplings) comprise J-SR closely apposed to the myoplasmic surface of the surface sarcolemma (Figs. 22, 29, 30, and 37–42). *Interior* couplings are formed between J-SR saccules and TATS elements, and thus occur at the deeper cell levels, as opposed to peripheral couplings that are consistently subsarcolemmal in location. In the adult mouse, interior couplings appear in significantly greater numbers than peripheral couplings (Bossen *et al.,* 1978). The fundamental characteristic of couplings is the collection of intervening densities—*junctional processes*—that occupy the interspace—the junctional gap—between the apposed J-SR and sarcolemmal surfaces. The substructure of junctional processes is considered at length in Section 2.2.3b.

2.2.3a. *Structural Variations of Couplings.* Peripheral couplings are quite similar in appearance in many mammalian hearts, being relatively simple appositions of SR and sarcolemma. Interior couplings are more varied, appearing in thin sections both as *dyads* ("diads," "dyadoids"), single J-SR saccules apposed to T- or axial tubules, and *triads,* each the combination of two J-SR saccules that abut on a single TATS element. In the mouse, the situation is not so clear-cut, however. In a "triad," SR does not necessarily terminate at the level of the T-tubule, but may extend around it in the form of either N-SR

Figure 27.   Guinea pig right ventricular wall. Longitudinal, 0.25-μm-thick section which grazes surface of SR-stained myocardial cell, revealing the complex expanse of subsarcolemmal SR, including network SR (N-SR) anastomosed with platelike distensions of peripheral junctional SR (PJ-SR). Scale bar represents 1.0 μm.

Figure 28.   Freeze–fracture replica of E face of mouse myocardial sarcolemma (long axis of the cell is in the horizontal plane), shown for the purpose of comparing the distribution of N-SR (its contours visible here as a system of replicated grooves) with that seen in the semithin section in Fig. 27. Z-line levels are indicated. Scale bar represents 1.0 μm.

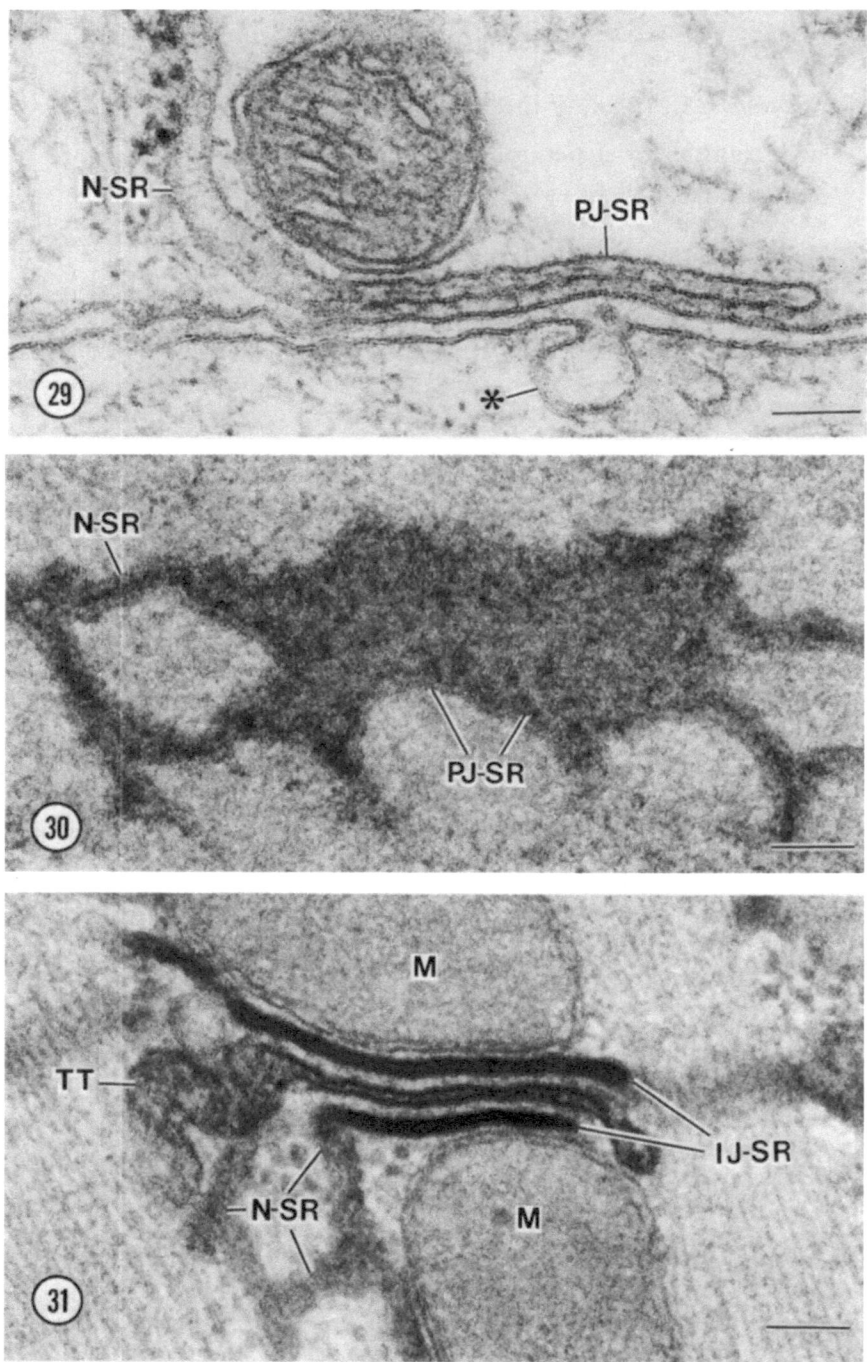

or J-SR; furthermore, this arrangement commonly occurs in both skeletal and cardiac muscle (e.g., Forbes and Sperelakis, 1977, 1980a; Forbes *et al.*, 1977; Sommer and Johnson, 1979). In thin sections of mouse heart, transversely cut TATS elements encircled by J-SR are common (Fig. 32), as are couplings formed by interposed, alternating membrane system elements (Figs. 33–35). Although the latter complexes may be interpreted in thin section as constituting "tetrads," "pentads," "hexads," or "heptads," certain micrographs indicate that such couplings are not composed of multiple individual saccules and tubules, but result rather from the intricate intertwining of the membrane systems (Figs. 35 and 36). "Bidirectional peripheral couplings" are present as well, produced by the entrapment of J-SR between apposed folds of the intercalated disc (Forbes and Sperelakis, 1977; Forbes *et al.*, 1977).

The relatively regular distribution, smooth contours, and large diameters of the "conventional" TATS found in hearts such as those of guinea pig, cat, and dog, are probably the factors that preclude the formation of stratified couplings similar to those found in the mouse. The highly caveolar nature of the mouse TATS, furthermore, imparts a multidirectional proliferation to its elements, exemplified in the architecture of the labyrinth (Fig. 17).

Multiplicity of J-SR is not limited to the mouse heart. "Lamellar" SR has been described in myocardial cells of *Didelphis* (opossum) by Waugh and Sommer (1974b). This particular variation consists of parallel arrays of extensive J-SR saccules, the majority of which give rise to junctional processes from both their appositional faces, as do the "trapped" SR elements of multipartite and bidirectional couplings of mouse heart (Figs. 33–36; also see Forbes and Sperelakis, 1977, 1980a; Forbes *et al.*, 1977).

2.2.3b. Junctional Processes. The hallmark of couplings in all types of muscle cells is the presence in the junctional gap of intervening densities, which have been known variously as "dimples" (Kelly, 1969), "SR feet" (Franzini-Armstrong, 1970), and "junctional processes" (Sommer and Johnson,

←——————————————————————————————————————

Figure 29.   Peripheral coupling in mouse ventricular cell, formed by apposition of junctional SR saccule (PJ-SR) to the inner sarcolemmal surface. The J-SR, with its associated intraluminal junctional granules (which in this section form an opaque line which bisects the luminal depth: the "coextensive density") and junctional processes, dense bodies of various configurations in the gap between SR and sarcolemma (also see Figs. 37–39), is clearly connected with the network SR (N-SR). Note the distinct flattening of the J-SR relative to the anastomosed N-SR; this is apparently brought about by the presence of junctional granules. The J-SR is caught, along part of its length, between the sarcolemma and a mitochondrion; such complexes are frequently encountered in the myocardium (cf. Fig. 11, 16, 31, and 42–44). A surface-connected caveola (*) is present in the adjacent cell. Scale bar represents 0.1 μm.

Figure 30.   SR-stained mouse ventricular myocardium. A portion of the SR system, equivalent to the elements shown in Fig. 29, but displayed *en face*, demonstrates the continuity of the peripheral J-SR saccule (PJ-SR) with N-SR tubules. Scale bar represents 0.1 μm.

Figure 31.   Interior coupling ("triad") in mouse ventricle, composed of two profiles of junctional SR (IJ-SR) separated by a T-tubule (TT). Two mitochondria (M) are closely applied to the nonjunctional surfaces of the J-SR (cf. Fig. 16). N-SR, network SR that leads into one of the J-SR saccules. Scale bar represents 0.1 μm.

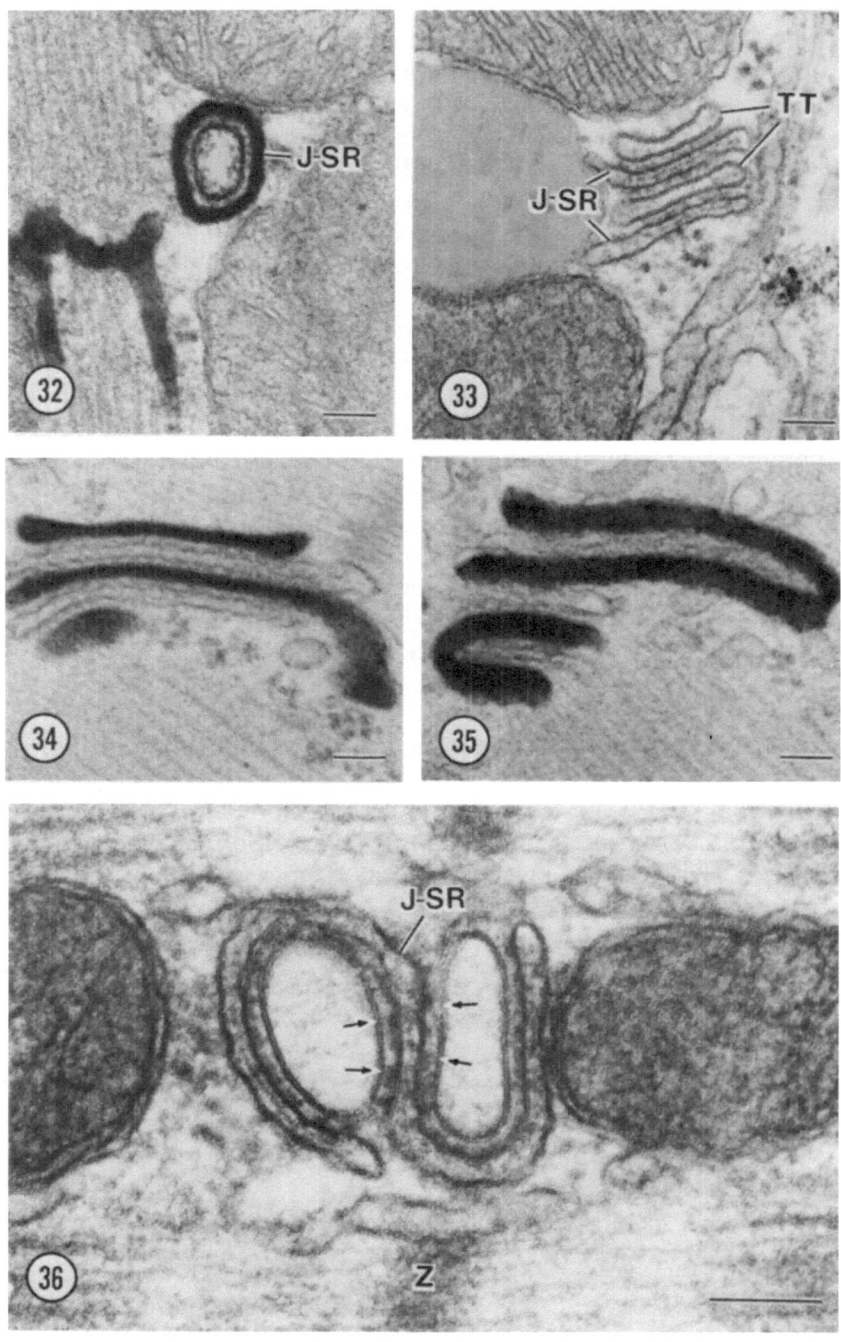

1968). This last term has stuck, and we have chosen in our work to use it to denote all discrete structures found in the junctional gap, whether they appear as clusters of opaque material or as membrane-based projections of the J-SR limiting membrane.

The basic structure of the periodically spaced junctional processes was interpreted by D. E. Kelly (1969)—from his examinations of vertical and *en face* thin sections of larval amphibian skeletal muscle—as being hemispherical eversions of the adsarcolemmal J-SR membrane face. Each hemisphere was thought not to make membrane-to-membrane contact with the T-tubule, but rather to connect via an amorphous adhesive substance. By means of infiltration of skinned frog sartorius fibers with ferritin particles, Franzini-Armstrong (1971) concluded, however, that such adhesive material did not fill the junctional gap. Freeze–fracture replication of couplings gave equivocal results: some workers found particles deemed correspondent with junctional processes (Rayns *et al.*, 1975), whereas others detected no such structural features in the membranes of either SR or T-tubules (Franzini-Armstrong, 1974, 1975, 1977). More recently, modifications to freeze–fracture regimens have produced replicas of skeletal muscle that display grouped intramembrane particles in the T-tubule membranes (Franzini-Armstrong, 1980) or aligned rows of bulges on the J-SR that are identical in position to junctional processes seen in *en face* thin sections (Kelly and Kuda, 1979). Therefore, in skeletal muscle, it seems that the structure of junctional processes is based on membranes.

Comparable freeze–fracture technique has not yet been applied to cardiac muscle. Nevertheless, our own investigations (Forbes and Sperelakis, 1982b) made on thin sections of myocardium confirm the observations of Somlyo (1979) and Eisenberg and Gilai (1979) on skeletal muscle and Somlyo (1979) on vascular smooth muscle, by the demonstration of distinct linear bridges ("pillars": Eisenberg and Gilai, 1979) between the J-SR and T-tubule (or surface sarcolemma). Pillars are unit membrane-like and are fused with

---

Figures 32–36. Structural variations of couplings in mouse ventricular myocardium.

Figure 32. SR-stained tissue. In this longitudinally cut cell, the transverse tubule is completely encircled by a profile of junctional SR (J-SR). Scale bar represents 0.1 μm.

Figure 33. Conventionally stained tissue. "Tetrad," composed of two profiles each of T-tubules (TT) and junctional SR (J-SR). Note that the upper J-SR saccule has produced junctional processes on both its faces that are apposed to T-tubule membrane. Scale bar represents 0.1 μm.

Figure 34. TATS-stained tissue. "Pentad" formed by interdigitation of three TATS profiles (opacified bodies) and two saccules of J-SR. Scale bar represents 0.1 μm.

Figure 35. TATS-stained tissue. Multilayered coupling with three J-SR saccules and two TATS profiles, each of the latter bent into a hairpin shape about J-SR. A section taken in another plane through this coupling could reveal as many as seven, apparently separate, coupling components, leading to its labeling as a "heptad." Scale bar represents 0.1 μm.

Figure 36. Conventionally prepared tissue. In this coupling, an extensive saccule of J-SR winds in a reverse "S" curve around two T-tubular profiles. Note junctional processes on both J-SR faces where the saccule is sandwiched between T-tubules. Z, Z line. Scale bar represents 0.1 μm.

the outermost membrane leaflets of the two coupled components. Such pillars can be made out in myocardial couplings of tissue contrasted *en bloc* with aqueous uranyl acetate solution (tannic acid treatment is not requisite to their demonstration, though recommended by Somlyo [1979]). Such interconnections (Figs. 37–39) may coexist with other forms of junctional processes whose profiles resemble inverted stirrups (Figs. 38 and 39), such as would be created by sections passing vertically through membranous eversions of the J-SR, similar to the "dimples" originally described by D. E. Kelly (1969).

In cardiac, skeletal, and vascular smooth muscle cells alike, pillars can be rendered visible in the transmission electron microscope by the use of a tilting specimen stage; for example, as little as 5–10° of tilt is sufficient to effect the resolution of a pillar, in a region in which in "flat" view there previously appeared only a poorly defined intermembranous density (Forbes and Sperelakis, 1982b). The coexistence of a variety of types of intervening processes between SR and sarcolemma, differing in intrinsic morphology, cannot at this point be ruled out.

In thin sections that encompass peripheral J-SR saccules, the superposition of junctional processes upon the J-SR adsarcolemmal membrane and the underlying junctional granules creates an ambiguous mélange of opaque material. In some micrographs there can be detected a pattern composed of interlocked punctations (Figs. 40 and 41); it is difficult to discern which structures represent the junctional processes, which the junctional granules, and which the "lattice fibers" (lateral connections between adjacent junctional processes: Sommer and Johnson, 1979). The judicious application of stereoscopy carried out on very thin sections, the use of hyperosmotic fixatives (Page and Upshaw-Earley, 1977), and improved methods of freeze–fracture replication (Kelly and Kuda, 1979) may eventually reveal the two- and three-dimensional geometry of myocardial junctional processes. The potential physiological sig-

-->

Figure 37.   Peripheral coupling in mouse heart. Long saccule of J-SR is connected to the inner sarcolemmal surface by a variety of electron-opaque bodies ("junctional processes": several indicated by arrows). Scale bar represents 0.1 μm.

Figures 38 and 39.   High-magnification details of coupling shown in Fig. 37. Junctional processes appear in varying configurations, including the extremes of amorphous densities and membranelike bodies. The latter may form U-shaped profiles (*) which extend from the J-SR, or connect the outer leaflets of J-SR and sarcolemma as linear "pillars" (between arrows). Scale bars represent 0.05 μm.

Figure 40.   Thin section that grazes the surfaces of two mouse ventricular myocardial cells. Connective tissue elements passing close to the sarcolemmata include collagen fibrils (Co) and a bundle of microfibrils (MF) (cf. Fig. 73). Saccules of peripheral junctional SR (PJ-SR), sectioned *en face*, can be distinguished from the associated network SR by their dilated profiles (cf. Fig. 30) and more opaque contents. MT, microtubules. Scale bar represents 0.2 μm.

Figure 41.   Detail of PJ-SR in Fig. 40. Its opacity is derived from the superposition of various of its components, including its intraluminal junctional granules, its adsarcolemmal limiting membrane and the junctional processes, these last probably accounting in large part for the pattern of interlocked punctate opacities (several indicated by arrows). Scale bar represents 0.05 μm.

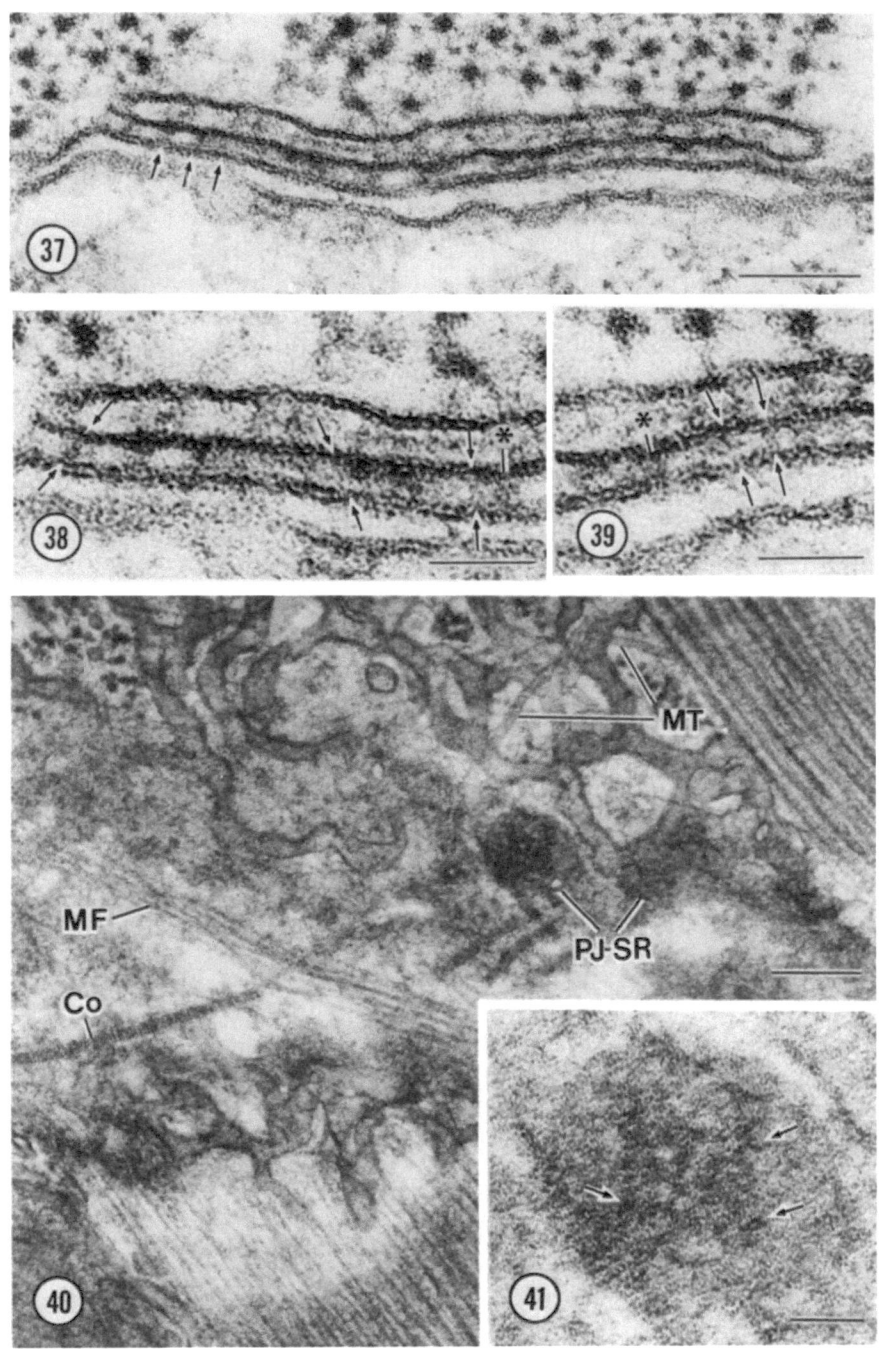

nificance of junctional processes in general, and of pillars in particular, is discussed in Section 2.2.7.

2.2.3c. Possible Specific Association of J-SR with Myocardial Mitochondria.   It is debatable to just what degree mitochondria contribute to the handling of Ca^{2+} in the normal myocardial cell. Yet mitochondria are the second most voluminous constituent of the cardiac muscle cell (e.g., Sommer and Johnson, 1979), and can be found in many mammalian hearts to lie in potentially strategic locations, for example, near the gap junctions (Forbes and Sperelakis, 1982a). It is inevitable, given the sizes and numbers of cardiac mitochondria, that they will by mere random topography appear to be specifically aligned with one or another component. The combination of two mitochondria with a "triad"-type coupling, in the longitudinal order of mitochondrion : J-SR : T tubule : J-SR : mitochondrion (Figs. 16 and 31) appears so often in dog heart as to have been deemed a "pentad" (Prasad and Singal, 1977; not to be confused with pentads composed purely of membrane-system elements). This juxtaposition may on some occasions be the result of nothing more significant than the lodging within the intermyofibrillar space of mitochondria against an interior coupling, the gravitation of which to the Z-line is in fact rather specific (Forbes and Sperelakis, 1980a). Likewise, peripheral J-SR saccules are frequently found sandwiched between mitochondria and the sarcolemma (Figs. 11 and 42). In some instances, prominent connections can be discerned between mitochondria and the "nonjunctional" surfaces of junctional SR (Figs. 43 and 44). Such structures are reminiscent in appearance of both the "pillars" in the junctional gap of couplings (see Section 2.2.3b) and the "intermembranous cross-bridges" described between adjacent membranes in a number of cells of nonmuscle origin (Franke *et al.*, 1971). The bridges between mitochondria and J-SR have been seen so far only in mouse heart, and it remains to be determined whether they occur in myocardial cells of other species. In this regard, it is pertinent to note the apparent connections formed with mitochondria by cisterns and tubules of SR seen in freeze–fracture replicas of dog heart (Scales, 1981).

2.2.4. Specialized, "Nonjunctional" Forms of SR

2.2.4a. Z-Tubules.   The network SR, far from being an undistinguished collection of randomly distributed tubules, is frequently disposed in sarcomere-oriented arrays. The so-called "M-retes" (Figs. 24 and 26) and fenestrated collars are prominent examples of specific segmentation.

→

Figure 42.   Mouse ventricle. In this field, a saccule of peripheral J-SR (PJ-SR) is sandwiched between the sarcolemma and a mitochondrion (M), a configuration commonly encountered in heart. A tubule of network SR (N-SR) is closely apposed to the mitochondrion as well. FA, *Fascia adherens* of intercalated disc. Scale bar represents 0.2 μm.

Figures 43 and 44.   Mitochondrion: J-SR apposition in mouse ventricular cell. Here the J-SR saccule is apposed to a transverse tubule (TT); distinct linear connections appear between the mitochondrial and J-SR limiting membranes; these are shown (arrows) at greater magnification in Fig. 44. Scale bar in Fig. 43 represents 0.2 μm; in Fig. 44, 0.05 μm.

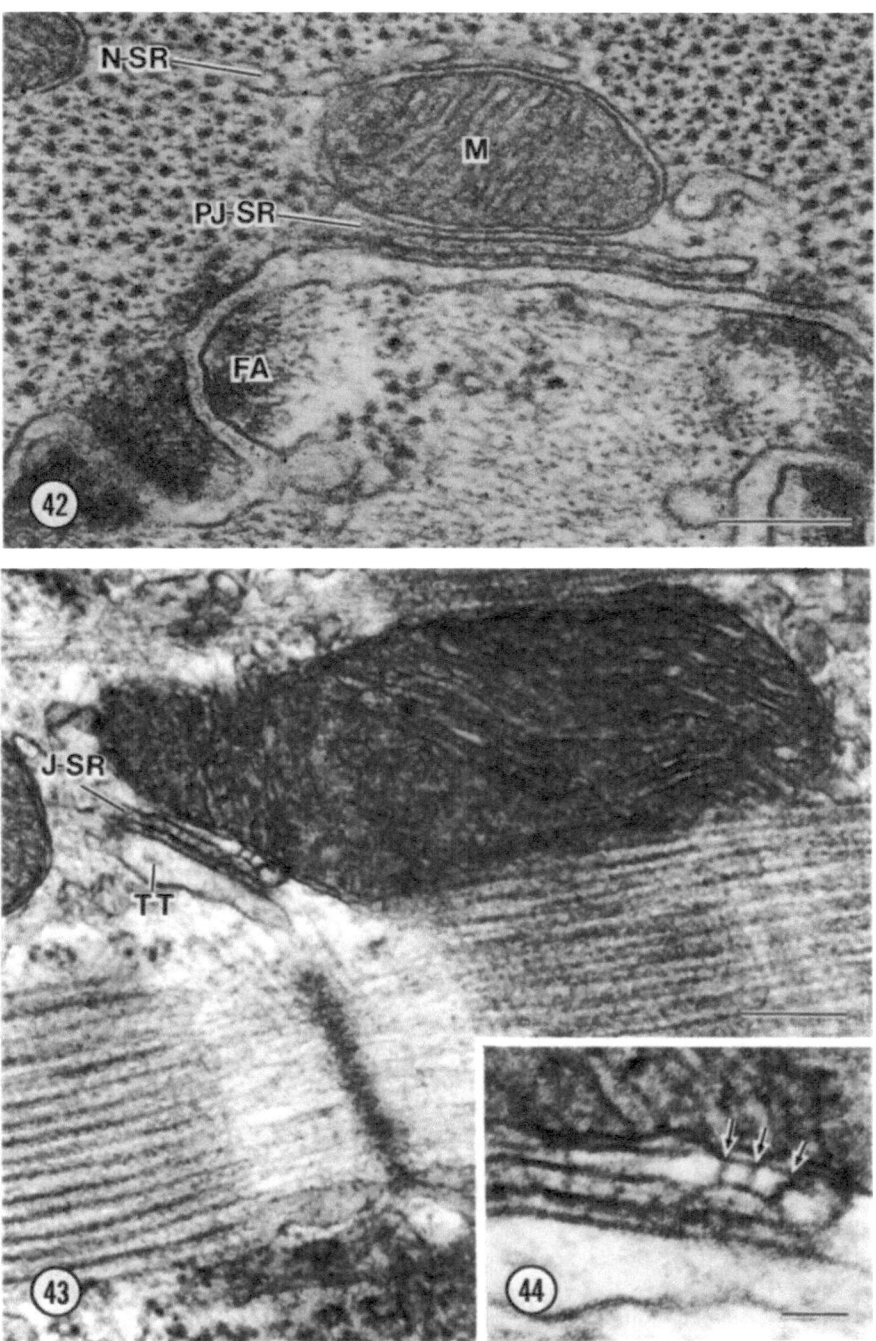

A noteworthy specialization of the SR is that portion which encircles Z-lines. This particular entity was named the "Z-tubule" (Simpson, 1965; Simpson and Rayns, 1968) (not to be confused with sarcolemma-connected invaginations of invertebrate heart [Sperelakis, 1970; Forbes *et al.*, 1972]). The density of myocardial SR is greatest around the Z-line levels of the sarcomeres, where double layers of SR often are present between adjacent myofibrils, whereas single layers appear along other levels of the sarcomere (Fig. 45). Smaller myofibrils in mouse heart may be completely surrounded by closely applied Z-tubules (Fig. 46), which at some points exhibit linear connecting structures between their membranes and the adjacent myofibril's surface (Fig. 47).

Since the first description of Z-tubules by Simpson (1965), there has ensued a fair amount of debate concerning their presence or absence in various myocardial cells. In the guinea pig, Simpson *et al.* (1973) were able to discern Z-tubules in transverse, but not in longitudinal, sections of ventricular cells. In conventionally prepared mouse heart, profiles of Z-tubules are more readily observed in transverse sections (Figs. 45 and 46). This is probably attributable to the relatively low contrast offered by SR tubules when they are superimposed upon the myofibrils (see Section 5.1). Our recognition of murine myocardial Z-tubules has been facilitated by the use of SR-stained preparations (Figs. 45 and 48). The occurrence of Z-tubules in some mammalian hearts has been denied (McNutt and Fawcett, 1974), but Z-tubules have been described in ox (Simpson, 1965), ferret (Simpson and Rayns, 1968), guinea pig (Simpson *et al.*, 1973), dog (Edge and Walker, 1970), mouse (Sommer and Waugh, 1976; Forbes and Sperelakis, 1980a), hamster (Sommer and Waugh, 1976) and opossum (Sommer and Johnson, 1979).

2.2.4b. Corbular ("Extended Junctional") SR. Various reviews and extensive primary studies directed toward the myocardial cells have included incidental observations on bodies referred to as "coated vesicles" (Fawcett and McNutt, 1969) and "coated dense vesicles" (Simpson *et al.*, 1973). These

---→

Figure 45. Transverse thin section of SR-stained mouse myocardium. The SR is most heavily concentrated about those myofibrils cut at their Z-line levels; transversely directed tubules are nearly continuous adjacent to the Z-line material, forming double layers at some points around its circumference. By contrast, the SR associated with the A- and I-bands is frequently discontinuous, its punctate profiles indicating that the section passes through longitudinally disposed tubules of the network characteristic at those sarcomere levels (cf. Figs. 22, 24, and 25). Scale bar represents 0.5 μm.

Figure 46. Transverse section of conventionally prepared mouse ventricle. A small myofibril is sectioned through one of its Z-discs (Z); at this level it is completely surrounded by the SR, which forms a girdling structure known as a "Z-tubule" (ZT). Scale bar represents 0.1 μm.

Figure 47. Detail of Fig. 46. The Z tubule is connected to the substance of the Z disc by opaque fibrils (arrows). Scale bar represents 0.05 μm.

Figure 48. SR-stained mouse ventricular myocardium. This longitudinal section has captured the interconnection of three SR categories: network SR (N-SR), leading into three examples of corbular SR (C-SR) and a prominent Z-tubule (ZT) that is so well aligned with the Z-line as to occlude it. Scale bar represents 0.2 μm.

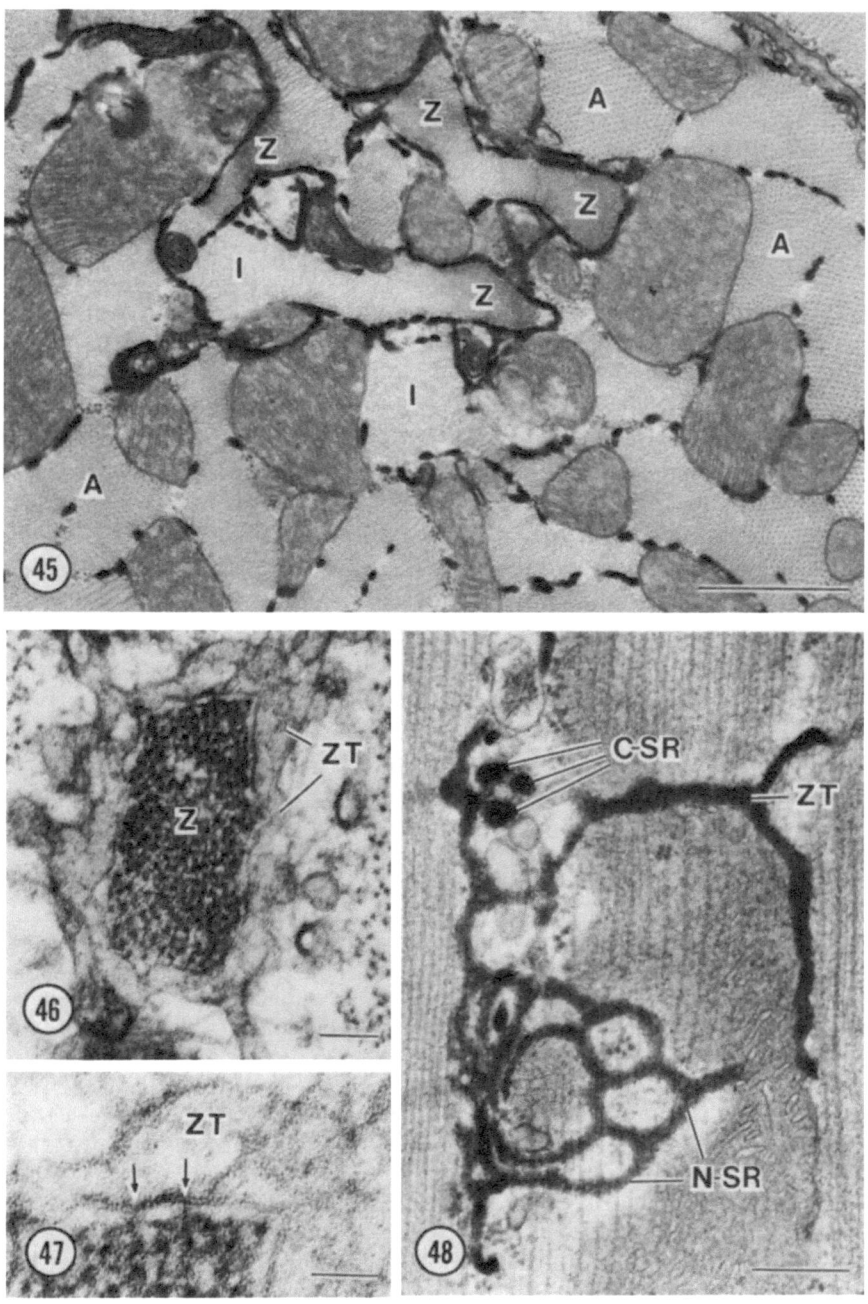

bodies, when encountered as isolated circular profiles, can be confused with "bristle-coated vesicles" or "acanthosomes"—a family of morphologically similar structures that may either exist free in the myoplasm or fused with surface sarcolemma or TATS elements. Many such bodies are, however, intrinsic segments of the SR, as is readily apparent from inspection of SR-stained material (Fig. 48). For such bulbous bodies, Sommer and Waugh (1976) introduced the term "corbular SR," denoting their resemblance to small baskets. It subsequently became obvious that corbular SR was an equivalent to the "extended junctional SR" (EJ-SR) described previously in studies of avian heart (which lacks the TATS: Jewett *et al.*, 1971, 1973; Sommer and Johnson, 1979). Such EJ-SR frequently appears at the Z-line levels of sarcomeres, where interior couplings would be formed in mammalian heart; avian EJ-SR resembles free-floating J-SR saccules and exhibits the salient features of true junctional SR, namely, junctional granules and processes. EJ-SR elements are freestanding, however, rather than abutted to another membrane as in the case of couplings. Likewise, corbular SR of mammalian heart often occupies Z-level myoplasm (Fig. 48; also see Forbes and Sperelakis, 1980a); it may, however, appear as interconnected vesicular chains disposed longitudinally toward other sarcomere levels (Fig. 49). The luminal contents of corbular SR exhibit a diffuse electron opacity (Figs. 49 and 50), and the coating of the limiting membranes can be resolved in side-view into semicircular or stirruplike profiles (Fig. 50), similar in appearance to junctional processes.

2.2.4c. Cisternal SR. There exist certain expansions of the SR that have come to be called "cisternal SR" (C-SR) (Dolber and Sommer, 1980; Scales, 1981). Studies of the C-SR have relied primarily on freeze–fracture (Figs. 49, 51, and 52). Frequently, cisternal SR appears as a saccular distension flattened in the myofibrillar plane and containing a collection of electron-opaque material, thus rendering such saccules quite similar in appearance to *en face* views of junctional SR—with which we previously (Forbes and Sperelakis, 1977) have confused them, as Scales (1981) rightly pointed out. However, it seems

→

Figure 49.   Longitudinal thin section of mouse ventricular myocardial cell. An extensive array of SR is present, network SR (N-SR) being most prominent. Specialized SR components appear near the Z-lines; these include numerous racemose bodies of corbular SR (C-SR) and a distended saccule with opaque contents ("cisternal" SR: Cs-SR). MT, microtubule directed obliquely across the SR. Scale bar represents 0.5 μm.

Figure 50.   Detail of the corbular SR in Fig. 49. The surfaces of such spherular bodies resemble those of junctional SR in that they bear structures (*arrows*) which in side view resemble U-shaped junctional processes (cf. Figs. 38 and 39), and in *en face* view appear punctate. Scale bar represents 0.1 μm.

Figure 51.   Cisternal SR saccule in Fig. 49. The opaque core of the saccule is made up of a collection of punctate densities similar to those which characterize *en face* views of peripheral junctional SR (cf. Figs. 40 and 41). Scale bar represents 0.1 μm.

Figure. 52.   Guinea pig ventricle, SR-stained. A saccule of SR, similar in overall form to the cisternal SR shown in Fig. 51, but not aligned with the Z-level, displays numerous punctate bodies, likely to be ribosomes, since OsFeCN staining of the SR would obscure the internal contents. Scale bar represents 0.2 μm.

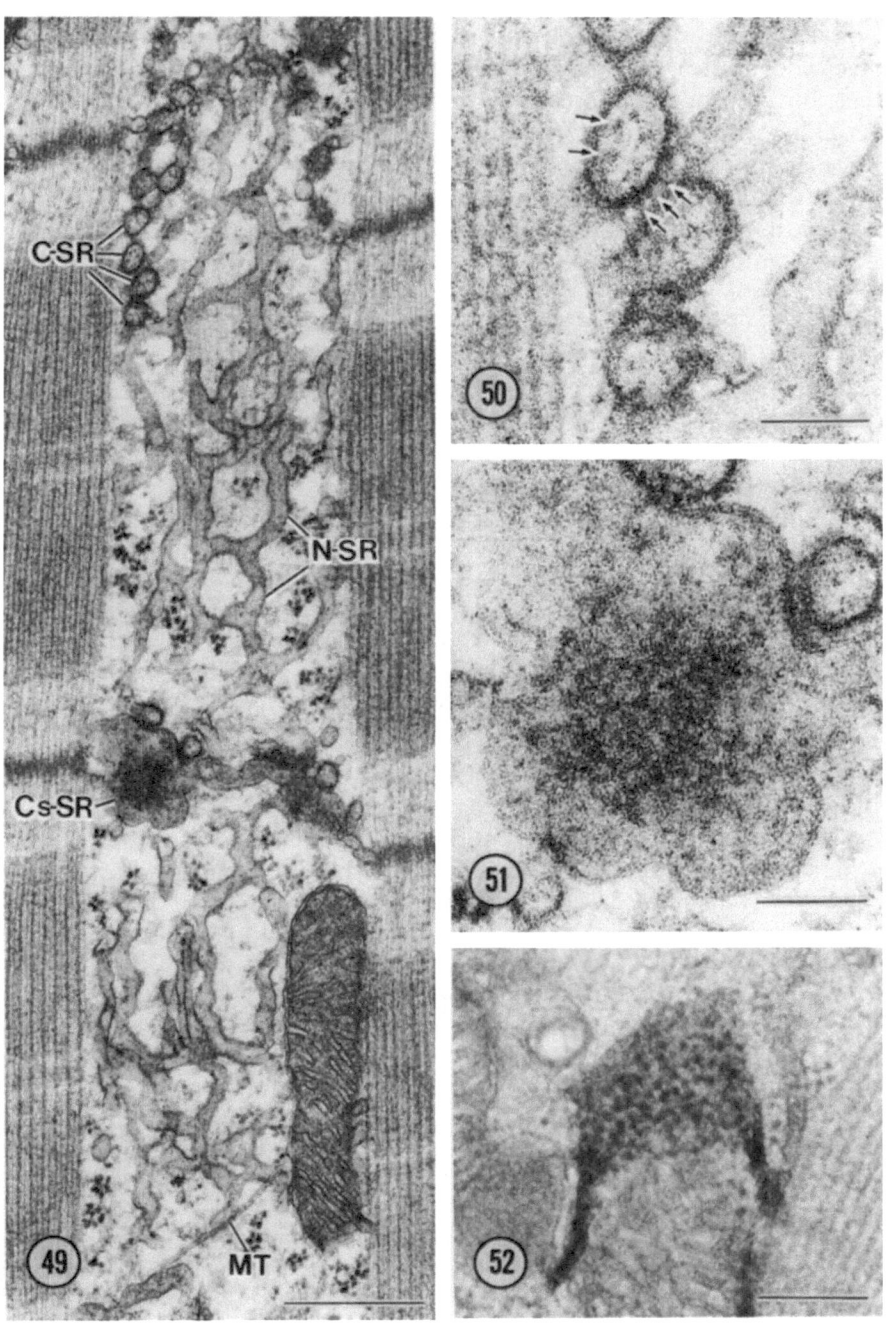

that such examples of cisternal SR constitute yet another form of *extended* junctional SR, especially in view of their location at the Z-lines of sarcomeres (Figs. 49 and 51).

A less structurally complex form of cisternal SR has been found by us in mouse and guinea pig heart (Fig. 52). Such bodies are studded with ribosomes, but do not seem to contain junctional granules or other opaque contents. The tendency for "ordinary" regions of N-SR to be decorated with ribosomes indicates that this second category of "cisternal SR" is merely a random dilatation of SR, rather than a particular specialization. Cisternal SR has been demonstrated at various sarcomeric levels by freeze–fracture (Dolber and Sommer, 1980; Scales, 1981), a procedure that would tend to impart the same appearance to both categories that we have described above. It seems that the "EJ-SR" category of cisternal SR is closely associated with the Z-line levels of the myofibrils (as are many other myocardial components: Forbes and Sperelakis, 1980a), and that this may be a criterion whereby these saccules can be distinguished both from chance expansions of N-SR and from peripheral J-SR, the latter of which also fails to observe sarcomere-specific localization (Figs. 29, 30, and 40).

2.2.4d. "Dense-Cored" SR.   Sommer and Johnson (1969) have described in the hearts of frog and chicken the association with the SR of dense-cored bodies that sometimes take the form of elongate tubules, clearly continuous with more electron-lucent SR portions (network SR). Such dense-cored SR is also found in the mouse, in which we have noted—during the preparation of this chapter—a number of similarly elongated SR segments, not specifically aligned with the Z-discs (see Section 2.2.4c, above), which contain moderately dense material (Figs. 53 and 54). These, as cisternal SR, may prove to be structural variations of the SR that will become more visible under the condition of specifically directed inspection. Still to be categorized are bodies, often ellipsoid in profile, which are identifiable on the basis of a distinct electron-opaque core, separated from the limiting membrane by a clear rim (Fig. 62).

2.2.4e. Proliferated SR.   We have discussed (Section 2.2.3a) the occurrence of multiple stacked J-SR elements and convoluted J-SR saccules that interdigitate with the TATS. The network SR of the mouse as well is capable of multiplication of its components (Fig. 55). We have encountered a number of examples of voluminous, subsarcolemmal collections of plicae and tubules, vaguely similar to the "columnar SR" of lizard skeletal muscle (Forbes and Sperelakis, 1980b), which is largely composed of N-SR derivatives. The N-SR has not been reported to undergo similar hypertrophy in other hearts, though SR may be the generator of elaborate arrays described in normal and

———————————————————————————→

Figure 53.   Mouse ventricle, longitudinal section. An elongate cisternal profile (*), filled with dense material, is fused with the network SR (N-SR). Note the microtubule (MT) whose profile curves to fit the surface contour of this "dense-cored SR." Scale bar represents 0.2 μm.

Figure 54.   Another, extensive example of dense-cored SR (*) that extends longitudinally for over half a sarcomere length. Note its anastomosis with the network SR, and the faint collection of transversely oriented intermediate filaments (F) at the Z-line level. Scale bar represents 0.2 μm.

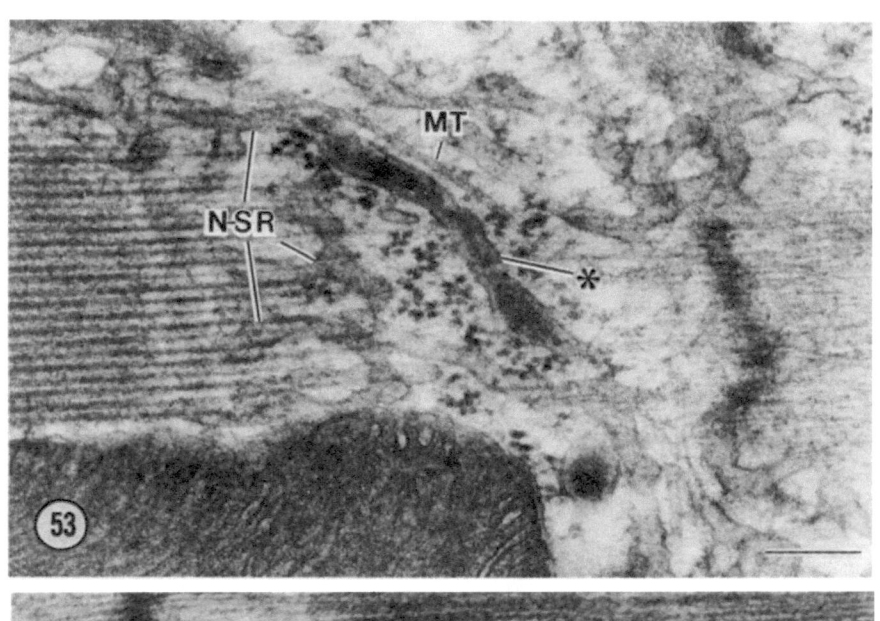

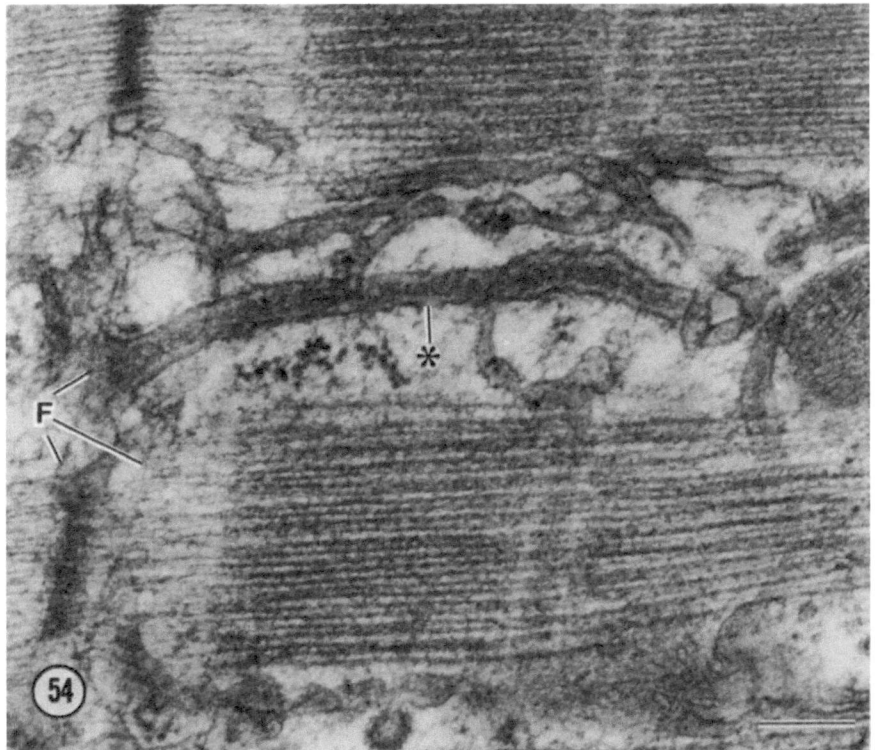

pathological skeletal muscles (Cho *et al.*, 1972; Mastaglia, 1973; Pachter and Breinin, 1977).

2.2.5. Development of the SR

Although the complexity of the mammalian myocardial SR has been demonstrated, surprisingly little effort to date has been expended toward the understanding of its generation. The research of Ishikawa and Yamada (1976) is particularly useful in that it focuses on the mouse, the heart of which has played a significant role in the study of myocardial embryony (e.g., Challice and Virágh, 1973), as well as in much of our own work (Forbes and Sperelakis, 1972a,b, 1973a,b, 1974, 1977, 1980a, 1982b; Forbes *et al.*, 1977; Sperelakis *et al.*, 1974). The inability of Ishikawa and Yamada (1976) to demonstrate the SR as "honeycomb networks" in the neonatal mouse is probably a consequence of the unavailability to them of regimens which lead to selective SR opacification, since OsFeCN-treated hearts of newborn mouse clearly display arrays of network SR in longitudinal section (Figs. 56 and 57), as well as examples of peripheral J-SR (Fig. 56) and corbular SR (Fig. 57). Necessarily, PJ-SR is the mainstay of the population of the couplings in the developing heart cell (Page and Buecker, 1981), statistically giving way in large measure (Bossen *et al.*, 1978; Page and Surdyk-Droske, 1979) to interior junctional SR both by means of induction by contact with the proliferating TATS and through the "sweeping in" of erstwhile PJ-SR elements by the active invagination of TATS components (for example, see Fig. 4 of Forbes and Sperelakis, 1976). It is likely that development of the SR in the embryonic myocardial cell will become better appreciated with the expanded utilization of the various contrasting procedures now available for the delineation of SR (Section 5.1).

2.2.6. Enzymatic Activities in SR

A myriad of conclusions concerning enzyme localization and ionic distributions has been drawn from studies on mammalian heart that have utilized cytochemical techniques (Table 2). It is obvious that a spectrum of

→

Figure 55. Adult mouse ventricle. In this OsFeCN-treated cell, normal SR components, such as Z-tubules (ZT), appeared through much of the cell, but there also appears a large subsarcolemmal accumulation of thin, closely packed tubules and plicae that stain in a manner similar to the rest of the SR and in some places incorporate junctional SR (at arrows), thus identifying the bulk of the structures as being derived from the proliferation of network SR. Scale bar represents 0.5 μm.

Figures 56. SR-stained ventricular cell of newborn mouse. This semithin section passes close to the cell surface, thus revealing caveolar profiles (*) and arrays of anastomosed N-SR and peripheral J-SR (PJ-SR) similar in general appearance to those seen in adult mammalian myocardium (cf. Figs. 22, 27, 30, and 40). Scale bar represents 0.5 μm.

Figure 57. Same tissue as Fig. 56. Toward the interior of this myocardial cell, SR configurations equivalent to those of adult mouse can be discerned, these including N-SR and—at the Z lines— examples of corbular SR (C-SR). Scale bar represents 0.5 μm.

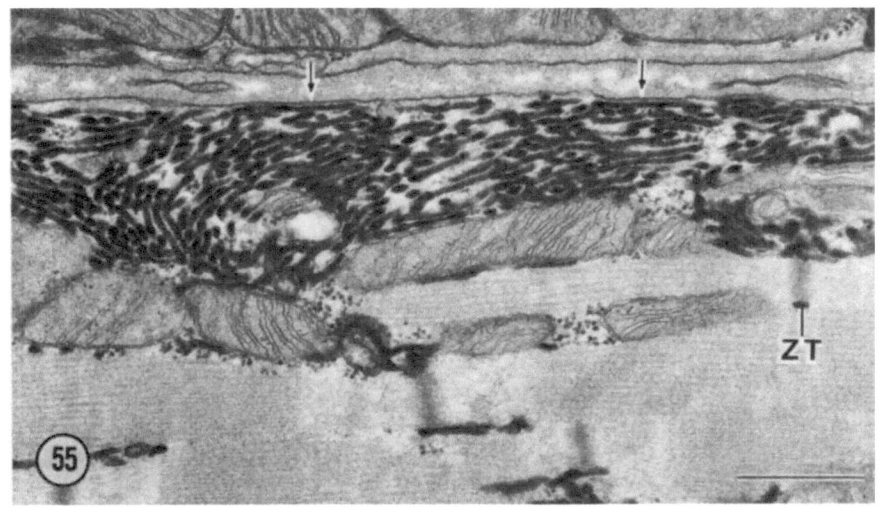

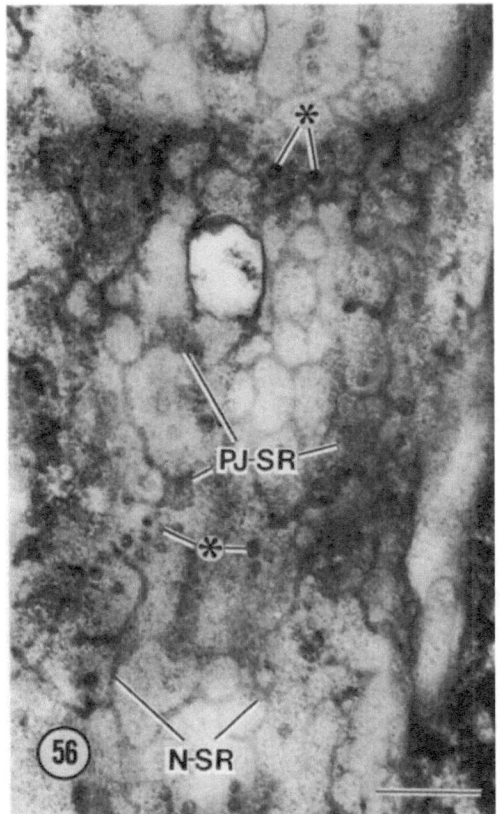

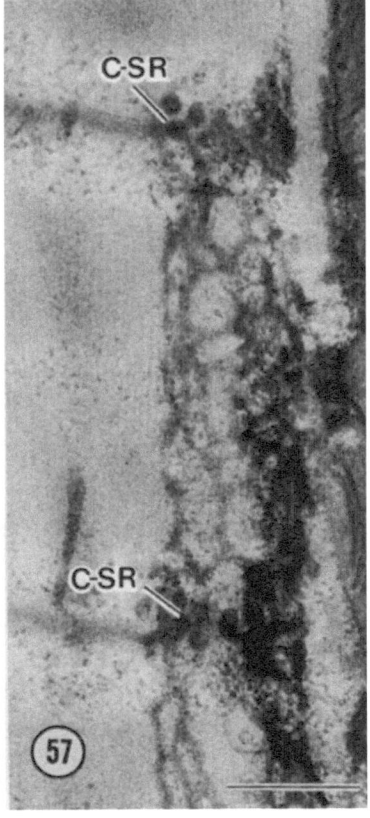

Table 2. Reported Cytochemical Studies Concerning Myocardial SR

Parameter	Localization		Reference
	N-SR[a]	J-SR[b]	
Acid phosphatase	+	+	Hoffstein et al., 1975
Alkaline phosphatase	−	+	Borgers et al., 1971
Aryl phosphatase	+	+	Hoffstein et al., 1975
ATPase (nonspecific)		+	Sommer and Spach, 1964; Borgers et al., 1971; Forbes and Sperelakis, 1972b; Ashraf et al., 1976;
ATPase (Ca^{2+}-specific)	+		Ashraf et al., 1976
ATPase (Ca^{2+}-specific		+	Forbes and Sperelakis, 1972b
ATPase (Ca^{2+}-specific)		+[c]	Strosberg et al., 1970
ATPase (Na^+,K^+)	+		Forbes and Sperelakis, 1972b
ADPase	−	+	Rostgaard and Behnke, 1965; Ferrans et al., 1969
UDPase, IDPase	+	+	Ferrans et al., 1969
5′-nucleotidase		+	Rubio et al., 1973
Cholinesterase	+	+	Karnovsky, 1964
Butyryl cholinesterase	+	+	Hagopian and Tennyson, 1971
Glucose 6-phosphatase	+	+	Borgers et al., 1971
Glutamic oxalacetic transaminase	+	+	Lee, 1969
Lactic dehydrogenase	+	−	Baba and Sharma, 1971
Thiamine pyrophosphatase	−	+	Ferrans et al., 1969, 1972
Oxalate precipitate	+	+	Diculescu et al., 1971
Pyroantimonate precipitate		+	Legato and Langer, 1969; Shiina et al., 1970
Ca^{2+} (calcium dihydroxytartrate osazone precipitate)	+	+	Shiina and Mizuhira, 1970
Ruthenium red accumulation	−	−	Forbes and Sperelakis, 1979
HRP accumulation	−	−	Sperelakis and Rubio, 1971
Osmium precipitation following OsFeCN postfixation	+	+	Waugh and Sommer, 1974; Forbes et al., 1977

[a] N-SR, that portion of the SR not specifically associated with surface sarcolemma or TATS elements.
[b] J-SR, SR portions apposed to surface sarcolemma or TATS.
[c] Specifically corbular SR.

enzymatic activities exists there and that the myocardial SR must be a complex entity. Debate continues as to the efficacy of various cytochemical techniques, especially those directed toward the demonstration of ATPase activity (e.g., Firth, 1980). Yet, the preponderance of observations leads to the general conclusion that phosphatases of one sort or another are localized in the SR (Sommer and Johnson, 1979) and that these enzymes are related to the process of excitation–contraction coupling.

The concept of a functional biochemical segmentation of myocardial SR is suggested by the results of various cytochemical studies, and is a tempting conclusion because of studies that indicate that the SR of skeletal muscle is divided into compartments (Sperelakis et al., 1978; Sommer et al., 1978; Sommer and Johnson, 1979), a division that may alter the total capacitance of the membrane system during certain phases of the contraction cycle. The precise location of various ATPases in myocardial SR (Table 2) obviously is debatable, but an Na:Ca exchange reaction (Baker *et al.*, 1969; Langer and Serena, 1970) is thought to occur across the SR wall (Forbes and Sperelakis, 1972b; Asano *et al.*, 1980). Studies continue to appear that purport definitive localization of various enzymes in the SR. In the following section, we engage in some physiological speculation that is based in part on certain cytochemical observations.

2.2.7. *Physiological Role of SR*

The classic concept of SR in all types of striated muscle is that of a uniformly distributed perimyofibrillar and subsarcolemmal meshwork that is involved in the metabolism of calcium ion (Ca^{2+}). This and other features of the myocardial cell that are involved with Ca^{2+} are diagrammed in Plate I. A central question concerns the directionality of Ca^{2+} movements relative to the SR. The myocardial SR actively pumps Ca^{2+} in from the myoplasm and myofibrils, and binds it internally. Langer (1974, 1976) has suggested that the SR may requilibrate Ca^{2+} with the interstitial fluid via the couplings. Langer also proposed that the entire source of Ca^{2+} for contraction was the extracellular coating (glycocalyx) of the sarcolemma and TATS (see Section 2.1.4). In his scheme, the SR is considered incapable of releasing Ca^{2+} to the myofilaments and therefore would function in a unidirectional manner (though this view has recently been modified considerably: Langer, 1980).

A second major school of thought contends that the SR of heart cells, like that of skeletal myocytes, is capable of cycling Ca^{2+}, pumping it in to effect relaxation by means of Ca^{2+}-ATPase and perhaps also by a Ca–Na exchange reaction. The SR releases Ca^{2+} in response either to a chemical event (e.g., Ca^{2+}-triggered Ca^{2+} release) or to an electrical signal from the sarcolemma/TATS that might invade the SR at the level of the couplings (see below). (For review, see: Schwartz, 1972; Reuter, 1974; Fabiato and Fabiato, 1977; Winegrad, 1979.) Since it appears that the concentration of SR tubules is greater at deeper levels within the myocardial cell than in the subsarcolemmal regions (see Figs. 25 and 26), this may play a role in excitation–contraction coupling.

Cytochemical experiments on intact tissue demonstrate Ca^{2+} sequestration in myocardial SR (Table 2), and biochemical experiments confirm the Ca^{2+}-sequestering ability of isolated SR vesicles. Mitochondria may also act as Ca^{2+}-accumulating bodies (Sordahl, 1979), and thus the J-SR–mitochondrion appositions described in Section 2.2.3c may prove to be of importance in Ca^{2+} metabolism in myocardial cells. Similarly, the localization of mitochondria near the gap junctions of intercalated discs may help to protect the cell junctions from cell-to-cell uncoupling because of their Ca^{2+}-sequestering abilities (see Plate I and Forbes and Sperelakis, 1982a).

The myocardial cell contains considerable extended junctional SR, including corbular SR (Section 2.2.4b) and certain elements of cisternal SR

ROLE OF Ca^{++} IN EXCITATION-CONTRACTION COUPLING

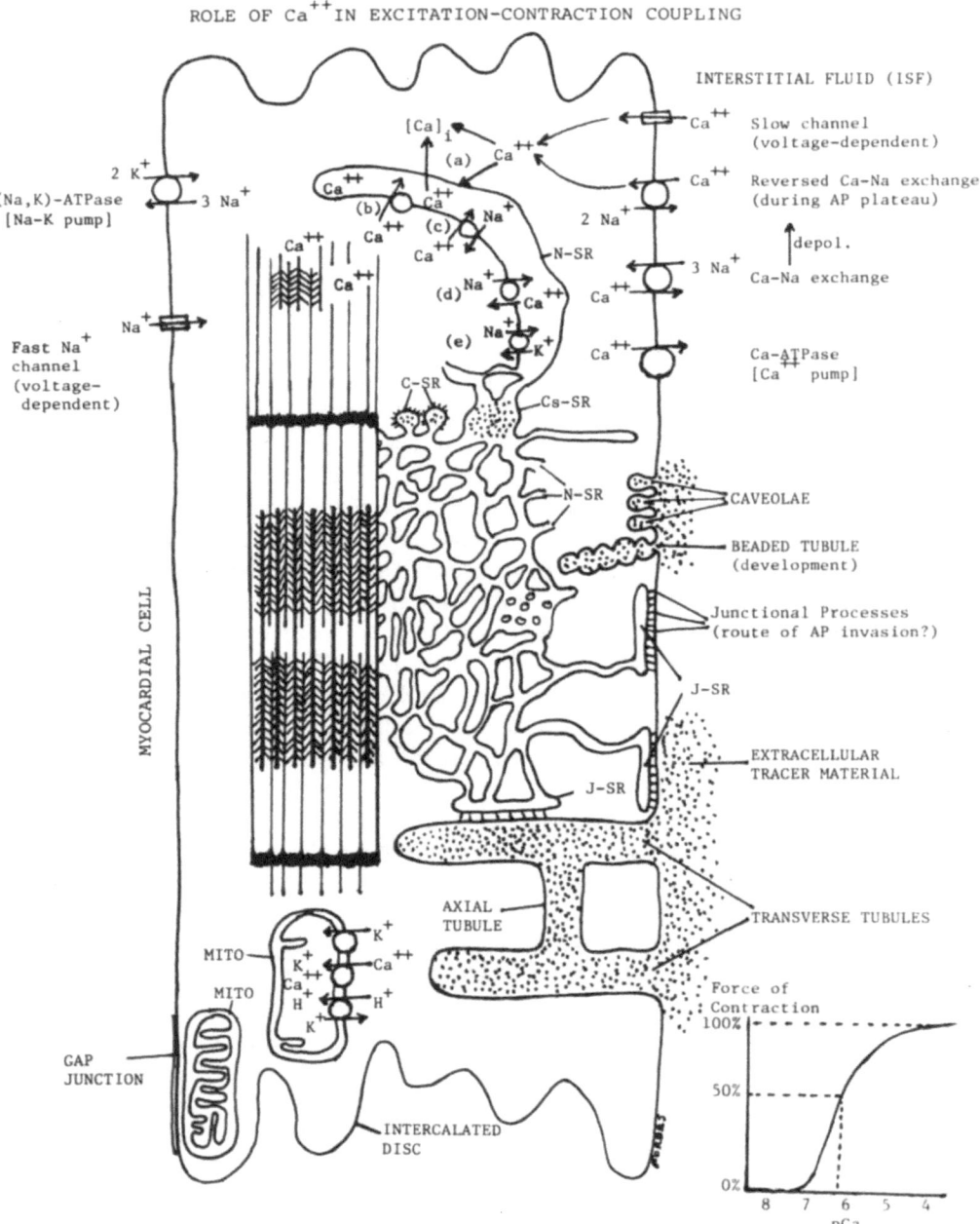

Plate I. Diagrammatic representation of a myocardial cell, illustrating structures and physiological processes likely to be involved in the metabolism of Ca^{2+}. The sarcolemma and its various invaginations, such as caveolae and T-axial tubules, are depicted filled with extracellular tracer material. A number of ionic channels, enzymes, and exchange mechanisms are associated with the sarcolemma. The interior of the cell is maintained low in Na$^+$ (ca. 15 mM) and high in K$^+$ (ca. 150 mM) by the action of a (Na,K)-ATPase and Na–K-linked pump in the sarcolemma, which

(Section 2.2.4c) that contain junctional processes but do not form couplings with the sarcolemma or TATS. If junctional granules are the matrix which supports Ca^{2+}-specific ATPase (Forbes and Sperelakis, 1974) or Ca^{2+}-binding proteins (Somlyo, 1979), it follows that all J-SR representatives may be calcium sinks, regardless of whether they are affiliated with the sarcolemma or its derivatives.

During excitation of myocardial cells, there is a Ca^{2+} influx (as part of the inward slow current) through the voltage-dependent slow channels located in the sarcolemma (see Plate I). This Ca^{2+} influx is from the interstitial fluid into the myoplasm. In addition, there is evidence that the Ca_i–Na_o exchange reaction across the sarcolemma reverses during the action potential plateau due to favorable energetics, thus giving rise to a Ca^{2+} influx (Ca_o–Na_i exchange) by this mechanism. The Ca^{2+} influx from both of these sources helps to raise $[Ca]_i$ to the level required to activate the contractile myofilaments (e.g., 10^{-5} M) (see inset graph of Plate I). In addition, the Ca^{2+} influx may act on the surface of the SR to trigger the release of more Ca^{2+} from the SR, the so-called Ca-triggered Ca mechanism of Fabiato and Fabiato (1977). This mechanism serves as a Ca^{2+} amplifier. For relaxation, $[Ca]_i$ must be lowered to the resting level of $< 1 \times 10^{-7}$ M. This is accomplished by Ca^{2+} sequestration (pumping) into the SR (primarily N-SR) by means of a Ca-ATPase. In addition, the Ca^{2+} uptake into the SR can occur by means of a Ca–Na exchange mechanism located in the SR. It is possible that $[Ca]_i$ may be lowered for relaxation also by Ca-ATPase activity and Ca_i–Na_o exchange in the sarcolemma. The diagram in Plate I and the accompanying legend summarize these processes.

The ubiquity and multiplicity of junctional couplings in cardiac muscle

---

pumps out 3 Na^+ ions to (usually) 2 K^+ ions pumped in per ATP hydrolyzed. The free Ca^{2+} concentration inside the resting cell ($[Ca]_i$) is maintained very low ($< 1 \times 10^{-7}$ M) by the action of a sarcolemmal Ca-ATPase and Ca pump which extrudes Ca^{2+} against a large electrochemical gradient (of about 210 mV). In addition, Ca^{2+} is effectively pumped out by the action of a Ca–Na exchange system in the sarcolemma that exchanges 1 Ca_i for (usually) 3 Na_o. Cl^- ion is low inside (ca. 6 mM) because of passive distribution across the membrane based on the membrane potential. Ca^{2+} enters the cell during excitation through the voltage-dependent slow channels, and possibly also by a reversal of the Ca_i–Na_o exchange reaction (Ca_o–Na_i). Ca^{2+} may then interact directly with the myofilaments and indirectly by way of the sarcoplasmic reticulum (SR) through Ca-triggered Ca^{2+} release (shown at [a]). The relationship between force of contraction and $[Ca]_i$ is shown in the inset graph. It is postulated that a reversed Ca–Na exchange (d) can also occur, which would constitute "Na-triggered Ca release" from the SR, the Na^+ having entered through the fast Na^+ channels during excitation. To produce relaxation, the SR may sequester Ca^{2+} by means of either the Ca-ATPase (Ca pump) (b) or the Ca-Na exchange reaction (c). Reactions (c) and (d) may be enhanced by the presence of a (Na,K)-ATPase and Na–K pump (e) shown to exist in the myocardial network SR (N-SR) (Forbes and Sperelakis, 1972b). Specialized portions of the SR, such as saccules of junctional SR (J-SR), corbular SR (C-SR), and cisternal SR (Cs-SR) may be additional sites which tape up and release Ca^{2+}. Furthermore, the junctional processes (pillars) of the J-SR may be the sites of electrical connection between the sarcolemma and SR, and involved in release of Ca^{2+} from the SR (ΔV_{SR}). Although mitochondria can accumulate Ca^{2+}, they probably play no role in normal excitation–contraction coupling.

cells suggests that they subserve some function in excitation–contraction (E–C) coupling. The characteristics of junctional processes among muscle cells from skeletal, cardiac, and vascular sources are similar in many respects (Forbes and Sperelakis, 1982b). The arrival of the action potential is somehow linked to the appearance of sufficient Ca^{2+} in the myoplasm to activate the contractile elements. The architecture of the couplings must be considered in studies aimed at elucidating the physiological function. For example, junctional processes within the J-SR couplings may serve as conduits for transfer of material or signals during E–C coupling. Though there exists a genuine possibility that patency exists between the J-SR and TATS of skeletal muscle (Birks and Davey, 1969; Rubio and Sperelakis, 1972; Kulczycky and Mainwood, 1972; Sperelakis *et al.*, 1978), there is no evidence to indicate such lumen-to-lumen connections in the heart. The "pillars" (Figs. 38 and 39) have been proposed (Somlyo, 1979; Eisenberg and Gilai, 1979; Franzini-Armstrong, 1980) as representing the "rigid-rod" model of Chandler *et al.* (1976), which was thought to be an electromechanical connection extending between the apposed membranes of J-SR and T-tubule in skeletal muscle. It was proposed that these linear bodies would insert into the J-SR and act as plugs; the arrival of the action potential into the J-SR coupling would lead to their dislodgment through the movement of some charged complex in the T-tubule membrane. It was further proposed (Schneider, 1981) that each rod–plug assembly blocks and unblocks one Ca^{2+} channel in the J-SR, and that each such assembly may correspond to a single junctional process. In view of the unit membrane-like appearance of pillars, an alternative interpretation has occurred to us, namely, that the action potential invades the SR by way of the pillars, depolarizing the SR membrane and opening up the slow channels there that are permeable to Ca^{2+} (see also Fabiato and Fabiato, 1977). Thus, the electrical signal would be conveyed by membranes that are continuous from TATS to J-SR to N-SR.

3. Cytoskeletal Elements

3.1. Microtubules

Among cytological structures, microtubules are among the best recognized in terms of their form and function. Numerous reviews consider their occurrence in a wide variety of cells and attest to their central role in intracellular frameworks. A number of articles in this series of volumes are concerned with microtubules, and several of them consider microtubules in muscle. We have not undertaken a study of cardiac microtubules that is nearly so detailed as that of Goldstein and Entman (1979), but previously we have presented some description of these organelles as they appear in mouse heart (Forbes and Sperelakis, 1980a).

Diameters of the microtubules encountered in mouse myocardium were, on the average, somewhat larger (28.8 ± 0.4 nm; $n = 20$) than those reported for the dog and guinea pig (26.2 ± 0.2 nm; $n = 108$: Goldstein and Entman,

1979), but still within the size ranges reported for microtubules (Tyson and Bulger, 1973). Transversely sectioned myocardial cells yield images of microtubules, also frequently displayed in cross section, and distributed rather evenly in the myoplasm at the periphery of nuclear profiles (Figs. 58–61), as well as between myofibrils and near the sarcolemmal borders (Figs. 22 and 71). Longitudinal sections of myocardium often exhibit longitudinal microtubule profiles (Figs. 22 and 65). Undulating profiles are frequently found, however (Fig. 62), and because of their obliquity these probably often escape notice in transversely cut cells. In mouse heart, we have found profiles of microtubules 3.5 μm and more in length.

We have found the microtubules in heart to be best preserved and most obvious in those tissues which have undergone "conventional" fixation with aldehyde solution and postfixation in cacodylate-buffered osmium tetroxide, as opposed to OsFeCN treatment (Section 5.1), which renders relatively low electron opacity to cellular components other than the elements of membrane systems. In transverse section, the typical microtubule presents a more-or-less rounded profile; the tubulin subunits are often poorly defined (Fig. 60), but photographic reinforcement by the technique of Markham rotation (Markham *et al.*, 1963) is effective in displaying the subunits, such as the 13 within the microtubule profile shown in Figs. 60 and 61. Microtubules occasionally can be discerned in freeze–fracture replicas (Fig. 63), but usually are not obvious there.

Microtubules often are found superimposed on the myofibrillar faces, along with SR tubules. Close association has been reported to exist between microtubules and SR elements of dog and guinea pig heart (Goldstein and Entman, 1979); we have found this to be true in the mouse as well (Figs. 24 and 53). Other close associations include that of transversely oriented microtubules with Z-tubules of SR or with T-tubules; such microtubules are therefore most often located in the myoplasm adjacent to the I-bands (Figs. 22 and 24).

As mentioned above, myocardial nuclei are generally accompanied by longitudinal microtubules; this has already given rise to the concept that microtubules are involved in the shaping of myocardial nuclei (Ferrans and Roberts, 1973a). The cytoskeleton of the elongate, bipolar smooth muscle cells of coronary arteries consists of prominent microtubules and intermediate filaments, all of which sweep longitudinally around the nucleus and project toward the cell ends within a myofilament-free region termed the "central sarcoplasmic core" (Forbes, 1982). The situation seems more complex in ventricular myocardial cells, probably as a result of their more complex architecture. According to Goldstein and Entman (1979), the microtubules form helical enwrapments of myofibrils and nuclei.

3.2. Intermediate Filaments

Intermediate filaments have been dealt with elsewhere in this series, and readers are referred in particular to the chapters by Fuseler *et al.*, 1981, Vol. 1; and Price and Sanger, 1983, Chapter 1 this volume for detailed consideration of these fibrillar elements.

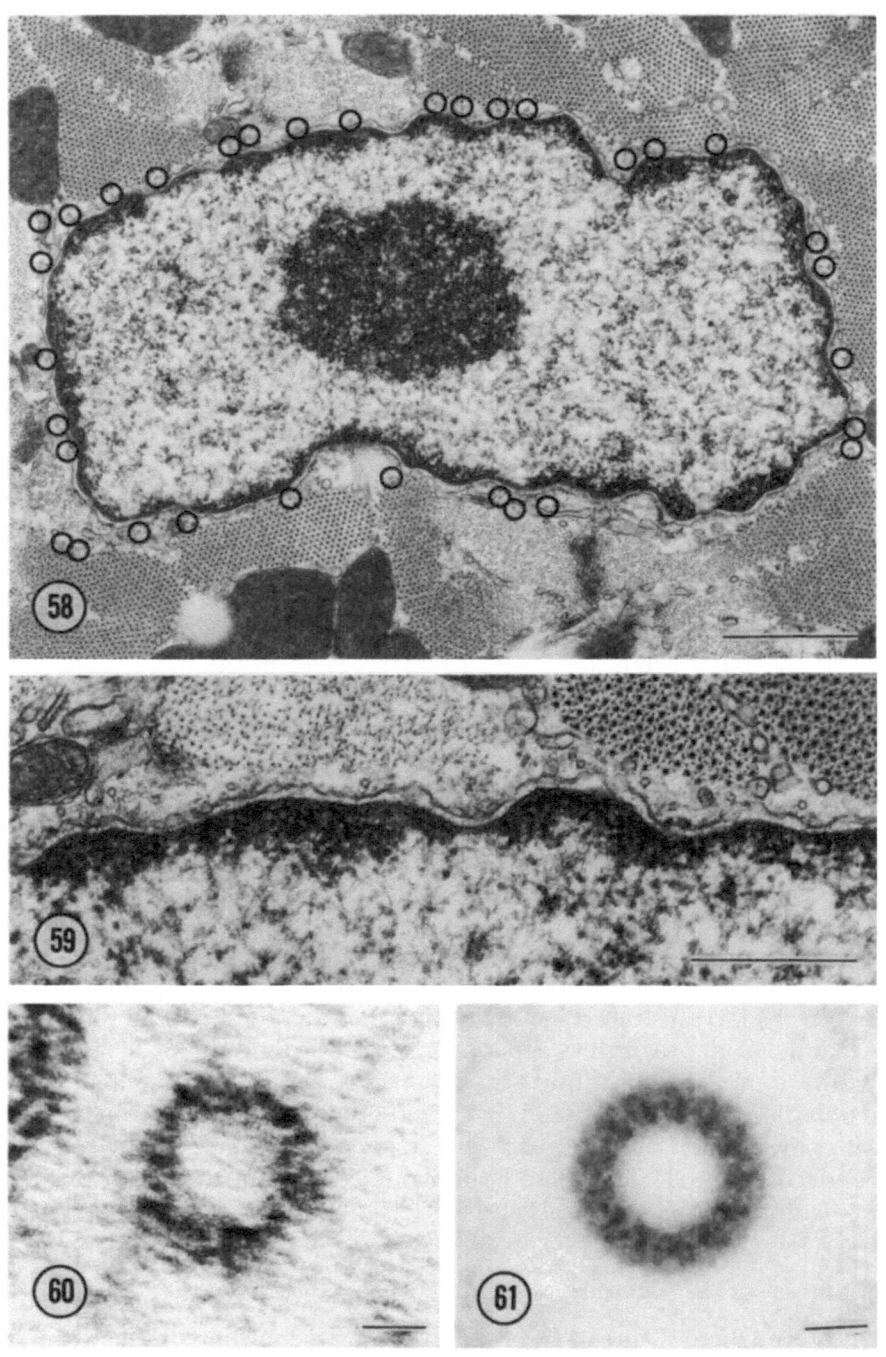

Because of the certain cytoskeletal function of intermediate filaments, they also have been called "skeletin" filaments (Eriksson and Thornell, 1979). The original term, which still is generally accepted, essentially resulted from a study by Heuson-Stiennon (1965), who found filaments in developing skeletal muscle of a diameter—approximately 10 nm—that was roughly intermediate between those of actin and myosin filaments. Intermediate filaments are now established as organelles that are virtually ubiquitous in animal cells (e.g., Ferrans and Roberts, 1973b; Lazarides, 1980; Anderton, 1981; Thornell and Eriksson, 1981). A rather wide range of diameters has been reported for intermediate filaments, the measurements by Eriksson and Thornell (1979) giving values in myocardial conducting fibers ranging from 7.0–9.5 nm. Our own calculations indicate a somewhat higher value for intermediate filaments in mouse ventricular myocardial cells (10.9 ± 0.1 nm; $n = 31$). The primary orientation of intermediate filaments in heart is one that is transverse with respect to the long axis of the cell (Figs. 64–68). The majority of such filaments is invested in bundles aligned with myofibrillar Z-lines (Figs. 54, 64, 65, and 68) (Forbes and Sperelakis, 1980a). As few as 2–3 filaments, and upwards of 50, can be found grouped at the Z-line levels of mouse heart (Fig. 65). In some cells a prominent transverse meshwork of intermediate filaments is intertwined with the myofibrils (Fig. 68) and in some instances can be traced to juxtaposition (and perhaps fusion) with the sarcolemma (Fig. 64) and nuclear membrane (see also Ferrans and Roberts [1973a] and Behrendt [1977]). The use of Markham rotation (Figs. 66 and 67) confirms the existence of four subunits which comprise the cross-sectional profile of such filaments, as previously demonstrated by Eriksson and Thornell (1979).

Longitudinally oriented intermediate filaments are in a definite minority in "working" cardiac muscle cells. Most of these take the form of bundles directed into the intracellular plaques of desmosomes (Figs. 69 and 70), and therefore such groups of filaments are restricted mainly to the myoplasmic regions near the cell tips, where the intercalated discs have formed. In mouse and guinea pig, however, individual examples or small groups of longitudinally oriented intermediate filaments can be found at points distal to the cell ends, either in the myofibrillar interstices or adjacent to other longitudinally disposed elements such as mitochondria and axial tubules (Fig. 14). It is not known how substantial a population is formed by these axial groups of inter-

---

Figure 58. Transversely sectioned myocardial cell, mouse left ventricle. Although microtubules cannot be easily discerned at this low magnification, their positions are circled, showing the affinity of microtubules for the perinuclear myoplasm. Scale bar represents 1.0 μm.

Figure 59. Detail of Fig. 58, showing transversely cut microtubules near the nuclear envelope. Scale bar represents 0.5 μm.

Figure 60. High magnification of a transversely cut microtubule in mouse myocardial cell. The circular array of subunits is clear, but the individual subunits are poorly defined. Scale bar represents 0.01 μm.

Figure 61. The microtubule in Fig. 60, photographically enhanced by Markham rotation (n of 13). Scale bar represents 0.01 μm.

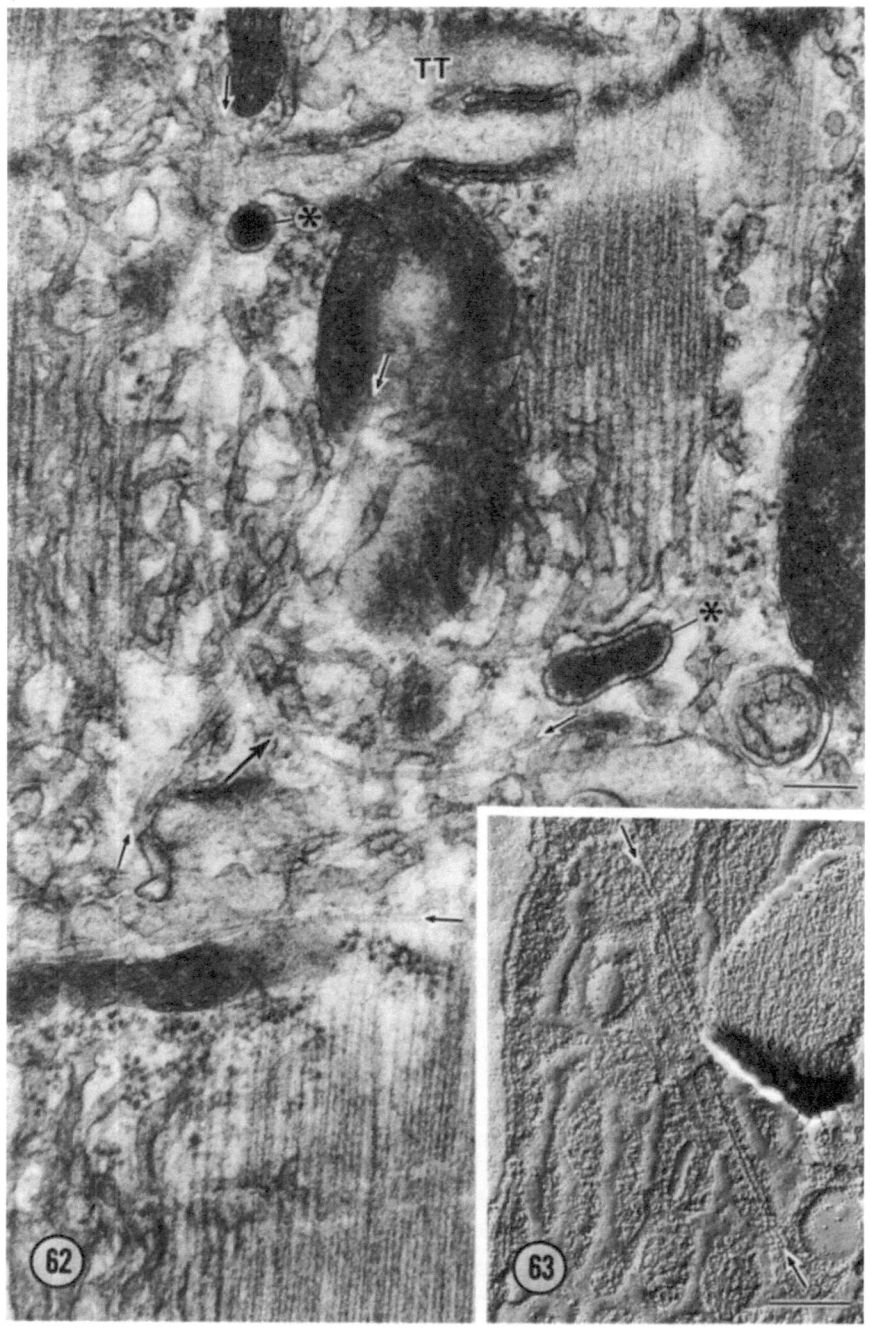

mediate filaments, nor whether they are interconnected with the Z-level frameworks. Neither has it been determined what relationship exists between intermediate filaments and microtubules in heart cells. It has become apparent in certain cell types—including cultured fibroblasts (Goldman, 1971), developing skeletal muscle (Ishikawa *et al.*, 1968), and certain secretory cells of the pituitary gland (Forbes and Dent, 1974)—that intermediate filaments depend on the structural integrity of microtubules for their own dispersion in the cytoplast. That is, when microtubules are disrupted, or disappear during certain phases of the cell cycle, the complement of intermediate filaments collapses into a perinuclear mass. Such comparative experimentation with microtubule-disrupting agents can now be readily carried out for heart, since a number of dependable methods for the isolation of adult myocardial cells are presently in vogue (e.g., Nag *et al.*, 1977; Vahouny *et al.*, 1979; Claycomb and Palazzo, 1980; Schwarzfeld and Jacobson, 1981). It has recently been shown that the desmin of isolated embryonic heart cells undergoes varying degrees of coalescence in response to relatively high concentrations of colchicine or vinblastine (Fuseler *et al.*, 1981).

It is thought that the intermediate filaments of adult cardiac muscle are primarily those of the "desmin" type, one of five morphologically similar categories of filaments in eukaryotic cells (Lazarides, 1980; Anderton, 1981). Some vimentin filaments may exist in muscle cells, however (Lazarides, 1980). We believe that the Z-level frameworks of intermediate filaments act in concert with Z-discs and transverse tubules to maintain an evenly spaced array of relatively rigid layers that traverse the cell and are interconnected between cells by collagenous struts (Caulfield and Borg, 1979; Forbes and Sperelakis, 1980a; see also Section 3.3). The intermediate filaments of heart may be instrumental in guiding the formation of the transversely arranged components of membrane systems (Lazarides, 1980). Such fibrils also appear sensitive to pathological conditions, responding by proliferation to varying degrees (Behrendt, 1977; Stoeckel *et al.*, 1981). The combination of transversely arrayed cytoskeletal components (the intermediate filaments) and longitudinally or helically arranged cytoskeletal elements (the microtubules, perhaps together with some intermediate filaments) is likely to be responsible, together with the compacted collections of myofilaments, for the maintenance of the overall shape of the individual myocardial cell, a form which is maintained when the cell is isolated intact by enzymatic dispersion.

←———

Figure 62.   Grazing longitudinal section of myofibrillar faces in mouse ventricle. Microtubules in various orientations are captured (indicated by small arrows), superimposed on SR elements and mitochondria. One microtubule (upper left of micrograph) runs parallel to the myofibrillar axis and then executes an extreme bend (at large arrow) to become perpendicularly oriented. Note dense-cored bodies (*) which may be components of the SR. TT, T-tubule complex. Scale bar represents 0.2 μm.

Figure 63.   Freeze-fracture replica of an area equivalent to that shown in Fig. 62. Network SR is clearly delineated, and an obliquely running linear structure, approximately 32 nm in width, is revealed (between arrows) which apparently represents a microtubule fractured through its wall. Scale bar represents 0.2 μm.

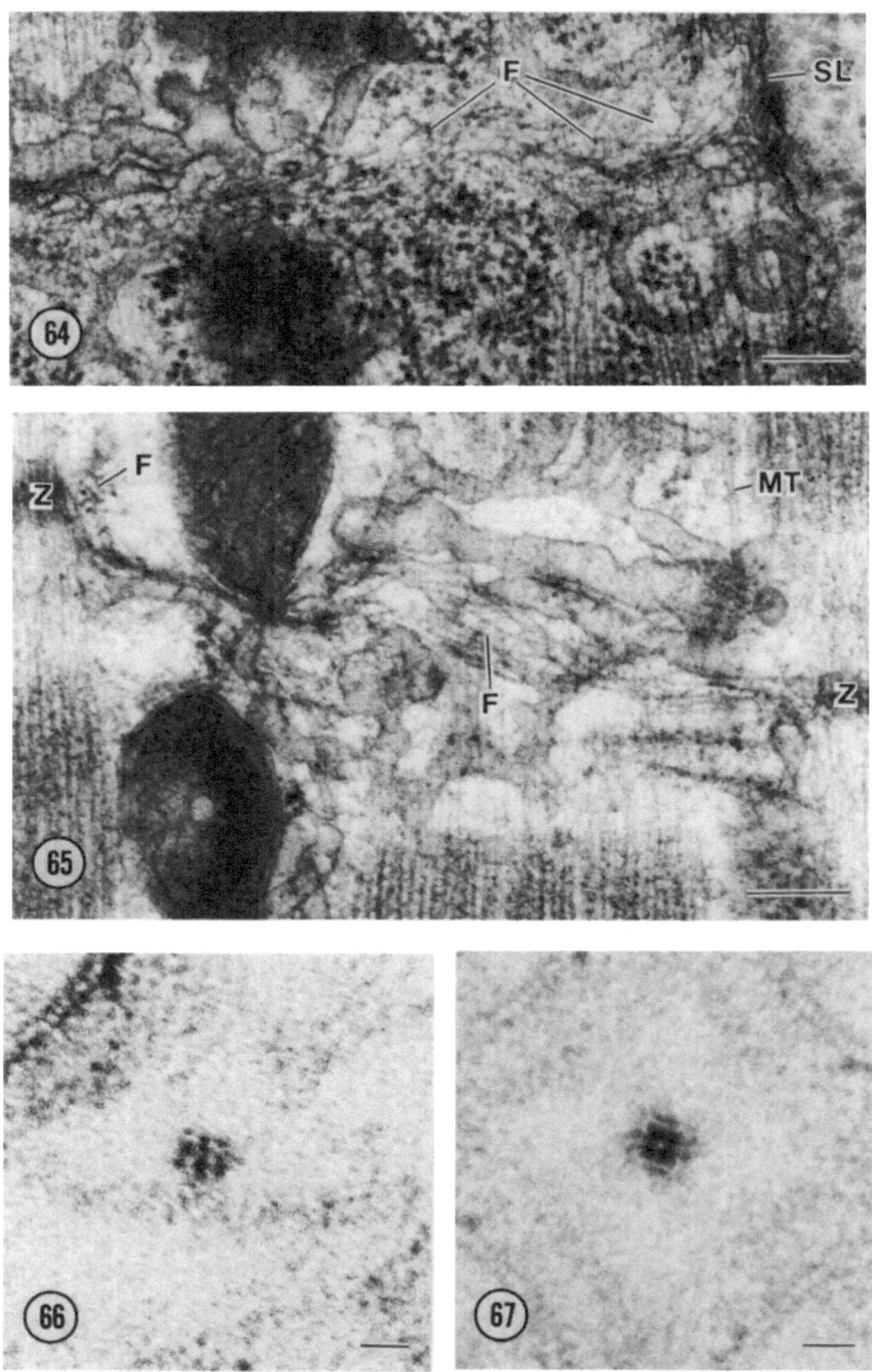

3.3. Extracellular Elements

In the consideration of skeletal elements of the heart, the "exoskeleton" associated with the various cellular elements has often been overlooked in the past. Recently, however, interest has been demonstrated in the extracellular components that enmesh the myocardial cells and circulatory components of the heart. It has been recognized that the generalized cell type known as "fibroblasts" constitutes a substantial occupant of the tissue space of the heart (Melax and Leeson, 1972; Nag, 1980). Generation of the connective tissue of this organ is probably attributable solely to this population of cells. Most prominent among the extracellular matrix components are the various collections of collagen fibrils; these are organized into strands and struts that form bridges between adjacent myocardial cells or among muscle cells and blood vessels (Caulfield and Borg, 1979; Borg and Caulfield, 1979). These elegant studies were carried out by means of scanning and high-voltage transmission electron microscopy. Examination of thin sections with conventional transmission electron microscopy also is useful in providing an appreciation of the variety of extracellular bodies that may be involved in the adhesivity and stabilization of working myocardial cells. In addition to the extracellular interconnections between muscle cells by the large collagen masses described by Caulfield and Borg (1979), there recently have been described a number of elaborate minuscule structures that form a maze of filamentous and granular profiles in the interspace between cardiocytes (Robinson, 1980; Robinson and Winegrad, 1981). Included in this pericellular embedment are the cell coats themselves, together with collagen fibrils, "microfibrils," "microthreads," granular bodies, and elastin (Robinson and Winegrad, 1981).

In mouse and guinea pig hearts, collagen fibrils 20–30 nm in diameter may be found even in the narrow clefts between the lateral surfaces of myocardial cells (Figs. 71–74), along with microfibrils 12–18 nm in diameter that are likely to be related to elastin (Figs. 71 and 73). These latter structures are closely associated with the cell coat, and perhaps are identical to the fibrils that line the TATS of guinea pig ventricle (Figs. 11–14). In addition there are

---

←—————————————————————————————————

Figure 64.   Mouse ventricle. A skein of intermediate filaments (F) weaves a sinuous course across the Z-line level of these longitudinally sectioned myofibrils and forms a reticular mat at the myoplasmic side of the sarcolemma (SL). Scale bar represents 0.2 μm.

Figure 65.   Mouse ventricle. As in Fig. 64, intermediate filaments (F) are aligned with the Z-lines (Z) of the associated myofibrils, and in this field can be found in both longitudinal (middle and right of micrograph) and transverse section (leftmost portion of micrograph). Note the superposition of filaments and membrane system elements, though in this conventionally prepared material the latter cannot be unequivocally identified. Scale bar represents 0.2 μm.

Figure 66.   Transversely cut intermediate filament, ca. 12 nm in diameter, in mouse ventricle. Like the microtubule in Fig. 60, this filament is poorly defined in terms of its wall structure. Scale bar represents 0.01 μm.

Figure 67.   The intermediate filament in Fig. 66 subjected to Markham rotation with an *n* value of 4, which reveals the four subunits and the lucent core characteristic of such filaments. Scale bar represents 0.01 μm.

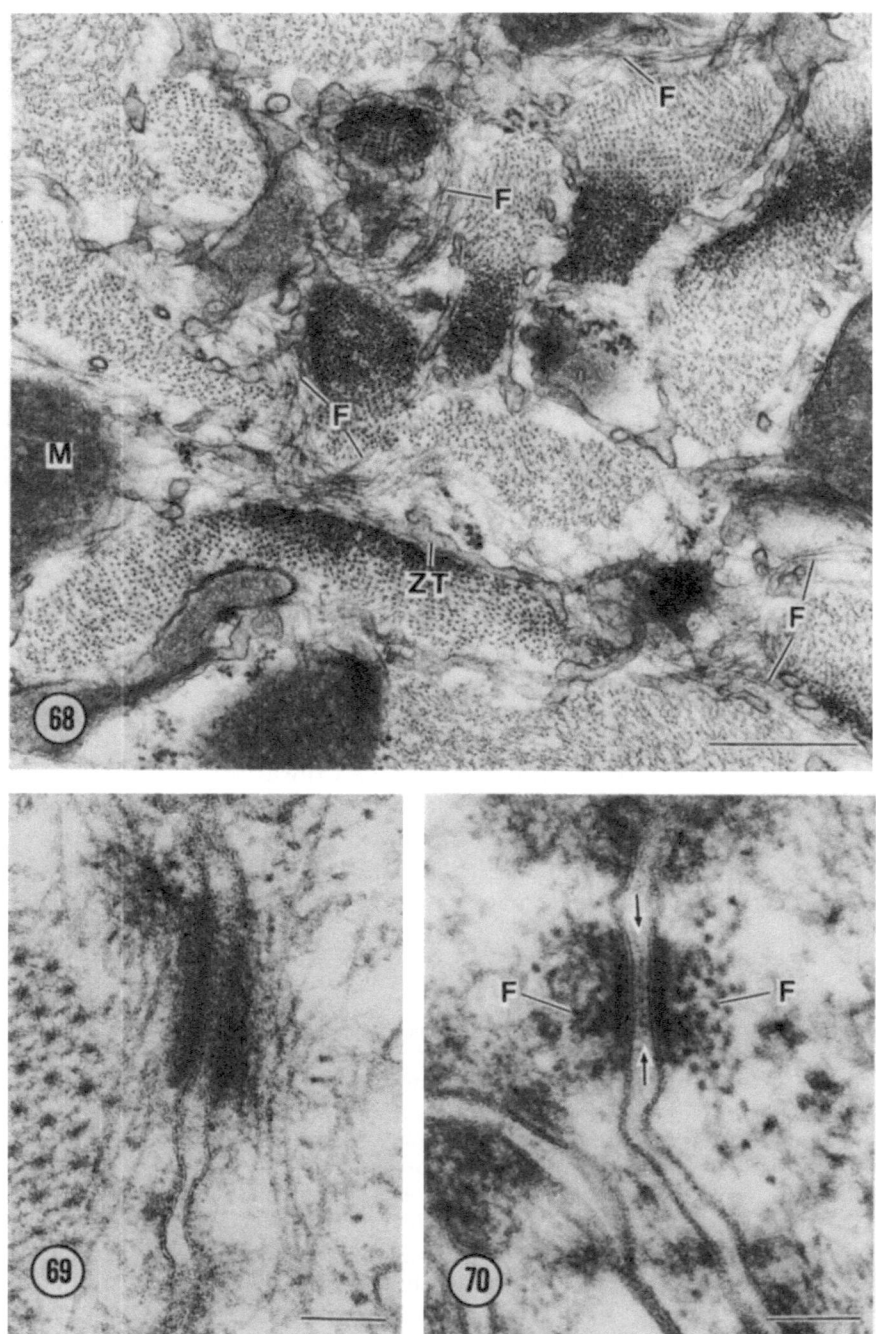

wisps of lightly opaque material, which seem to correspond to the interconnected microthreads and granules described by Robinson and Winegrad (1981); these adhere to the cell coat and at various points may radiate outward to form cohesive strands between cells (Figs. 72–74) or alternately interdigitate with other extracellular elements, thereby bridging the intercellular gap (Figs. 71–73).

The exploitation of "ultracytochemical" techniques that selectively amplify the opacification of connective tissue components such as glycosaminoglycans (e.g., Coltoff-Schiller and Goldfischer, 1981) will prove useful to further studies, along with other intricate techniques such as the combination of HVEM and "deembedment" procedures (Robinson, 1980), which demonstrate the existence of substantial lateral interconnections between rat ventricular myocardial cells. Such attachment, together with the transverse intracellular skeletal plates we have described, is in all probability intrinsic to the purely mechanical properties of the mammalian myocardium.

4. Summary

In this chapter, we have described structural and functional features of the membrane systems and cytoskeletal fibrils of mammalian myocardial cells. The organization and interrelation of these two particular features underscores the complex overall architecture of the myocardial cell. It is clear that all types of sarcolemmal invaginations are homologous, since caveolar proliferation gives rise to beaded tubules, which give rise to definitive TATS elements. In the adult myocardial cell, many of the TATS elements are physically interconnected, and—particularly in mouse and guinea pig—form elaborate latticeworks of tubules that are aligned transversely, longitudinally, and obliquely with respect to the long axis of the cell. TATS of mouse and guinea pig ventricular cells represent extremes of development, that of the mouse being more primitive and that of the guinea pig being quite highly advanced.

The SR of heart is highly structured; elements of N-SR are arranged in different patterns that correlate with zones of the sarcomere and that may vary according to the particular depth at which they are located. N-SR "Z-

←───

Figure 68.   Transverse view of mouse ventricular cell. Much of the myofibrillar material is captured at the level of the Z-line, and accordingly bundles of intermediate filaments (F) are much in evidence, forming an intermyofibrillar web composed of varying numbers of transversely oriented fibrils that cross over SR elements (note Z-tubules: ZT) and come into close association with mitochondria (M). Scale bar represents 0.5 μm.

Figure 69.   Junctional region between two mouse ventricular myocardial cells, in which a desmosome is flanked by numerous intermediate filaments, many of which are directed into the intracellular dense material (plaques) of the desmosome. Scale bar represents 0.1 μm.

Figure 70.   Myocardial desmosome in which the central lamina (between arrows) is clearly visible, and intermediate filaments (F) in transverse section can be discerned in close association with the plaque material. Scale bar represents 0.1 μm.

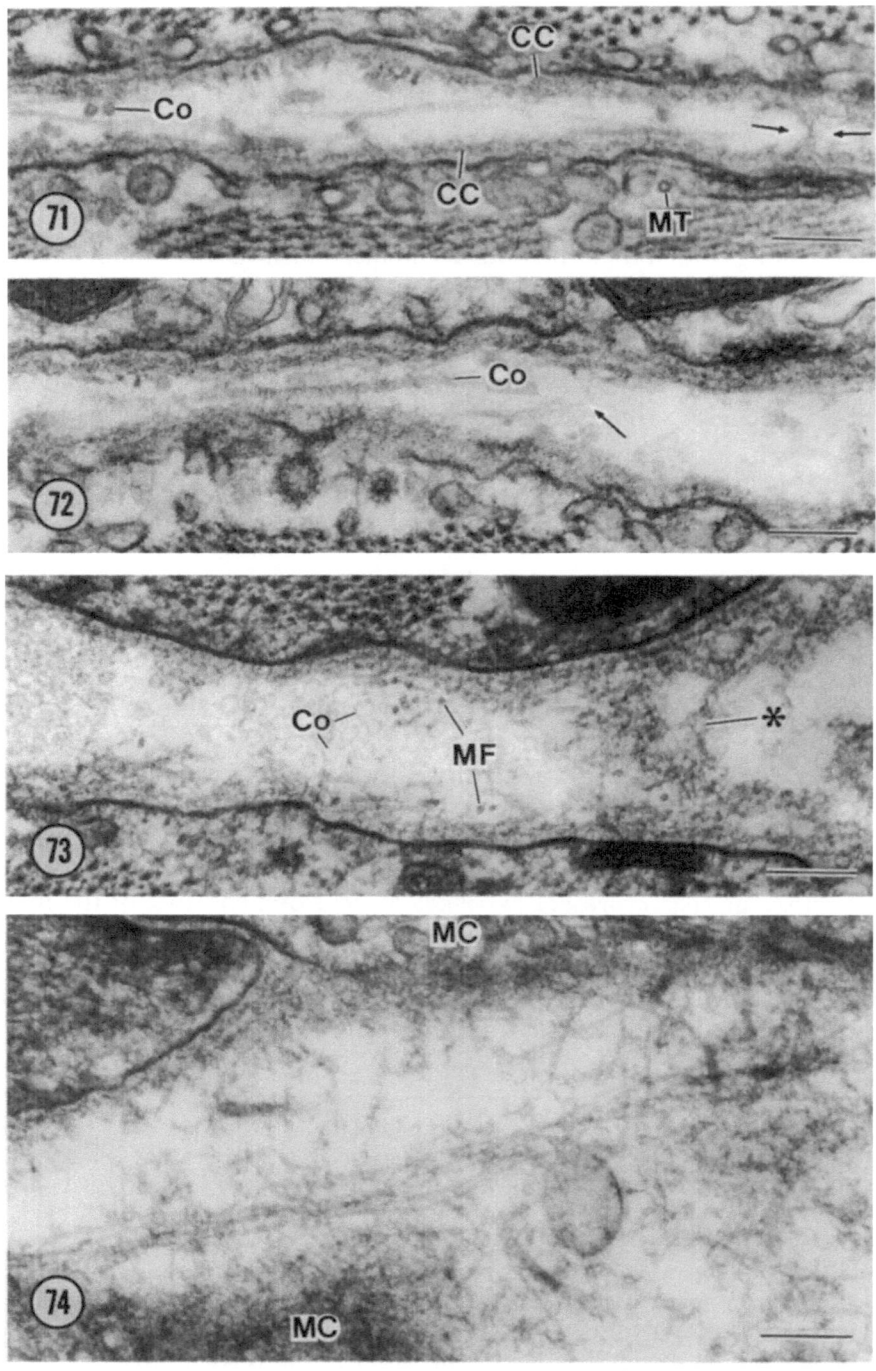

tubules" appear to be connected to the substance of sarcomere Z-discs. Junctional SR (J-SR) may function to effect electrical connection between the surface sarcolemma or TATS and the entire system of SR membranes. In many couplings, the junctional processes of the J-SR appear in the form of membranelike "pillars," which are fused with the apposed outer leaflets of the J-SR and sarcolemma/TATS and may serve as connecting structures between the two lumina. Corbular and cisternal SR, two major categories of "extended junctional SR" (which does not form junctions with the sarcolemma/TATS), are present. Additional specializations of SR occur, including dense-cored tubules and proliferated N-SR and J-SR. In some instances, the J-SR makes specialized contact with mitochondria. In the neonatal mouse heart, TATS development is beginning, whereas the SR is already well organized around myofibrils, and resembles the adult SR in many respects.

The major cytoskeletal entities in mammalian myocardial cells, the microtubules and intermediate filaments, form intricate arrays. Microtubules gravitate especially to the nucleus and the membrane systems, whereas intermediate filaments usually form prominent transverse meshworks around the myofibrils at their Z-line levels. Recently, there has arisen the concept that there exists an extracellular system that supports and interconnects myocardial cells. This is embodied in bundles of collagen, as well as components which form a complex, delicate matrix in which the muscle cells are embedded.

The membranes of both the sarcolemma/TATS complex and the SR are involved in regulation of the force of contraction of the heart by exercise of control over the myoplasmic calcium concentration ($[Ca]_i$). Both sets of membranes exhibit permeability changes and voltage changes during excitation. The Na–K pump, Ca pump, and Ca–Na exchange systems maintain steady-state ion distributions across the sarcolemma. In E–C coupling, Ca^{2+} is the key messenger. The influx of Ca^{2+} across the sarcolemma, through voltage-

---

Figures 71–74.   Transversely sectioned ventricular myocardial cells from mouse and guinea pig, illustrating extracellular "connective tissue" components which lie between and connect the myocardial cells.

Figure 71.   Mouse. A variety of extracellular structures is present, including the cell coats of the myocytes (CC), which are fused at one point (between arrows), probable collagen fibrils (Co), and other filamentous bodies. MT, peripheral myocardial microtubule. Scale bar represents 0.2 μm.

Figure 72.   Mouse. A banded collagen fibril is seen in longitudinal section (Co), and filamentous material (at arrow) forms a faint bridge across the interspace to connect adjacent myocardial cells. Scale bar represents 0.2 μm.

Figure 73.   Guinea pig. Though the gap between myocardial cells is rather wide, it is filled with extracellular substances, including the cell coat material, which merges with a farther-removed, similar-appearing filamentous matrix to create a broad intercellular connection (*). Discrete microfibrils (MF), probably of elastin, and numerous profiles of collagen (Co), are present as well. Scale bar represents 0.2 μm.

Figure 74.   Mouse. Delicate filamentous material forms a complex connection between two widely separated myocardial cells (MC). Scale bar represents 0.2 μm.

dependent slow Ca channels, occurs during excitation, and is part of the inward slow current that flows during most of the action potential plateau. This influx of Ca^{2+} directly helps to raise $[Ca]_i$ to the level required to activate the myofilaments, and in addition acts to bring about further release of Ca^{2+} from the SR.

Relaxation of the myofibrils is effected largely by means of sequestration of Ca^{2+} into the SR, through the action of its Ca pump, and also perhaps by the Ca–Na exchange mechanism coupled with the Na–K pump thought to exist in the SR. In addition, Ca^{2+} may be pumped out of the cell by a sarcolemmal Ca pump, in addition to the Ca_i–Na_o exchange reaction known to be located in the sarcolemma.

5. Appendix

Certain of the technical procedures which have been used in our research are described below.

5.1. Selective Staining of Membrane Systems

Specific visualization of the SR or TATS is difficult in tissues prepared for electron microscopy in a "conventional" manner, i.e., by means of fixation and staining procedures that impart a uniform overall contrasting to most myocardial components. For example, in a section of myocardium which has been fixed with aldehyde solution and then exposed to an osmium solution made up with an ordinary buffer vehicle (e.g., phosphate or cacodylate), the "signal-to-noise ratio" of the SR to underlying myofibrils is so low as to preclude ready identification of the SR tubules, especially when they are exposed *en face*, superimposed on myofilaments. Even T-axial tubules can be difficult to identify unequivocally in many instances, when encountered in "typical" thin sections. For this reason, a great amount of effort has been expended in search for preparatory techniques which result in opacification of one or another membrane system. We have in the past utilized a variety of tracer or marker substances (Forbes and Sperelakis, 1971; Forbes and Sperelakis, 1972a; Sperelakis *et al.*, 1974), and fairly recently have developed a technique that appears "bifunctional," i..e, applicable to one or the other membrane system upon proper handling. This procedure entails the incorporation, into the embedding sequence, of a postfixation step incorporating ferrocyanide-reduced osmium tetroxide (osmium ferrocyanide: OsFeCN). The opacification of SR bestowed by this procedural modification was first noted by Waugh and his co-workers (Peachey *et al.*, 1974; Waugh and Sommer, 1974a). This finding apparently resulted from experiments that sought endogenous peroxidatic sites (Sommer and Waugh, 1976), and it originally was thought that diaminobenzidine (DAB) was requisite to the staining. Our further experimentation, however, revealed that DAB was unnecessary and therefore that the postfixation step alone was responsible for the deposition of opaque material (Forbes *et al.*, 1977). Further, the choice of primary fixative was found to be critical, in that the use of $CaCl_2$-containing, phosphate-lacking vehicles tended to produce SR staining, whereas the omission of calcium from the fixative solution or the utilization of phosphate-containing buffer would result in the specific filling of the system of extracellular spaces (including the TATS). Energy-dispersive X-ray analysis of mouse myocardial tissue postfixed with the OsFeCN solution indicates that the opaque material is primarily composed of osmium, with only a little iron sometimes detectable. The mechanism of staining with mixtures of osmium tetroxide and potassium ferrocyanide has been dealt with in depth by White *et al.* (1979), who conclude that Os^{VII} and Os^{VI} oxidation states are primarily responsible for immobilization and deposition of the elemental osmium visualized in most embedded tissue. The staining of SR membrane vesicles is likely attributable though to the presence of Os^{IV}, this owing to the presence of associated proteins (such as calsequestrin) that are capable of capturing metal ions.

As with nearly any cytochemical technique, there appear to be some drawbacks to the "OsFeCN method." Though the staining and filling of the various constituents of SR or TATS give quite a clean picture, with very little discernible background, the procedure for SR identification is subject to occasional anomalous staining of the associated TATS elements. In such cases, one's own knowledge of myocardium will dictate which element is which (Forbes *et al.*, 1977).

We have successfully utilized the OsFeCN technique in studies of mouse, guinea pig, rat, and shrew myocardium. It should be noted, however, that in heart cells having a significant glycogen content, such as those of dog, cat, pig, or monkey, the propensity of osmium tetroxide-potassium ferri/ferrocyanide mixtures to confer specific opacification to glycogen particles takes the fore, often to the exclusion of delineation of particular membrane systems (though on occasion conjoined cells—one glycogen-stained, one membrane system-stained—are encountered). In such carbohydrate-rich hearts, it appears prudent to apply exogenous materials such as horseradish peroxidase or lanthanum colloid for demonstration of the TATS. Specific presentation of the SR remains a problem, though the "Ur–Pb–Cu" impregnation procedure utilized successfully in rat skeletal (Rambourg and Segretain, 1980) and heart muscle (Segretain *et al.*, 1981) may prove to be of some use, particularly because of its intense deposition of electron particles within the SR.

5.2. "Pseudo High Voltage" Electron Microscopy

"High voltage electron microscopy" (HVEM) is a relative term. The resolution theoretically achievable in a 0.5-μm thick section, viewed at an accelerating voltage of 100 keV, is 7.5 nm (Cosslett, 1974). Above 500 keV, there results severely decreased contrast (Glauert, 1974). Hence, one should match available accelerating voltage to the nature of the specimen to be examined, rather than overpowering it with an excessive setting.

With the advent of HVEM techniques (depending primarily on voltage above 200 keV) utilizing section thicknesses of several micrometers, there has been a general neglect of the midrange of section thickness. In transmission electron microscopy, the extremes—"ultrathin" and "thick" plastic sections—are popularly utilized. Often the "semi-thin" sections are relegated to the level of the light microscope for the purposes of confirmation or determination of the origin, state of preservation, and orientation of tissues under examination, or are discarded during trimming of the block face. The neglect into which the 0.1- to 0.5-μm-thick plastic section thus has fallen has long deprived the typical investigator, ensconced in "conventional" TEM facilities, of a valuable source of information.

The fact has been noted—but not widely promulgated—that an officially "high voltage" electron microscope (200 keV or more) is not always requisite to the effective practice of HVEM. Accelerating voltages of 60, 80, or 100 keV are quite adequate for the penetration of epoxy-embedded tissue sections up to a micrometer or more in thickness (see: Yamada and Ishikawa, 1976; Kelly, 1980; Rambourg and Segretain, 1980; Forbes and Sperelakis, 1980b; Segretain *et al.*, 1981—all of whom have examined membrane systems of muscle). Our own work incorporates routine examination of 0.1 to 2.0-μm-thick sections of OsFeCN-stained heart and skeletal muscle with a Zeiss EM-9A instrument that possesses a single accelerating voltage of 60 keV; this method has afforded us excellent compromise between contrast and resolution in thin, semithin, and "thick" plastic sections.

In the case of myocardial SR, the thinness of the "torn sleeve" of SR that surrounds myofibrils allows conclusions to be drawn from sections that are only barely "semithin" (i.e., in the range 0.1–0.3 μm). Such specimens in longitudinally sectioned, SR-stained myocardial cells will include, in many instances, the entire thickness of the SR layer, superimposed only on comparatively electron-lucent myofilaments and mitochondria (as discussed in Section 5.1. above). Furthermore, the largely planar nature of the SR sleeve in myocardium acts to optimize the obtainment of useful data from studies of semithin sections, presuming that such studies are performed on sections that capture the SR sleeves in a plane passing tangential to the surface of the associated myofibril.

For study of the myocardial TATS, the substantially greater diameter and pronounced three-dimensionality of many of its elements (as opposed to those of the SR) dictates that commensurately thicker sections be studied in order to capture its interconnected arrays. Because of the three-dimensional nature of the TATS arrays, it is often essential that stereoscopic pairs of micrographs be prepared for proper appreciation of the various spatial relationships. The use of sections 1–2 μm in thickness for surveys of the TATS has been quite successful (Figs. 4 and 6), and tilting of the sections illustrated here was accomplished via the single-tilt stage in the Zeiss EM-9A, which allows a total convergence angle of 12°.

5.3. Scanning Electron Microscopy

The major point of contention emerging from the use of scanning electron microscopy (SEM) in the study of myocardial membrane systems has been the validity of the reported observations, since interpretations are being made from the observation of rather rudely exposed surfaces. It has been demonstrated by Sommer and his associates (Sommer and Waugh, 1976; Sommer and Johnson, 1979) that the myofibrillar Z-discs tend to become topographically pronounced upon critical-point drying, thus creating structures that could be construed as representing transverse tubules; this is likely the result of shrinkage of the remainder of each sarcomere relative to the Z-discs. In any event, this phenomenon can create images in bird heart (which lacks T-tubules) equivalent in many respects to the published micrographs that purport to depict T-tubules in mammalian hearts. One criterion for identification of such vermiform bodies as T-tubules, it seems, is their continuity between adjacent myofibrils, a condition which would not be expected to occur consistently with Z-discs, even in the case of severe shrinkage of the remaining myofibrillar segments. This condition appears to be met in the case of micrographs appearing in several publications to date (Crissman *et al.*, 1978; Myklebust *et al.*, 1975; Dalen *et al.*, 1978), and in our own SEM preparations as well (Figs. 9 and 10). Proof of this interpretation will require the reembedment and TEM examination of such exposed surfaces in order to confirm or deny the presence of transverse (and axial) tubules there. At this point in time, however, it seems reasonable to conclude that we are sometimes able to demonstrate elements of the TATS by means of SEM. Further, it appears that the SR sleeves about the collections of contractile material also can be made patent by such examination.

Our own success with this technique has come primarily by our association with Mr. Myles D. Brager, who some years ago recommended to us the efficacy of immersion fixation of myocardial tissue previous to its dehydration and critical-point drying, followed by its mechanical disruption by tearing between pairs of forceps. We have not yet rigorously examined by TEM such preparations, but thus far can attest that there is considerable extraction of "ground substance" from cells thus treated, which circumvents objections that such a matrix obscures the formed elements of the myocardial cell (Sommer and Waugh, 1976). A valid objection to this preparatory procedure is that one is resorting to suboptimal fixation in order to present one's case. Certainly there is increased opportunity for artifact with our regimen. Improvement of such techniques may be made, for example by the use of thiocarbohydrazide-osmium fixation (Kelley *et al.*, 1973), which imparts an intrinsic conductive coating to the material, thus obviating the need for deposition of gold or other extraneous coverings which obscure fine details of such structures as the SR and TATS elements.

ACKNOWLEDGMENTS. The research summarized in this chapter was carried out with funding from the American Heart Association (Grant-in-Aid 78-753 to M. S. F.) and from the Public Health Service (HL-28329 to M. S. F. and HL-18711 to N. S.). Some of the research was conducted while Dr. Forbes was a postdoctoral fellow (1-FO2-HL-51147) of the Public Health Service. Dr. Forbes is currently the recipient of a Research Career Development Award (5 KO4 HL00550) from the National Institutes of Health.
Miss Barbara A. Plantholt (Department of Pathology, the Johns Hopkins

Hospital) has contributed greatly by her advice and assistance in the formulation of procedures involving the use of ferrocyanide-reduced osmium. We thank the following investigators for donation of myocardial tissue: monkey heart from Drs. S. K. Jirge and K. R. Brizzee of the Delta Regional Primate Research Center, Covington, Louisiana; dog heart from Dr. Rafael Rubio of the University of Virginia Department of Physiology; guinea pig heart from Dr. Rubio and Dr. Elaine Zelcer of the Department of Physiology; shrew heart from Dr. Orin B. Mock, Department of Anatomy, Kirksville College of Osteopathic Medicine, Kirksville, Missouri; cat heart from Drs. Marshall L. Rennels, Juanita A. Anders, and Katsukuni Fujimoto of Neurology and Anatomy, University of Maryland School of Medicine. Thanks also go to Dr. Margaret Ann Goldstein (Baylor College of Medicine) for making available to us a preprint of her manuscript on microtubules of muscle cells.

The freeze–fracture replicas were prepared by Ms. Margaretta Alietta (Department of Pathology) and Ms. Susie Sellers (Central Electron Microscope Facility of the University of Virginia School of Medicine). We thank Messrs. Rafa Rubio, James D. Ennis, and Lawrence A. Hawkey for their technical contributions, and Ms. Susan I. Purdy-Ramos for her efficient overseeing of our electron microscopy laboratory.

References

Anderton, B. H., 1981, Intermediate filaments: a family of homologous structures, *J. Muscle Res. Cell Motility* **2**:141–166.

Asano, G., Ashraf, M., and Schwartz, A., 1980, Localization of Na-K-ATPase in guinea-pig myocardium, *J. Molec. Cellular Cardiol.* **12**:257–266.

Ashraf, M., Jones, H. M., and Livingston, L. H., 1976, Localization of ATPase activity in the sarcoplasmic reticulum of myocardium, *Thirty-fourth Annu. EMSA Meeting* (C. J. Arceneaux, ed.), pp. 84–85, Claitor's Pub. Div., Baton Rouge.

Ayettey, A. S., and Navaratnam, V., 1980, The fine structure of myocardial cells in the grey seal, *J. Anat.* **131**:748.

Ayettey, A. S., and Navaratnam, V., 1981, The ultrastructure of myocardial cells in the golden hamster *Cricetus auratus*, *J. Anat.* **132**:519–524.

Baba, N., and Sharma, H. M., 1971, Histochemistry of lactic dehydrogenase in heart and pectoralis muscles of rat, *J. Cell Biol.* **51**:621–635.

Baker, P. F., Blaustein, M. P., Hodgkin, A. L., and Steinhardt, R. A., 1969, The influence of calcium on sodium efflux in squid axons, *J. Physiol.* **200**:431–458.

Behrendt, H., 1977, Effect of anabolic steroids on rat heart muscle cells. I. Intermediate filaments, *Cell and Tissue Res.* **180**:303–315.

Birks, R. I., and Davey, D. F., 1969, Osmotic responses demonstrating the extracellular character of sarcoplasmic reticulum, *J. Physiol. (Lond.)* **202**:171–188.

Borg, T. K., and Caulfield, J. B., 1979, Collagen in the heart, *Tex. Rep. Biol. Med.* **39**:321–333.

Borgers, M., Schaper, J., and Schaper, W., 1971, Localization of specific phosphatase activities in canine coronary blood vessels and heart muscle, *J. Histochem. Cytochem.* **19**:526–539.

Bossen, E. H., Sommer, J. R., and Waugh, R. A., 1978, Comparative stereology of the mouse and finch left ventricle, *Tissue and Cell* **10**:773–784.

Carrascal, E., Leon, L., and Alexandre, C., 1981, Sistema de caveolas en miocardio de rata, *Morfol. Normal y Patol., Secc. A* **5**:121–128.

Caulfield, J. B., and Borg, T. K., 1979, The collagen network of the heart, *Lab. Invest.* **40**:364–372.

Challice, C. E., and Virágh, S., 1973, The embryologic development of the mammalian heart, in: *Ultrastructure of the Mammalian Heart* (C. E. Challice and S. Virágh, eds.), pp. 91–126, Academic Press, New York.

Chandler, W. K., Rakowski, R. F., and Schneider, M. F., 1976, Effects of glycerol treatment and maintained depolarization on charge movement in skeletal muscle, *J. Physiol. (Lond.)* **254:**285–316.

Cho, Y., Sidie, J. M., and DeBruyn, P. P. H., 1972, Electron microscopic studies on a tubulofilamentous fasciculus in the bat cricothyroid muscle, *J. Ultrastruct. Res.* **41:**344–357.

Claycomb, W. C., and Palazzo, M. C., 1980, Culture of the terminally differentiated adult cardiac muscle cell: a light and scanning electron microscope study, *Dev. Biol.* **80:**466–482.

Coltoff-Schiller, B., and Goldfischer, S., 1981, Glycosaminoglycans in the rat aorta. Ultrastructural localization with toluidine blue O and osmium-ferrocyanide procedures, *Amer. J. Pathol.* **105:**232–240.

Cosslett, V. E., 1974, Current developments in high voltage electron microscopy, *J. Microscopy* **100:**233–246.

Crissman, R. S., Lane, R. D., DiDio, L. J. A., and Johnson, R. C., 1978, The relationship of myofilaments and T-tubules to the nuclei of canine myocardium studied by scanning and transmission electron microscopy, *J. Submicr. Cytol.* **10:**155–162.

Dalen, H., Myklebust, R., and Saetersdal, T. S., 1978, Cryofracture of paraffin-embedded heart muscle cells, *J. Microscopy* **112:**139–151.

Diculescu, I., Popescu, L. M., Ionescu, N., and Butucescu, N., 1971, Ultrastructural study of calcium distribution in cardiac muscle cells, *Z. Zellforsch. mikrosk. Anat.* **121:**181–198.

Dolber, P. C., and Sommer, J. R., 1980, Freeze-fracture appearance of rabbit cardiac sarcoplasmic reticulum. *Thirty-eighth Annu. EMSA Meeting* (G. Bailey, ed.), pp. 630–631, Claitor's Pub. Div., Baton Rouge.

Edge, M. B., and Walker, S. M., 1970, Evidence for a structural relationship between sarcoplasmic reticulum and Z-lines in dog papillary muscle, *Anat. Rec.* **166:**51–66.

Eisenberg, B. R., and Gilai, A., 1979, Structural changes in single muscle fibers after stimulation at a low frequency, *J. Gen. Physiol.* **74:**1–16.

Eriksson, A., and Thornell, L.-E., 1979, Intermediate (skeletin) filaments in heart Purkinje fibers. A correlative morphological and biochemical identification with evidence of a cytoskeletal function, *J. Cell Biol.* **80:**231–247.

Fabiato, A., and Fabiato, F., 1977, Calcium release from the sarcoplasmic reticulum, *Circ. Res.* **40:**119–129.

Fawcett, D. W., and McNutt, N. S., 1969, The ultrastructure of the cat myocardium. I. Ventricular papillary muscle, *J. Cell Biol.* **42:**1–45.

Ferrans, V. J., and Roberts, W. C., 1973a, Intermyofibrillar and nuclear-myofibrillar connections in human and canine myocardium. An ultrastructural study, *J. Molec. Cellular Cardiol.* **5:**247–257.

Ferrans, V. J., and Roberts, W. C., 1973b, Structural features of cardiac myxomas. Histology, histochemistry and electron microscopy, *Human Pathology* **4:**111–146.

Ferrans, V. J., Hibbs, R. G., and Buja, L. M., 1969, Nucleoside phosphatase activity in atrial and ventricular myocardium of the rat: a light and electron microscopic study, *Amer. J. Anat.* **125:**47–86.

Ferrans, V. J., Hibbs, R. G., Cipriano, P. R., and Buja, L. M., 1972, Histochemical and electron microscopic studies of norepinephrine-induced myocardial necrosis in rats, in: *Recent Advances in Studies on Cardiac Structure and Metabolism*, Vol. 1, *Myocardiology* (E. Bajusz and G. Rona, eds.), pp. 495–525, University Park Press, Baltimore.

Firth, J. A., 1980, Reliability and specificity of membrane adenosine triphosphatase localizations, *J. Histochem. Cytochem.* **28:**69–71.

Forbes, M. S., 1982, Ultrastructure of vascular smooth-muscle cells in mammalian heart, in: *The Coronary Artery* (S. Kalsner, ed.), pp. 3–58, Croom-Helm, Ltd., London.

Forbes, M. S., and Dent, J. N., 1974, Filaments and microtubules in the gonadotrophic cells of the lizard, *Anolis carolinensis*, *J. Morphol.* **143:**409–434.

Forbes, M. S., and Sperelakis, N., 1971, Ultrastructure of lizard ventricular muscle, *J. Ultrastruct. Res.* **34:**439–451.

Forbes, M. S., and Sperelakis, N., 1972a, Ultrastructure of cardiac muscle from dystrophic mice, *Amer. J. Anat.* **134:**271–290.

Forbes, M. S., and Sperelakis, N., 1972b, (Na^+, K^+)-ATPase activity in tubular systems of mouse cardiac and skeletal muscles, *Z. Zellforsch. mikrosk. Anat.* **134:**1–11.

Forbes, M. S., and Sperelakis, N., 1973a, A labyrinthine structure formed from a transverse tubule of mouse ventricular myocardium, *J. Cell Biol.* **56:**865–869.

Forbes, M. S., and Sperelakis, N., 1973b, Cardiomyopathy in the dystrophic mouse, in: *Recent Advances in Studies on Cardiac Structure and Metabolism,* Vol. 3, *Myocardial Metabolism,* (N. S. Dhalla, ed.), pp. 455–466, University Park Press, Baltimore.

Forbes, M. S., and Sperelakis, N., 1974, Spheroidal bodies in the junctional sarcoplasmic reticulum of lizard myocardial cells, *J. Cell Biol.* **60:**602–615.

Forbes, M. S., and Sperelakis, N., 1976, The presence of transverse and axial tubules in the ventricular myocardium of embryonic and neonatal guinea pigs, *Cell and Tissue Research* **166:**83–90.

Forbes, M. S., and Sperelakis, N., 1977, Myocardial couplings: their structural variations in the mouse, *J. Ultrastruct. Res.* **58:**50–65.

Forbes, M. S., and Sperelakis, N., 1979, Ruthenium-red staining of skeletal and cardiac muscles, *Cell and Tissue Research* **200:**367–382.

Forbes, M. S., and Sperelakis, N., 1980a, Structures located at the level of the Z bands in mouse ventricular myocardial cells, *Tissue and Cell* **12:**467–489.

Forbes, M. S., and Sperelakis, N., 1980b, Membrane systems in skeletal muscle of the lizard *Anolis carolinensis, J. Ultrastruct. Res.* **73:**245–261.

Forbes, M. S., and Sperelakis, N., 1982a, Association between gap junctions and mitochondria in mammalian myocardial cells, *Tissue and Cell* **14:**25–37.

Forbes, M. S., and Sperelakis, N., 1982b, Bridging junctional processes in couplings of striated, cardiac, and smooth muscle cells, *Muscle and Nerve* (in press).

Forbes, M. S., Rubio, R., and Sperelakis, N., 1972, Tubular systems of *Limulus* myocardial cells investigated by use of electron-opaque tracers and hypertonicity, *J. Ultrastruct. Res.* **39:**580–597.

Forbes, M. S., Plantholt, B. A., and Sperelakis, N., 1977, Cytochemical staining procedures selective for sarcotubular systems of muscle: applications and modifications, *J. Ultrastruct. Res.* **60:**306–327.

Forbes, M. S., Rennels, M. L., and Nelson, E., 1979, Caveolar systems and sarcoplasmic reticulum in coronary smooth muscle cells of the mouse, *J. Ultrastruct. Res.* **67:**325–339.

Forssmann, W. G., and Girardier, L., 1966, Untersuchungen zur Ultrastruktur des Rattenherzmuskels mit besonderer Berücksichtigung des Sarcoplasmatischen Retikulum, *Zeit. Zellforsch. mikrosk. Anat.* **72:**249–275.

Forssmann, W. G., and Girardier, L., 1970, A study of the T system in rat heart, *J. Cell Biol.* **44:**1–19.

Frank, J. S., and Langer, G. A., 1974, The myocardial interstitium: its structure and its role in ionic exchange, *J. Cell Biol.* **60:**586–601.

Frank, J. S., Langer, G. A., Nudd, L. M., and Seraydarian, K., 1977, The myocardial cell surface, its histochemistry, and the effect of sialic acid and calcium removal on its structure and cellular ionic exchange, *Circ. Res.* **41:**702–714.

Franke, W. W., Kartenbeck, J., Zentgraf, H., Scheer, U., and Falk, H., 1971, Membrane-to-membrane cross-bridges. A means to orientation and interaction of membrane faces, *J. Cell Biol.* **51:**881–888.

Franzini-Armstrong, C., 1970, Studies of the triad. I. Structure of the junction in frog twitch fibers, *J. Cell Biol.* **47:**488–499.

Franzini-Armstrong, C., 1971, Studies of the triad. II. Penetration of tracers into the junctional gap, *J. Cell Biol.* **49:**196–203.

Franzini-Armstrong, C., 1974, Freeze fracture of skeletal muscle from the tarantula spider. Structural differentiations of sarcoplasmic reticulum and transverse tubular system membranes, *J. Cell Biol.* **61:**501–513.

Franzini-Armstrong, C., 1975, Membrane particles and transmission at the triad, *Fed. Proc.* **34:**1382–1389.

Franzini-Armstrong, C., 1977, The comparative structure of intracellular junctions in striated muscle fibers, in: *Pathogenesis of Human Muscular Dystrophies*, (L. P. Rowland, ed.), pp. 612–625, Excerpta Medica, Amsterdam.

Franzini-Armstrong, C., 1980, Structure of sarcoplasmic reticulum, *Fed. Proc.* **39:**2403–2409.

Fuseler, J. W., Shay, J. W., and Feit, H., 1981, The role of intermediate (10-nm) filaments in the development and integration of the myofibrillar contractile apparatus in the embryonic mammalian heart, in: *Cell and Muscle Motility*, Vol. 1 (R. M. Dowben and J. W. Shay, eds.), pp. 205–259, Plenum Press, New York.

Gabella, G., 1978, Inpocketings of the cell membrane (caveolae) in the rat myocardium, *J. Ultrastruct. Res.* **65:**135–147.

Geddes, L. A., 1970, *The Direct and Indirect Measurement of Blood Pressure*, Year Book Medical Publishers, Chicago.

Glauert, A. M., 1974, The high voltage electron microscope in biology, *J. Cell Biol.* **63:**717–748.

Goldman, R. D., 1971, The role of three cytoplasmic fibers in BHK-21 cell motility. I. Microtubules and the effects of colchicine, *J. Cell Biol.* **51:**752–762.

Goldstein, M. A., and Entman, M. L., 1979, Microtubules in mammalian heart muscle, *J. Cell Biol.* **80:**183–195.

Hagopian, M., and Tennyson, V. M., 1971, Cytochemical localization of cholinesterase activity in adult rabbit heart, *J. Histochem. Cytochem.* **19:**376–381.

Heuson-Stiennon, J. A., 1965, Morphogènese de la cellule musculaire striée au microscope électronique. Formation des structures fibrillaires, *J. Microscopie (Paris)* **4:**657–678.

Hirakow, R., and Gotoh, T., 1980, Quantitative studies on the ultrastructural differentiation and growth of mammalian cardiac muscle cells. II. The atria and ventricles of the guinea pig, *Acta Anat.* **108:**230–237.

Hirakow, R., and Krause, W. J., 1980, Postnatal differentiation of ventricular myocardial cells of the opossum (*Didelphis virginiana* Kerr) and T-tubule formation, *Cell and Tissue Research* **210:**95–100.

Hoffstein, S., Gennaro, D. E., Weissmann, G., Hirsch, J., Streuli, F., and Fox, A. C., 1975, Cytochemical localization of lysosomal enzyme activity in normal and ischemic dog myocardium, *Amer. J. Pathol.* **79:**193–206.

Isenberg, G., and Klöckner, U., 1980, Glycocalyx is not required for slow inward calcium current in isolated rat heart myocytes, *Nature* **284:**358–360.

Ishikawa, H., Bischoff, R., and Holtzer, H., 1968, Mitosis and intermediate-sized filaments in developing skeletal muscle, *J. Cell Biol.* **38:**538–555.

Ishikawa, H., and Yamada, E., 1976, Differentiation of the sarcoplasmic reticulum and T-system in developing mouse cardiac muscle, in: *Developmental and Physiological Correlates of Cardiac Muscle* (M. Lieberman and T. Sano, eds.), pp. 21–35, Raven Press, New York.

Jewett, P. H., Sommer, J. R., and Johnson, E. A., 1971, Cardiac muscle. Its ultrastructure in the finch and hummingbird with special reference to the sarcoplasmic reticulum, *J. Cell Biol.* **49:**50–65.

Jewett, P. H., Leonard, S. D., and Sommer, J. R., 1973, Chicken cardiac muscle. Its elusive extended junctional sarcoplasmic reticulum and sarcoplasmic reticulum fenestrations, *J. Cell Biol.* **56:**595–600.

Johnson, E. A., and Sommer, J. R., 1967, A strand of cardiac muscle: its ultrastructure and the electrophysiologic implications of its geometry, *J. Cell Biol.* **33:**103–129.

Karnovsky, M. J., 1964, The localization of cholinesterase activity in rat cardiac muscle by electron microscopy, *J. Cell Biol.* **23:**217–232.

Kelley, R. O., Decker, R. A. F., and Bluemink, J. G., 1973, Ligand-mediated osmium binding: its application, in coating biological specimens for scanning electron microscopy, *J. Ultrastruct. Res.* **45:**254–258.

Kelly, A. M., 1980, T-tubules in neonatal rat soleus and extensor digitorum longus muscles, *Dev. Biol.* **80:**501–505.

Kelly, D. E., 1969, The fine structure of skeletal muscle triad junctions, *J. Ultrastruct. Res.* **29**:37–49.

Kelly, D. E., and Kuda, A. M., 1979, Subunits of the triadic junction in fast skeletal muscle as revealed by freeze-fracture, *J. Ultrastruct. Res.* **68**:220–233.

Kerr, L. M., and Sperelakis, N., 1982, Effects of the calcium antagonists bepridil (CERM-1978) and verapamil on Ca^{++}-dependent slow action potentials in frog skeletal muscle, *J. Pharm. Exp. Therap.* (in press).

Kulczycky, S., and Mainwood, G. W., 1972, Evidence for a functional connection between the sarcoplasmic reticulum and the extracellular space in frog sartorius muscle, *Can. J. Physiol. Pharmacol.* **50**:87–98.

Langer, G. A., 1974, Ionic movements and the control of contraction, in: *The Mammalian Myocardium*, (G. A. Langer and A. J. Brady, eds.), pp. 193–217, John Wiley & Sons, New York.

Langer, G. A., 1976, Events at the cardiac sarcolemma: localization and movement of contractile-dependent calcium, *Fed. Proc.* **35**:1274–1278.

Langer, G. A., 1980, The role of calcium in the control of myocardial contractility: an update, *J. Molec. Cellular Cardiol.* **12**:231–239.

Langer, G. A., and Serena, S. D., 1970, Effects of strophanthidin upon contraction and ionic exchange in rabbit ventricular myocardium: relation to control active state, *J. Molec. Cellular Cardiol.* **1**:65–90.

Lazarides, E., 1980, Intermediate filaments as mechanical integrators of cellular space, *Nature* **283**:249–256.

Lee, S. H., 1969, Ultrastructural localization of glutamic oxalacetic transaminase activity in cardiac muscle fiber and cardiac mitochondrial fraction of the rat, *Histochemie* **19**:99–109.

Leeson, T. S., 1978, The transverse tubular (T) system of rat cardiac muscle fibers as demonstrated by tannic acid mordanting, *Can. J. Zool.* **56**:1906–1916.

Leeson, T. S., 1980, T-tubules, couplings and myofibrillar arrangements in rat atrial myocardium, *Acta Anat.* **108**:374–388.

Legato, M. J., and Langer, G. A., 1969, The subcellular localization of calcium ion in mammalian myocardium, *J. Cell Biol.* **41**:401–423.

Markham, R., Frey, S., and Hills, G. J., 1963, Methods for the enhancement of image detail and accentuation of structure in electron microscopy, *Virology* **20**:88–102.

Masson-Pévet, M., Gros, D., and Besselsen, E., 1980, The caveolae in rabbit sinus node and atrium, *Cell and Tissue Research* **208**:183–196.

Mastaglia, F. L., 1973, Pathological changes in skeletal muscle in acromegaly, *Acta Neuropath.* (Berlin) **24**:273–286.

McNutt, N. S., and Fawcett, D. W., 1969, The ultrastructure of the cat myocardium. II. Atrial muscle, *J. Cell Biol.* **42**:46–67.

McNutt, N. S., and Fawcett, D. W., 1974, Myocardial ultrastructure, in: *The Mammalian Myocardium* (G. A. Langer and A. J. Brady, eds.), pp. 1–49, John Wiley & Sons, New York.

Melax, H., and Leeson, T. S., 1972, Electron microscope study of myocardial tissue space contents in rat heart, *Cardiovasc. Res.* **6**:89–94.

Myklebust, R., Dalen, H., and Saetersdal, T. S., 1975, A comparative study in the transmission and scanning electron microscope of intracellular structures in sheep heart muscle cell, *J. Microscopy* **105**:57–65.

Myklebust, R., Saetersdal, T. S., and Engedal, H., 1978, The T-tubule system in the myocardia of the sand rat and mouse as demonstrated by horseradish peroxidase, *Cell and Tissue Research* **192**:205–213.

Nag, A. C., 1980, Study of non-muscle cells of the adult mammalian heart: a fine structural analysis and distribution, *Cytobios* **28**:41–61.

Nag, A. C., Fischman, D. A., Aumont, M. C., and Zak, R., 1977, Studies of isolated adult rat heart cells: the surface morphology and the influence of extracellular calcium ion concentration on cellular viability, *Tissue and Cell* **9**:419–436.

Osculati, F., Amati, S., Petrini, E., Marelli, M., and Gazzanelli, G., 1978, A study on the organization of the tubular endoplasmic system in the rat heart conduction fibres, *J. Submicr. Cytol.* **10**:371–380.

Pachter, B. R., and Breinin, G. M., 1977, Double-membrane arrays in Type II fibers of mouse extraocular muscle, *Invest. Ophthalmol. Visual Sci.* **16**:666–668.

Page, E., and Buecker, J. L., 1981, Development of dyadic junctional complexes between sarcoplasmic reticulum and plasmalemma in rabbit left ventricular myocardial cells. Morphometric analysis, *Circ. Res.* **48**:519–522.

Page, E., and Surdyk-Droske, M., 1979, Distribution, surface density, and membrane area of diadic junctional contacts between plasma membrane and terminal cisterns in mammalian ventricle, *Circ. Res.* **45**:260–267.

Page, E., and Upshaw-Earley, J., 1977, Volume changes in sarcoplasmic reticulum of rat hearts perfused with hypertonic solutions, *Circ. Res.* **40**:355–366.

Peachey, L. D., 1965, The sarcoplasmic reticulum and transverse tubules of the frog's sartorius, *J. Cell Biol.* **25**:209–231.

Peachey, L. D., Waugh, R. A., and Sommer, J. R., 1974, High voltage electron microscopy of sarcoplasmic reticulum, *J. Cell Biol.* **63**:262a.

Porter, K. R., 1961, The sarcoplasmic reticulum. Its recent history and present status, *J. Biophys. Biochem. Cytol.* **10**:219–226.

Prasad, K., and Singal, P. K., 1977, Ultrastructure of failing myocardium due to induced chronic mitral insufficiency in dogs, *Br. J. Exp. Path.* **58**:289–300.

Rambourg, A., and Segretain, D., 1980, Three-dimensional electron microscopy of mitochondria and endoplasmic reticulum in the red muscle fiber of the rat diaphragm, *Anat. Rec.* **197**:33–48.

Rayns, D. G., Simpson, F. O., and Bertaud, W. S., 1967, Transverse tubule apertures in mammalian myocardial cells: surface array, *Science* **156**:656–657.

Rayns, D. G., Simpson, F. O., and Bertaud, W. S., 1968, Surface features of striated muscle. I. Guinea-pig cardiac muscle, *J. Cell Sci.* **3**:467–474.

Rayns, D. G., Devine, C. E., and Sutherland, C. L., 1975, Freeze fracture studies of membrane systems in vertebrate muscle. I. Striated muscle, *J. Ultrastruct. Res.* **50**:306–321.

Reuter, H., 1974, Exchange of calcium ions in the mammalian myocardium. Mechanisms and physiological significance, *Circ. Res.* **34**:599–605.

Robinson, T. F., 1980, Lateral connections between heart muscle cells as revealed by conventional and high voltage transmission electron microscopy, *Cell and Tissue Research* **211**:353–359.

Robinson, T. F., and Winegrad, S., 1981, A variety of intercellular connections in heart muscle, *J. Molec. Cellular Cardiol.* **13**:185–195.

Rostgaard, J., and Behnke, O., 1965, Fine structural localization of adenine nucleoside phosphatase activity in the sarcoplasmic reticulum and the T system of rat myocardium, *J. Ultrastruct. Res.* **12**:579–591.

Rubio, R., and Sperelakis, N., 1971, Entrance of colloidal ThO_2 tracer into the T tubules and longitudinal tubules of the guinea pig heart, *Z. Zellforsch. mikrosk. Anat.* **116**:20–36.

Rubio, R., and Sperelakis, N., 1972, Penetration of horseradish peroxidase into the terminal cisternae of frog skeletal muscle fibers and blockade of caffeine contracture by Ca^{++} depletion, *Z. Zellforsch. mikrosk. Anat.* **124**:57–71.

Rubio, R., Berne, R. M., and Dobson, J. G., Jr., 1973, Sites of adenosine production in cardiac and skeletal muscle, *Amer. J. Physiol.* **255**:938–953.

Scales, D. J., 1981, Aspects of the mammalian cardiac sarcotubular system revealed by freeze fracture electron microscopy, *J. Molec. Cellular Cardiol.* **13**:373–380.

Schneider, M. F., 1981, Membrane charge movement and depolarization-contraction coupling, *Ann. Rev. Physiol.* **43**:507–517.

Schwartz, A., 1972, Calcium metabolism, *Cardiology* **57**:16–23.

Schwarzfeld, T. A., and Jacobson, S. L., 1981, Isolation and development in cell culture of myocardial cells of the adult rat, *J. Molec. Cellular Cardiol.* **13**:563–575.

Segretain, D., Rambourg, A., and Clermont, Y., 1981, Three dimensional arrangement of mitochondria and endoplasmic reticulum in the heart muscle fiber of the rat, *Anat. Rec.* **200**:139–151.

Shiina, S.-I., Mizuhira, V., Uchida, K., and Amakawa, T., 1969, Electronmicroscopic study of sodium ion distribution in cardiac ventricle cells, *Jpn. Circ. J.* **33**:601–605.

Shiina, S.-I., and Mizuhira, V., 1970, Calcium ion localization in cardiac ventricle muscle on electron microscopic level, *Jpn. Circ. J.* **34:**1047–1051.

Simpson, F. O., 1965, The transverse tubular system in mammalian myocardial cells, *Amer. J. Anat.* **117:**1–18.

Simpson, F. O., and Rayns, D. G., 1968, The relationship between the transverse tubular system and other tubules at the Z-disc levels of myocardial cells in the ferret, *Amer. J. Anat.* **122:**193–208.

Simpson, F. O., Rayns, D. G., and Ledingham, J. M., 1973, The ultrastructure of ventricular and atrial myocardium, in: *Ultrastructure of the Mammalian Heart,* (C. E. Challice and S. Virágh, eds.), pp. 1–41, Academic Press, New York.

Simpson, F. O., Rayns, D. G., and Ledingham, J. M., 1974, Fine structure of mammalian myocardial cells, *Adv. Cardiol.* **12:**15–33.

Slautterback, D. B., 1966, The ultrastructure of cardiac and skeletal muscle, in: *The Physiology and Biochemistry of Muscle as a Food,* (E. J. Briskey, R. G. Cassens, and J. C. Trautman, eds.), pp. 39–68, U. Wisconsin Press, Madison.

Somlyo, A. V., 1979, Bridging structures spanning the junctional gap at the triad of skeletal muscle, *J. Cell Biol.* **80:**743–750.

Sommer, J. R., and Johnson, E. A., 1968, Cardiac muscle. A comparative study of Purkinje fibers and ventricular fibers, *J. Cell Biol.* **36:**497–526.

Sommer, J. R., and Johnson, E. A., 1969, Cardiac muscle. A comparative ultrastructural study with special reference to frog and chicken hearts, *Z. Zellforsch. mikrosk. Anat.* **98:**437–468.

Sommer, J. R., and Johnson, E. A., 1979, Ultrastructure of cardiac muscle, in: *Handbook of Physiology. Section 2: The Cardiovascular System,* Vol. I: *The Heart,* (R. M. Berne, N. Sperelakis, and S. R. Geiger, eds.), pp. 113–186, Amer. Physiol. Soc., Bethesda.

Sommer, J. R., and Spach, M. S., 1964, Electron microscopic demonstration of adenosinetriphosphatase in myofibrils and sarcoplasmic membranes of cardiac muscle of normal and abnormal dogs, *Amer. J. Pathol.* **44:**491–505.

Sommer, J. R., and Waugh, R. A., 1976, The ultrastructure of the mammalian cardiac muscle cell—with special emphasis on the tubular membrane systems, *Amer. J. Pathol.* **82:**191–232.

Sommer, J. R., Wallace, N. R., and Hasselbach, W., 1978, The collapse of the sarcoplasmic reticulum in skeletal muscle, *Z. Naturforsch.* **33:**561–573.

Sordahl, L. A., 1979, Role of mitochondria in heart cell function, *Tex. Rep. Biol. Med.* **39:**5–18.

Sperelakis, N., 1970, Ultrastructure of the neurogenic heart of *Limulus polyphemus, Z. Zellforsch. mikrosk. Anat.* **116:**443–463.

Sperelakis, N., and Rubio, R., 1971, An orderly lattice of axial tubules which interconnect adjacent transverse tubules in guinea-pig ventricular myocardium, *J. Molec. Cellular Cardiol.* **2:**211–220.

Sperelakis, N., Forbes, M. S., and Rubio, R., 1974, The tubular systems of myocardial cells: ultrastructure and possible function, in: *Recent Advances in Studies on Cardiac Structure and Metabolism,* Volume 4, *Myocardial Biology,* (N. S. Dhalla, ed.), pp. 163–194, University Park Press, Baltimore.

Sperelakis, N., Shigenobu, K., and Rubio, R., 1978, [^{3}H]sucrose compartments in frog skeletal muscle relative to sarcoplasmic reticulum, *Amer. J. Physiol.* **234:**C181–C190.

Stenger, R. J., and Spiro, D., 1961, Structure of the cardiac muscle cell, *Amer. J. Med.* **30:**653–665.

Stoeckel, M.-E., Osborn, M., Porte, A., Sacrez, A., Batzenschlager, A., and Weber, K., 1981, An unusual familial cardiomyopathy characterized by aberrant accumulations of desmin-type intermediate filaments, *Virchows Arch. (Pathol. Anat.)* **393:**53–60.

Strosberg, A. M., Katzung, B. G., and Lee, J. C., 1970, Demonstration of adenosine triphosphatase activity in coated dense vesicles and membranes of specific granules in mammalian myocardium, *Lab. Invest.* **23:**386–391.

Sybers, H. D., and Ashraf, M., 1973, Preparation of cardiac muscle for SEM, in: *Scanning Electron Microscopy: Proceedings of the Workshop on Scanning Electron Microscopy in Pathology* (O. Johari, ed.), pp. 342–348, Illinois Inst. of Technol. Res. Inst., Chicago.

Sybers, H. D., and Gann, M. K., 1975, Non-specificity of a "selective" stain for sarcoplasmic reticulum in cardiac cells. *Thirty-third Annu. EMSA Meeting* (C. J. Arceneaux, ed.), pp. 539–540, Claitor's Pub. Div., Baton Rouge.

Thornell, L.-E., and Eriksson, A., 1981, Filament systems in the Purkinje fibers of the heart, *Amer. J. Physiol.* **241**:H291–H305.

Tyson, G. E., and Bulger, R. E., 1973, Vinblastine-induced paracrystals and unusually large microtubules (macrotubules) in rat renal cells, *Z. Zellforsch. mikrosk. Anat.* **141**:443–458.

Vahouny, G. V., Wei, R. W., Tamboli, A., and Albert, E. N., 1979, Adult canine myocytes: isolation, morphology and biochemical characterizations, *J. Molec. Cellular Cardiol.* **11**:339–357.

Van Winkle, W. B., 1977, The fenestrated collar of mammalian cardiac sarcoplasmic reticulum: a freeze-fracture study, *Amer. J. Anat.* **149**:277–282.

Waugh, R. A., and Sommer, J. R., 1974a, A reproducible selective stain for cardiac sarcoplasmic reticulum. *Thirty-second Annu. Meeting EMSA* (C. J. Arceneaux, ed.), pp. 214–215, Claitor's Pub. Div., Baton Rouge.

Waugh, R. A., and Sommer, J. R., 1974b, Lamellar junctional sarcoplasmic reticulum. A specialization of cardiac sarcoplasmic reticulum, *J. Cell Biol.* **63**:337–343.

White, D. L., Mazurkiewicz, J. E., and Barrnett, R. J., 1979, A chemical mechanism for tissue staining by osmium tetroxide-ferrocyanide mixtures, *J. Histochem. Cytochem.* **27**:1084–1091.

Winegrad, S., 1979, Electromechanical coupling in heart muscle, in: *Handbook of Physiology. Section 2: The Cardiovascular System*, Vol. I: *The Heart*, (R. M. Berne, N. Sperelakis, and S. R. Geiger, eds.), pp. 393–428, Amer. Physiol. Soc., Bethesda.

Yamada, E., and Ishikawa, H., 1976, High voltage electron microscopy combined with molecular tracers: observations on the cardiac muscle T-system and renal epitheliocyte, in: *Recent Progress in Electron Microscopy of Cells and Tissues*, (E. Yamada, V. Mizuhira, K. Kurosumi, and T. Nagano, eds.), pp. 354–362, University Park Press, Baltimore.

6

Control of Gene Expression in Muscle Development

S. M. Heywood, M. C. Thibault, and E. Siegel

1. Introduction

This review does not attempt to exhaustively summarize all of the recent studies concerning the mechanisms of gene expression during muscle development, but rather attempts to critically analyze recent studies describing which regulatory mechanisms are utilized by muscles during terminal differentiation. Terminal differentiation of muscles involves the transition from dividing myoblasts to multinucleate fibers, or myotubes, and leads to the ultimate appearance of muscle-specific proteins and their assembly into functional contractile apparatus. In addition, the determination of those isoforms of the contractile proteins to be synthesized by different fibers is an integral part of this differentiative process. For an excellent review of muscle proteins and their synthesis, readers are referred to Buckingham (1977).

Cellular differentiation is generally thought to involve the coordinate synthesis of specific proteins. In muscle, this likely involves both the synthesis as well as the assembly of the contractile proteins into a functional unit. This process is as yet poorly understood and, even further, the coordinate synthesis of these proteins has yet to be stringently demonstrated. Coordinate control must involve a simultaneous gene activation leading to the qualitative and quantitative synthesis of proteins finally resulting in the assembly of the contractile apparatus. This synthesis must account for the proper stoichiometric amounts of newly synthesized protein subunits. Alternatively,

S. M. Heywood, M. C. Thibault, and E. Siegel • Department of Biology, University of Connecticut, Storrs, Connecticut 06268.

the process may rely on excessive synthesis of some proteins or protein sub-units which are eventually discarded.

The identification of the control mechanisms that eukaryotic cells utilize to differentially express genes has been a major emphasis of developmental biology. It is expected that muscle differentiation is not unique and in addition to utilizing control devices involved in mRNA synthesis (transcription) also makes use of posttranscriptional controls in mRNA processing and transport, translational controls involving mRNA utilization and mRNA half-life, and posttranslational modifications of newly synthesized proteins. It is clear that eukaryotic cells possess the means to control gene expression at many points between the level of gene structure and organization and the final assembled protein(s). What is necessary is to define these mechanisms and to determine to what degree each cell system utilizes the different mechanisms during various stages of its development. Modern technology, utilizing recombinant DNA technology, is just beginning to yield information on muscle differentiation and undoubtedly will aid in clarifying many unanswered questions (see Chapter 7).

2. Gene Expression during the Transition from Dividing Myoblasts to Myotubes

Myoblasts can be cultured *in vitro* and undergo cell multiplication. Once the proper conditions of cell density and cell media are met, the cells subsequently fuse into multinucleated myotubes (Herrmann *et al.,* 1970). This *in vitro* process is thought to closely resemble the formation of multinucleated myofibers *in vivo*. The process of cell fusion has been the basis of most studies that attempt to determine the timing and processes involved in differential gene expression in muscle. Generally, three types of cell systems have been used: (1) established rat muscle cell lines, (2) primary cell cultures in which cell fusion has been synchronized by conditioned media or by Ca^{2+} deficiency, and (3) primary cell cultures which are less synchronized and therefore have a more extended time frame for the fusion process. In addition, a number of studies have utilized embryonic muscle tissue for the large scale preparations of substances not feasibly isolated from cell cultures. These differences in experimental conditions are most likely important in the interpretation of results and will be referred to in the following discussion.

2.1. Analysis of Evidence for Transcriptional Control

Transcriptional control in any differentiating system is of primary importance. All macromolecules (RNA and proteins) must originally arise from gene activation. These molecules may be either structural or enzymatic components of muscle or regulatory elements involved in the synthesis of other components. A large number of reports have been concerned with the appearance of muscle specific mRNAs in relation to the subsequent translation of these mRNAs. Depending on cell culture conditions, muscle cell fusion normally begins at about 40 h after plating; and is followed by a lag period after which a burst of myosin (and in general, myofibrillar protein) synthesis occurs due to an increase in the rate of production. This increased rate presumably results from an increased amount of available or newly synthesized mRNAs (Emerson and Beckner, 1975; Paterson and Strohman, 1972; Coleman and Coleman, 1968; Devlin and Emerson, 1979, 1980; Strohman *et al.*, 1977; Strohman *et al.*, 1980). What the rate-limiting control step is, and whether mRNA transcripts are available for translation in proliferative myoblasts, postmitotic myoblasts, or appear only in myotubes after cell fusion, is the major concern of these studies.

It is generally accepted that after cell fusion there are pronounced changes in the pattern of protein synthesis with a several 100-fold increase in the synthesis of the myofibrillar proteins actin, myosin, troponin, tropomyosin and others (Devlin and Emerson, 1979; Buckingham, 1977). This is thought to be an example of coordinate control of gene expression; however, as previously mentioned, little information is available on the precise quantity of proteins produced in relation to the number of individual mRNA transcripts present in the cell. For example, myosin heavy chain (MHC) is approximately ten times the size of myosin light chains (MLC). Does this mean that equal numbers of mRNAs are transcribed and translated with MHC-mRNA being ten times as efficient as the MLC-mRNAs, or are variable amounts of mRNAs produced so as to reflect a balanced synthesis of proteins? Alternatively, mRNAs may utilize mechanisms which alter their translational efficiency (see below). Similar questions can be raised in regard to the stoichiometric relationships in the synthesis of the other myofibrillar proteins. Indeed, the quantitation of the myofibrillar protein mRNAs as reported by Bowman and Emerson (1977, 1980) indicate the mRNAs exist in equimolar concentrations. If this is correct, it would indicate either a significant contribution of translational control to obtain a balanced synthesis of proteins (assuming elongation rates for protein synthesis are similar for all mRNAs) or a noncoordinate control for the quantitative aspects of myofibrillar protein synthesis.

It has been reported (Devlin and Emerson, 1979), however, that there is a coordinate accumulation of contractile protein mRNAs when dividing myoblasts are shifted to culture conditions which support cell fusion. The shift to conditioned media was performed to obtain a more synchronous culture. Differentiation is known to occur over a less extended period of time in cell culture than *in vivo* (Herrmann *et al.*, 1970). In addition, while the synchronization of cells in culture is beneficial in defining two distinct muscle cell types (myoblasts vs. myotubes) it most probably shortens the period in which cells prepare to fuse more than normally occurs under *in vitro* conditions. It is necessary to keep these distinctions in mind when comparing results from different laboratories as well as attempting to relate these observations to those occurring *in vivo*.

If it is accepted that cell culture conditions adequately reflect *in vivo* differentiation, the evidence that the formation of mRNA transcripts is the sole contributor of gene expression during myofibrillar formation is still questionable. A detailed study by Devlin and Emerson (1979), utilizing the translation of total mRNA from differentiating muscle cells, claims that the mRNAs for the contractile proteins are present in stoichiometric amounts. Although in some cases this appears to be the case, myosin LC_2 is in excess (three- to fourfold) while β-tropomyosin is twofold less and MHC is present in less than one-tenth of the stoichiometric amount. As indicated by the authors, translational assays are difficult to use for quantitative information, particularly with relation to MHC-mRNA, emphasizing the difficulty in using translational assays to determine the quantity of mRNA present in a cell system. MHC-mRNA has been effectively translated by a number of laboratories (Bragg *et al.*, 1980; Benoff and Nadal-Ginard, 1979a; Bag and Sarkar, 1976; Umeda *et al.*, 1980; Strohman *et al.*, 1980) and negative results espousing "mRNAs coding for large proteins are translated inefficiently" should be viewed with caution. More likely, an exact quantitation of specific mRNAs present in the cytoplasm of differentiating muscle cells will require much more detailed studies utilizing specific DNA probes obtained from cloned DNA representing a sizeable portion of a particular gene.

Translational assays for specific mRNAs have also been utilized to determine when, in relation to myotube formation, myofibrillar protein mRNAs appear in the cytoplasm (Devlin and Emerson, 1979; Strohman *et al.*, 1977; Strohman *et al.*, 1980; Bragg *et al.*, 1980; Benoff and Nadal-Ginard, 1979b). Although some reports are conflicting, it is generally accepted that in the avian systems, dividing myoblasts contain little if any myofibrillar mRNA (MHC chain has been particularly studied) (Dym *et al.*, 1979; Strohman *et al.*, 1980; Devlin and Emerson, 1979). The data collected on postmitotic myoblasts containing MHC transcripts is even more difficult to analyze because of the different culture conditions employed (see above). In addition, measuring only translatable mRNA does not account for inactive cytoplasmic forms. Strohman *et al.* (1977), utilizing a 26 S mRNA isolated under denaturing conditions (most reports agree MHC-mRNA is 32–33 S under denaturing conditions) (Havaranis and Heywood, 1981; Dym *et al.*, 1979; Benoff and Nadal-Ginard, 1979b; Sarkar *et al.*, 1973) as well as total cell mRNA, found no translatable MHC-mRNA transcripts in dividing myoblasts but did find translatable MHC-mRNA in fused muscle cell cultures. More recently the same authors (Strohman *et al.*, 1980) have analyzed for the presence of MHC immune precipitates in cell-free systems in which total mRNA extracted from muscle cell cultures at various times after plating had been translated. The results of these studies when analyzed under conditions in which similar RNA concentrations were added to cell-free systems revealed the presence of MHC-mRNA from 22 h of culture to 72 h. The authors point out that these results are difficult to interpret for it was not known if the MHC-mRNA was present in a few contaminating myotubes (22 h) or was present in nondividing myoblasts and myoblasts as well.

Neither the Emerson group nor the Strohman group have attempted to localize the muscle-specific mRNAs using either sucrose density gradient or buoyant density gradient centrifugation to separate active (polysomal mRNAs) from inactive or mRNP (stored) mRNAs. Both groups (Devlin and Emerson, 1979; Strohman *et al.,* 1980) have claimed that no inhibitory RNAs are present in their RNA preparations based on experiments in which myoblast RNA is mixed with myotube RNA. Under these experimental conditions, the translation of myotube mRNA is unaffected; however, this experimental design fails to take into account the published reports which have indicated that mRNP-associated tcRNA is a stoichiometric inhibitor (inhibits only that mRNA with which it is associated) and does not act catalytically (Heywood and Kennedy, 1976).

Benoff and Nadal-Ginard (1979b), utilizing a rat muscle cell line, have observed that translatable MHC-mRNA is present prior to cell fusion and that these mRNAs have different lengths of poly-A tails than those of MHC-mRNA isolated from later appearing MHC-mRNAs. Interestingly these MHC-mRNAs, isolated at two stages of differentiation, have different translational efficiencies, thus suggesting that posttranscriptional alterations in mRNAs may effect their detectability by translational assays as well as their detectability by virtue of oligo (dT)-cellulose binding.

In one case, where muscle cell lysates have been fractionated into polysomal and non-ribosome bound mRNAs (mRNPs), translatable MHC-mRNA has been found in cell cultures where the majority of cells were postmitotic and at early fusion (Bragg *et al.,* 1980). In this case however, 20–30% of the cells were already fused and it is unknown whether the stored MHC-mRNPs were actually present in postmitotic myoblasts or newly formed myotubes. The difficulties in identifying the cell types involved, the use of culture conditions that may alter the timing of events occurring after myoblasts stop dividing, and the capability of assaying only for translatable transcripts, makes it difficult to determine with any certainty when during differentiation muscle-specific mRNAs are synthesized. In addition, neither the quantities of mRNAs present in the cell nor whether transcriptional controls alone can account for a coordinated accumulation of contractile proteins has been determined with any precision.

RNA–DNA hybridizations, utilizing specific cDNA probes synthesized from either total cell mRNA or specific mRNAs, has also been a method of choice in attempting to analyze when specific transcripts appear during muscle differentiation (Affara *et al.,* 1980a; Affara *et al.,* 1980b; Dym *et al.,* 1979; Benoff and Nadal-Ginard, 1980). In order to elucidate the mechanisms responsible for the increased levels of MHC, Benoff and Nadal-Ginard (1979b) isolated total cytoplasmic RNA from cells (rat L_6E_9 muscle cell line) at various stages of myogenesis and hybridized it to MHC-cDNA. These investigators found, using this particular system, that small amounts of MHC transcripts were present in dividing myoblasts followed by a fourfold increase within 24 h prior to the morphological differentiation. Interestingly, this increase appears to be correlated with a transient induction of MHC-mRNAs containing relatively short poly-A tails. Subsequently, a 12-fold increase in the amount of

MHC transcripts is observed in the differentiated myotubes followed by a reduction occuring in 10-day cultures. This later increase of MHC-mRNA transcripts is associated with mRNAs containing longer poly-A tails. The meaning of the variation in poly-A tail length is unclear. Bragg *et al.* (1980) observed that poly-A tails on MHC-mRNA tended to be considerably shorter when isolated from cultured cells as opposed to MHC-mRNA isolated from embryonic muscle tissue that contained poly-A tails of greater length. This indicates that *in vitro* differentiation of muscle may not actually represent *in vivo* differentiation in all details. Nevertheless, the results upon utilizing a highly specific probe suggest that muscle-specific transcripts are present prior to myoblast fusion and that a substantial increase occurs as morphological differentiation proceeds. The less sensitive assays detecting the presence of a particular protein cannot be compared to specific assays for nucleic acids and therefore one cannot say with assurance whether these early transcripts are utilized with similar efficiency as those found in myotubes.

In two recent reports from the laboratories of Buckingham and Gros (Affara *et al.*, 1980a; Affara *et al.*, 1980b), the importance of transcriptional controls in muscle differentiation has been investigated utilizing cDNA hybridizations to mRNA from three very distinct cell stages found during mammalian myogenesis. These included an undifferentiated carcinoma cell line, teratocarcinoma-derived myoblasts, and mouse myotubes formed *in vitro*. A comparison was made of both the complexity of nuclear RNAs in the three stages of differentiation as well as the amount of sequences shared in common by hybridization to single-copy mouse DNA. In addition cDNA, synthesized from polysomal mRNA from each of these stages, was used to compare the actual mRNA populations being translated. Nuclear RNA from myoblasts and myotubes showed a 30% increase in the extent of hybridization over the undifferentiated cells, while the polysomal mRNA populations of the undifferentiated cells were present in myoblasts as well as myotubes. Also noted were new sets of mRNAs entering the polysomes as differentiation proceeded from myoblasts to myotube. These results suggest that mRNAs associated with polysomes from differentiated cells are absent from the nuclear RNA of the undifferentiated cells. This conclusion was further supported by the finding that chromatin of the undifferentiated carcinoma cells was resistant to nuclease digestion at specific gene sites which showed nuclease susceptibility in the myoblast and myotubes (Affara *et al.*, 1980a). These sites were tested by use of myoblast- and myotube-specific cDNA hybridizations. From these studies, it would seem that new mRNAs appearing in myoblasts and myotubes result from transcriptional control. The fact that some RNA sequences, common to both myoblasts and myotubes, are present in identical amounts in nuclei but in differing amounts on polysomes increases the likelihood that posttranscriptional controls also play a significant role in muscle differentiation. Translation of specific mRNAs from myotubes purified by DNA hybridization, has shown the presence of myosin light chains, troponin and actin (Affara *et al.*, 1980b). Although the experiments reported by these authors clearly demonstrate that transcriptional controls are important for the appearance of new mRNAs in the nuclei and ultimately into polysomes, the importance of additional control mechanisms or of stored cytoplasmic mRNA cannot be ruled out. Disregarding the fact that the teratocarcinoma cell line may not be the best model system, the following reservations can be made: (1) the cDNA probes are not pure and sequence similarities certainly exist between isoforms of proteins involved with muscle differentiation; (2) the results do not distinguish between dividing myoblasts, which a number of laboratories have demonstrated do not contain mRNAs for the muscle specific contractile proteins (see above), and postmitotic myoblasts which have been reported to contain muscle specific transcripts (Dym *et al.*, 1979); (3) no attempt was made to determine if there were inactive (nonpolysomal) mRNAs present in the cells at the various stages.

To date, only Heywood's laboratory has attempted to localize cytoplasmic muscle specific transcripts during differentiation (Dym *et al.*, 1979; see below). Dym *et al.* (1979) have made a MHC-cDNA from mRNA isolated from 14-day embryonic leg muscle MHC-mRNP-mRNA. This cDNA preparation was used for the localization of transcripts in muscle cell cultures at various stages of differentiation with the following observations: (1) dividing myoblasts (24 h of culture) contain little if any MHC-mRNA; (2) postmitotic cell cultures (40 h of culture with 20–30% of cells in small myotubes) contained 80% of the MHC-mRNA transcripts in the non-ribosomal bound mRNP fraction; (3) myotubes (70 h of culture) contained the majority of MHC transcripts in the polysomal fraction. These results suggest that MHC-mRNA transcription may proceed MHC-mRNA translation by a definable time period which occurs in the extended G1 period prior to cell fusion. These results do not contradict those results previously discussed. Apparent differences most likely exist as a result of the utilization of different culture conditions and cell systems. Rigorous objections have been made to these studies however, because the Rot ½ values of the back-hybridization to MHC-mRNA is higher than expected (Dym *et al.*, 1979). Although these are not completely unfounded, it should be kept in mind that the MHC-mRNA used for the synthesis of the cDNA is one size under denaturing conditions, synthesizes only MHC, and most likely contains mRNAs for 4–7 isoforms of MHC which have been isolated from muscle tissue. Nevertheless it is thought by the authors that the results demonstrate a pool of stored MHC-mRNA existing prior to its appearance in polysomes. The purity of the cDNA and the cell type actually containing the MHC transcripts remain an open question. As discussed in Chapter 7, it is likely that many of these questions will be resolved by careful utilization of specific cDNA or genomic clones in conjunction with cell culture conditions that reflect, as much as possible, *in vivo* embryogenesis. At this time, it can be concluded that transcriptional controls must play a significant role in muscle differentiation but it is not at all clear if they act alone or in concert with posttranscriptional and translational controls.

2.2. *Mechanisms of Transcriptional Control*

Little is known concerning the mechanisms for transcriptional control in eukaryotes. By utilizing recombinant DNA technology, rapid advances are being made in our understanding of gene organization, nucleotide sequence of genes, and the mechanisms involved in gene splicing. RNA polymerase binding sites and regions upstream to the 5′ capping sites involved in control of transcription rates are being determined in many systems including the histones, adenovirus, globin, ovalbumin, immunoglobulins, collagen, 5 S ribosomal RNA (see the following for a partial list of references: Grosschedl and Birnsteil, 1980; Konkel *et al.*, 1978; Wasylyk *et al.*, 1980; Engelke *et al.*, 1980; Korn and Brown, 1978; Bogenhagen *et al.*, 1980; Luse and Roeder, Roeder, 1980; Pelham and Brown, 1980; Tsai *et al.*, 1981; Hu and Manley, 1981; Ford, 1980; Flavell, 1980).

With regard to differentiating muscle, a number of laboratories have obtained specific DNA clones which undoubtedly will yield information of a similar nature concerning the structure and organization of muscle specific genes (see Chapter 7). New technical breakthroughs and insights will be required before a precise knowledge is gained about the mechanisms by which developmental signals lead to a recognition of specific sequences in packaged super-coiled DNA. These signals must alternately lead to the production of specific RNA sequences which both qualitatively and possibly quantitatively reflect the proteins synthesized for a developmental program.

2.3. *Analysis of Evidence for Posttranscriptional and Translational Controls*

It is well established that eukaryotic cell systems utilize posttranscriptional and translational controls (Tolstoshev *et al.*, 1981; Kistler *et al.*, 1981; Hill *et al.*, 1981; Savouret *et al.*, 1980; Rosenthal *et al.*, 1980; Jagus and Kay, 1979; Jenkins *et al.*, 1978; Northemann *et al.*, 1980; Civelli *et al.*, 1980; Rogers *et al.*, 1980; Early *et al.*, 1980) and it is very unlikely that muscle is unique and makes no use of such controls in its developmental program of gene expression. To estimate the relative importance of transcriptional and translational control mechanism is a difficult task and no attempt will be made to do so in this paper. Indeed, one does not know whether differentiation or chaos would ensue if all posttranscriptional controls were eliminated. For a previous review of translational controls in muscle, see Buckingham (1977).

Probably the most decisive evidence that translational controls are involved in developing muscle is the fact that stored mRNP particles are actually found and isolated from embryonic muscle tissue as well as muscle cells in culture (Buckingham *et al.*, 1976; Heywood and Kennedy, 1976; Jain and Sarkar, 1979; Bester *et al.*, 1980; Havaranis and Heywood, 1981). Nonribosomal bound mRNA, associated with proteins, may exist in cells as a result of either the backup of the protein synthetic machinery due to the inefficient translation of particular mRNAs (Lodish, 1974) or are actively repressed as "stored" mRNPs (Civelli *et al.*, 1980; Northemann *et al.*, 1980; Havaranis and

Heywood, 1981). Goeghegan *et al.* (1979) have demonstrated that both types of mRNPs exist in mouse sarcoma-180 ascites cells.

Utilizing a modification of the procedure of Buckingham and Gros (1975), MHC-mRNPs have been purified from both embryonic muscle tissue and muscle cell cultures (Bester *et al.*, 1980; Havaranis and Heywood, 1981; Siegel, 1981). These procedures utilize a combination of sucrose density gradient centrifugation and buoyant density gradient centrifugation and yield a native particle incapable of translation *in vitro*. The isolated MHC-mRNA derived from these mRNPs show no differences from polysomal derived MHC-mRNA in their ability to direct the synthesis of MHC in a variety of *in vitro* cell-free systems (Heywood and Kennedy, 1976; Jain and Sarkar, 1979; Bragg *et al.*, 1980). It therefore seems unlikely that any major modification and the mRNA itself is involved in the transfer from mRNPs to polysomes. Other components must be involved in actively repressing the mRNA. There are reports which claim that isolated muscle mRNPs are as translationally active as deproteinized poly A^+-mRNA (Bag and Sarkar, 1976), but these results must be viewed with caution. Jenkins *et al.* (1978) have demonstrated that the isolation of mRNPs must be carried out under conditions which preserve mRNP stability. This includes utilization of the proper salts, salt concentrations, and the avoidance of chelating agents. Scherrer's laboratory has demonstrated that even at 0.5 M salt duck erthyroblast mRNPs remain inactive (Civelli *et al.*, 1976; Vincent *et al.*, 1980). In studying the distribution of MHC mRNPs, Jain and Sarkar (1979) report that the distribution of MHC-mRNP between mRNP and polysomes as well as actin mRNP and polysomes is unchanged from 11 to 17 days of development. These results argue that the presence of these mRNAs in mRNPs is a result of inefficient translation. In contrast, Doetschman *et al.* (1980) observe a rapid increase in the population of MHC-mRNPs from 12 to 13 days of development followed by a drop in the amount of MHC-mRNPs after 14 days. These latter results agree with what is observed in muscle cell culture (Dym *et al.*, 1979; Doetschman *et al.*, 1980). It is likely that the preparative procedure used by Jain and Sarkar (1979) in isolating polysomes and mRNPs is fundamentally inadequate for separating the large 80–120 S MHC-mRNPs from polysomes. In addition, our laboratory has been unable to identify mRNPs containing actin mRNAs while other small mRNPs including those for the myosin light chains have been isolated (McCarthy and Heywood, manuscript in preparation; Mroczkowski, Ph.D. thesis in progress). It therefore seems established that MHC-mRNA is present in developing muscle as a nonribosomal bound mRNP. Its function in MHC synthesis however, is unclear.

Havaranis and Heywood (1981) have purified MHC-mRNPs using mild conditions and metrizimide buoyant density centrifugation from 40- to 44-h muscle cell cultures containing postmitotic myoblasts and 25–30% newly formed myotubes. The protein components of these mRNPs contain 9–10 major proteins including the 78,000- and 53,000-dalton proteins commonly found on mRNPs (see review by Greenberg, 1975; Jain and Sarkar, 1979). In addition to the protein components, MHC-mRNA is present as a single RNA band migrating at 6500 nucleotides under denaturing conditions as well as three low-molecular-weight RNA components (from 70–102 nucleotides in length). Interestingly, those mRNPs obtained from cultures of chick embryonic pectoralis muscle, contain MHC-mRNA coding for both fast and slow isoforms of MHC as determined by *in vitro* protein synthesis followed by specific antibody precipitations for fast and slow MHC. One of the low-molecular-weight RNA molecules, containing 102 nucleotides, has been identified as containing tcRNA activity (Bester *et al.*, 1975; Heywood *et al.*, 1975; Heywood and Kennedy, 1976) and will be further discussed.

In an effort to determine whether the isolated MHC-mRNPs are translatable in intact cells, Havaranis and Heywood (1981) encapsulated the purified MHC-mRNPs in liposomes and delivered them to cell cultures consisting either of 24-h dividing myoblasts or differentiated myotubes. In the first case, the ^{3}H-labeled MHC-mRNA remained as nonribosomal bound mRNPs migrating between 80–120 S on sucrose density gradients while myotubes were able to "activate" the mRNPs and transfer the ^{3}H]MHC-mRNA to polysomes. In both cases the reextracted ^{3}H]-mRNA was functional in various cell-free systems suggesting that the mRNA was still intact. The use of RNA and protein synthesis inhibitors indicated that a reutilization of isotope was not involved and that the transfer of MHC-mRNA to polysomes in myotubes required protein synthesis. These results suggest that MHC-mRNPs contain a repressor molecule(s) that must maintain them in an inactive state (otherwise a distribution of mRNA in mRNP and polysomes would be observed) in dividing myoblasts receiving liposome-encapsulated mRNPs. In addition myotubes must contain the mechanisms for the activation of these mRNPs that is absent in the myoblasts. This latter observation indicates that posttranscriptional controls are developmentally programmed in differentiating muscle. Although there are obvious difficulties in interpreting the results using liposome insertion of macromolecules into cells, this technology should allow for a varied approach and comparison with cell-free systems. The utilization of liposomes to deliver inactive mRNPs to myogenic cells at distinct stages of differentiation should allow for both the identification of the repressor molecule(s) as well as the mechanism of activation.

A central question, as pointed out above, concerns the relationship between cytoplasmic appearance and utilization of MHC-mRNA and other muscle-specific mRNAs. At least some of these transcripts are inactive and can be found as stored mRNPs in committed cells (postmitotic myoblasts and early myotubes). Buckingham *et al.* (1976) reported the putative existence of a 26 S MHC-mRNA in muscle cells prior to fusion which showed a precursor–product relationship to a 26 S MHC-mRNA species present on myotube polysomes. A more recent report (Doetschman *et al.*, 1980) has confirmed this finding and included corrections for the reutilization of isotope to determine the amount new synthesis of this mRNA contributes to its appearance in the polysomes of myotubes. These reports emphasize that this 26 S mRNA is found in postmitotic and early fusing myotubes and is transferred to polysomes in developing myotubes. It is thought that the stored mRNA contributes to the early burst of MHC synthesis and that new synthesis of mRNA contributes to the increased pool of mRNA thus replacing mRNAs which are being degraded. Due to the high Rot ½ values of this poly A$^+$-mRNA, these results have been criticized with regard to their validity in describing MHC-mRNA metabolism. It was demonstrated however, that this RNA species directed the synthesis of MHC *in vitro*. Even if the mRNA is somewhat impure, the results, at a minimum, indicate that a poly A$^+$-mRNA, found soon after cell committment, is found in mRNPs and shows a precursor-product relationship to similar mRNAs in polysomes of myotubes 10–30 h later. As mentioned previously, these results mimic those found *in vivo* (Doetschman *et al.*, 1980), suggesting that they are not an artifact of tissue culture conditions. Other laboratories have not observed similar mRNPs but the appropriate experiments analyzing cytoplasmic compartmentalization of mRNAs and using culture conditions that do not force excessively rapid cell fusion have not been employed. It is unlikely that results concluding the existence of a precursor–product relationship between mRNP and polysomes can be explained by rRNA contamination. If this was the case, a coordinate loss of label in mRNPs, new synthesis of polysomal mRNA, and transfer would not be observable. In addition the mRNAs are highly purified and show no rRNA upon electrophoretic analysis. Undoubtedly, the use of specific DNA probes from cloned DNA will further clarify these results. Emphasis should also be given to establishing cell culture systems which best reflect the events occuring *in vivo*. Nevertheless, it seems well established from both *in vivo* and *in vitro* findings that the role of nonribosomal bound mRNPs must be included when deciperhing gene expression in differentiating muscle.

In eukaryotic cells, the initiation steps in protein synthesis are thought to be rate limiting (Revel, 1977). According to the model proposed by Lodish (1974), competition between mRNAs for initiation factors is the sole mechanism used by eukaryotic cells to regulate the relative rate of protein synthesis. This model predicts that a reduction in the rate of initiation will result in a preferential inhibition of those mRNAs with low rate constants for initiation. Initiation rate constants are a measure of the affinity of mRNAs for initiation factors resulting only from the primary structure of the initiation regions of mRNAs. Such a mechanism likely exists and may be sufficient in itself for cells translating only a limited number of mRNAs. However, in cells containing tens of thousands of mRNA species, the cells may possess alternative mechanisms which can override that of mRNA competition. Indeed, a paradox exists when analyzing MHC translation. It is obvious that *in vivo* MHC-mRNA is efficiently initiated in that it is found on polysomes consisting of 50–60 ribosomes, a maximum for an mRNA of this size, whereas *in vitro* it is a poor competitor (Heywood *et al.*, 1967; Gette and Heywood, 1979). It is therefore unlikely that MHC-mRNA would be found in untranslatable mRNPs (Havaranis and Heywood, 1981) or as inefficiently translated naked mRNA in muscle if it were in fact a supercompetitive mRNA. We have therefore suggested that in addition to the purely competitive model for mRNA utilization, eukaryotic initiation factor 3 (eIF3) can be modulated by associated polypeptides and thereby discriminate between mRNAs making certain mRNAs more efficient for translation (Heywood and Kennedy, 1979; Gette and Heywood, 1979). In this manner eIF3, containing discriminating polypeptides, would have a higher affinity for particular mRNAs and increase their rate of initiation. In addition to other reports, suggesting that mRNA discrimination occurs (DePhilip *et al.,* 1980; Jagus and Kay, 1979; Ilan and Ilan, 1976; Kabat and Chappell, 1977), we have attempted to determine whether such mechanisms occur in muscle utilizing a variety of *in vitro* systems including reticulocyte lysates (Heywood and Kennedy, 1976), wheat germ lysates (Heywood and Kennedy, 1979), and a reconstituted muscle cell-free system that most closely resembles the *in vivo* situation (Gette and Heywood, 1979). Given that basic competition exists between mRNAs and the difficulties in interpreting results from *in vitro* systems, this has been a difficult question in which to get an unambiguous answer. However, the *in vitro* results have continually implicated that factors are involved and have encouraged further pursuit of these factors. Utilizing a test system involving two easily obtainable mRNAs, globin and MHC, we were able to demonstrate that a preparation of active eIF3 obtained from muscle would preferentially enhance MHC synthesis while a similar preparation of reticulocyte eIF3 had no effect on the ability of MHC-mRNA to compete with globin mRNA (Heywood and Kennedy, 1976; Heywood and Kennedy, 1979). Subsequently, utilizing a MHC-mRNA affinity column we have been able to subfractionate eIF3 into "core" eIF3 and discriminatory components (Heywood and Kennedy, 1979; Gette and Heywood, 1979). The "core" eIF3, although active in initiation complex formation, did not enhance MHC synthesis while the high affinity fraction (HAF) (obtained from muscle eIF3) when reconstituted to either reticulocyte or muscle "core" eIF3 was subsequently able to alter the recognition properties of "core" eIF3 and specifically stimulate MHC synthesis. A similar fraction from reticulocyte eIF3 either isolated from a globin mRNA or a MHC-mRNA affinity column was not found to have specific stimulating activity (Heywood and Kennedy, 1979). Indeed the muscle eIF3-HAF was not capable of binding globin mRNA, suggesting that the HAF recognizes specific sequences of MHC-mRNA.

To further test the validity of these cell-free experiments, we have recently undertaken an analysis of the regulatory activity of these macromolecules in intact cells (O'Loughlin *et al.,* 1981). In these studies cytoplasmic control elements, as well as mRNAs, were introduced into differentiating muscle cells utilizing liposomes. When unfractionated muscle eIF3 is taken up by the cells, it was found to both associate with 40 S ribosomal subunits indicating retention of biological activity, and the MHC-mRNPs in the cells. Alternatively, "core" eIF3 devoid of HAF sediments primarily with the 40 S ribosomes. The HAF, separated from eIF3 by MHC-mRNA affinity chromatography, was found to mainly associate with MHC-mRNPs, however, a limited amount was associated with the 40 S ribosomal subunits as well—presumably resulting from association with eIF3 on these ribosomal subunits. Beside the association with subcellular particles, the discriminatory nature of

HAF containing eIF3 on endogenous mRNA translation was investigated (O'Loughlin *et al.*, 1981). Unfractionated eIF3, prepared from 14-day embryonic muscle tissue, was found to have no effect on total protein synthesis; however, a rather specific stimulatory effect on the synthesis of myofibrillar proteins was observed. This suggests that the increased translatability conferred by unfractionated eIF3 is not limited to MHC-mRNA but encompases a class of related mRNAs. It cannot be discerned from these experiments, whether the HAF isolated by MHC-mRNA affinity columns chromatography is solely responsible for the effects observed on other myofibrillar proteins; in addition other HAF proteins may be isolated utilizing different mRNA affinity columns which may give a cummulative effect. The greatest effect of added eIF3 to myogenic cells is observed when the cells are just beginning to fuse at 40 h. This is likely a result of the absence of myofibrillar mRNAs in dividing myoblasts, and the presence of discriminatory eIF3 molecules already present in differentiated myotubes. If this is the case, it would follow that the HAF proteins associated with eIF3 are developmentally programmed regulatory molecules. While unfractionated eIF3 can be shown to stimulate myofibrillar protein synthesis, liposome delivered purified HAF had no effect on total protein synthesis or myofibrillar protein synthesis. This may be due to the sequestration of HAF as a result of binding to inactive mRNPs. When HAF was introduced into the cells with MHC-mRNA however, a significant increase in polysome size for the translation of the labeled mRNA was observed over that observed when MHC-mRNA was inserted by itself. This preassociation of HAF with mRNA has also been shown to be required for maximal activity in cell extracts (Gette and Heywood, 1979; Heywood and Kennedy, 1979). Although it is difficult to determine if the results observed in intact cells are direct or indirect, these "*in vivo*" experiments lend support to our previous *in vitro* studies which suggest mRNA discriminatory elements play a role in the cytoplasmic utilization of mRNA during muscle differentiation.

The stability of mRNA is an important contributing factor to overall gene expression in eukaryotic cells. Although it has not been proven in all studies, polyadenylation of mRNA is thought to be involved with mRNA stability as well as other aspects of mRNA metabolism (for review see Brawerman, 1981). Although there have been varying reports (see review by Buckingham, 1977) it is generally accepted that muscle is not very different in its polyadenylation pattern of mRNAs as compared to other eukaryotic cells. It is possible that variations in the length of poly-A tails which have been reported reflect cell culture conditions (Benoff and Nadal-Ginard, 1979b; Medford *et al.*, 1980; Bragg *et al.*, 1980). This again emphasizes that cell cultures may not reflect every aspect of *in vivo* muscle differentiation.

Buckingham and Gros (1975) originally observed that 26 S mRNA (putatively MHC-mRNA) might differ in its half-life at two different developmental stages. A more recent study by Bowman and Emerson (1980) confirms the fact that mRNA half-lives do change during muscle differentiation. They observed that mitotic myoblasts and myotubes both have short- and long-lived mRNAs. The half-life of the short-lived mRNA is always 2–4 h. The long-lived mRNAs in myoblasts have a half-life of 60–100 h while in myotubes they have a decreased half-life of 20 h. Significantly, some of the myofibrillar proteins were found to be encoded for by the long-lived myotube mRNAs. Unfortunately, the relationship between the long-lived myoblast mRNAs and the long-lived myotube mRNAs was not examined. The manner by which mRNA half-life is regulated and, in fact, whether it can be altered or is inherent in the structure of the mRNAs themselves is not known.

2.4. Mechanisms of Translation Control

2.4.1. mRNP Storage

As previously discussed, mRNPs isolated from a variety of sources appear to contain a variable set of proteins, mRNA, and low-molecular-weight RNAs. It is unclear as to which molecules present have the repressor function, as it is also unclear as to the function for any of the associated proteins. This laboratory presented a model (Bester *et al.*, 1975) with subsequent supporting evidence (Heywood and Kennedy, 1976; Heywood *et al.*, 1975; Kennedy *et al.*, 1978) suggesting that a small RNA [translational control RNA (tcRNA)] was involved with the inactivation and storage of MHC-mRNA as an mRNP particle. The model suggested that a U-rich region of tcRNA interacted with the poly-A tail of the mRNA while another portion of tcRNA interacted with a region of the mRNA distal to the poly-A tail. In so doing, the mRNA was folded in such a way as to make it unavailable for protein synthesis. It was also suggested that a degree of specificity existed in the RNA–RNA interaction. Associated proteins most likely perform a function stabilizing these complexes. The basic evidence in support of this model has involved the demonstration of the inhibition of translation of mRNA with stoichiometric quantities of tcRNA as well as changes in the structure of mRNA as measured by increased RNase resistance upon hybridization to tcRNA. The association of the molecules in isolated mRNPs, (Havaranis and Heywood, 1981; Siegel, 1981) is also evidence for this function *in vivo*. It is also conceivable that tcRNA may also play a role in the transport of mRNA from the nucleus to the cytoplasm. Until recently, a major criticism of this model has been a lack of a well defined tcRNA molecule and a demonstration of its purity. Recent advances in purification and sequence analysis have allowed us to isolate tcRNA from MHC-mRNPs as well as other muscle mRNPs (Siegel, 1981). Previous studies utilized oligo d(A) binding to isolate tcRNA. Recent observations now confirm that this method of purification yielded a highly purified preparation of tcRNA; however, a sizeable portion of cellular tcRNA was not retained by the column. Nevertheless, our previous studies are valid in the manner in which they were performed. The sequence of tcRNA is:

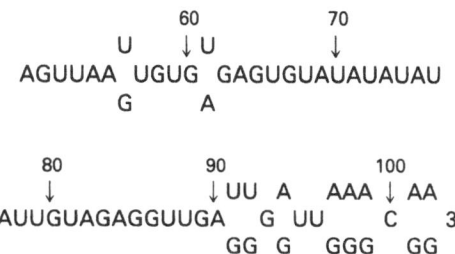

As can be readily seen, the molecule is unique in its sequence, is G-rich as well as U-rich, and has both a G-U- and an A-U-rich region. Purified tcRNA inhibits MHC-mRNA translation as well as the translation of mRNAs found in less than 40 S mRNPs including myosin light chain mRNAs. The tcRNA isolated from the small mRNPs has an identical sequence to that isolated from MHC-mRNPs and also inhibits MHC-mRNA translation (Siegel, 1981; Siegel *et al.*, 1982). In addition, the tcRNAs can complex with the muscle mRNAs; however, tcRNA does not effect globin mRNA or TMV-mRNA translation nor will it form complexes with these mRNAs or ribosomal RNA. Although we do not discount the importance of the mRNP proteins in mRNA metabolism, this evidence strongly supports our original model suggesting that tcRNA is very much involved in mRNP inactivation.

Recently, several reports have indicated that a small RNA, inhibitory to protein synthesis is found as a 10 S RNP in muscle (Bag and Sells, 1979; Sarkar *et al.*, 1981). It is not known if these investigators have also isolated tcRNA (they disclaim this interpretation of their result) or if they have isolated a different cytoplasmic small RNA. Until these small RNAs are sequenced and their function understood, it will not be possible to either claim or disclaim a relation of these RNAs to tcRNA. Nevertheless, it is now becoming well accepted that cytoplasmic low-molecular-weight RNAs play a role in gene expression during muscle differentiation. To establish the relationship between the genes coding for these regulatory RNAs and the genes coding for the structural proteins will be of value in understanding the interrelationship and interdependence of transcriptional and translational controls.

2.4.2. mRNA Utilization

It is generally thought that different mRNAs are translated with different efficiencies (see Lodish, 1974). Utilizing this model in itself, one could suppose that muscle-specific mRNAs were generally inefficient and that upon their translation some limiting factor, required by all mRNAs, would become available at a higher concentration, thus allowing a burst of myofibrillar protein synthesis as seen when myoblasts fuse. At first this seems an adequate explanation and may indeed be correct; however, it is difficult to explain (1) the varying efficiencies of translation of the contractile protein mRNAs (Bowman and Emerson, 1980), (2) the fact that MHC-mRNA loads up to maximal capacity with ribosomes (Heywood *et al.*, 1967) after being inactive as an mRNP (Havaranis and Heywood, 1981), and (3) the stoichiometric production of proteins that vary tenfold in size and that most likely have similar elongation rates during synthesis. We have suggested, while it is difficult to unequivocally demonstrate, that factors may bind to nondiscriminatory eIF3 and subsequently help specific mRNAs to compete more favorably for this limiting initiation factor. Although not completely purified, a 33,000-dalton protein has been identified as a candidate for such a factor (O'Loughlin *et al.*, 1981). This factor, isolated from the HAF of muscle eIF3, was found to bind MHC-mRNPs present in myogenic cells after delivery to the cells via liposomes. We would like to suggest that the 33,000-dalton polypeptide is originally present on mRNPs. Upon activation of the mRNPs, the first initiation event, allows a transfer of HAF to "core" eIF3. Subsequently, the muscle specific mRNAs could now compete with all other cellular mRNAs for "core" eIF3 and in addition would preferentially utilize the eIF3 molecules modulated with the HAF protein(s). Such a mechanism would allow a temporary and rapid synthesis of specific proteins. Recent studies suggest that the appearance of HAF for MHC-mRNA may coincide with the appearance of MHC-mRNPs (Dym *et al.*, 1979) and that the 33,000-dalton component of HAF can be isolated from purified MHC-mRNPs (Heywood, unpublished). Therefore, the model is obtaining some experimental support and at least presents a working hypothesis which may be valuable in explaining how mRNAs are differentially utilized during development.

3. Gene Expression in the Formation of Differentiated Fiber Types

The differentiation of muscle cells involves the developmental progression of mesodermal cells to myoblasts and the subsequent fusion of these cells into multinucleated myotubes capable of synthesizing the contractile proteins. Terminal differention that includes the appearance of distinct muscle fiber types apparently occurs as a result of exposure of multinucleated muscle cells to exogenous factors such as innervation, hormones, or increased levels of activity. The continued presence of these exogenous factors appears necessary for maintaining the stability of the fiber type. Even once the muscle phenotype has been established, it can be changed simply by modifying the patterns of innervation, by altering the level of muscular activity, or by exposing the muscle to different hormonal levels. Any such switch in muscle fiber type must result, in part, from an alteration in the expression of the structural genes encoding the contractile proteins. In order to understand the regulatory mechanisms involved in muscle specialization and the plasticity of this process, it is first necessary to determine the number of contractile protein phenotypes which are expressed at different stages of development. Studying and cataloging the different isoforms of myosin, actin, tropomyosin, and troponin, which appear and are subsequently replaced as development proceeds, allows an estimation of the number of structural genes involved in the developmental program. This, in turn, should help investigators understand the signals responsible for changes in phenotype, as well as determine gene organization, and the mechanisms involved in the expression of the various contractile protein genes.

3.1. Myosin

Within any single organism, various types of myosin can be distinguished on the basis of their electrophoretic mobility under nondenaturing conditions. The isoforms of native myosin may result from either the assembly of homologous subunits of myosin heavy chains (MHC) and myosin light chains (MLC) as in the case of fast muscle where all isoforms are of the fast variety, or alternatively, from the assembly of heterologous subunits, in which case, fast MHC associates with slow MLC. Such hybrid forms of native myosin have in fact been identified by d'Albis *et al.* (1979).

In any individual species, both the light and heavy subunits of myosin are found in a number of differing forms. As shown in Table 1, at least 11 different forms of MLC have been identified in chicken, nine in rabbit and 13 in other mammals. These include three fast skeletal muscle forms (MLC_{1F}, MLC_{2F}, MLC_{3F}), two slow skeletal muscle forms (MLC_{1S}, MLC_{2S}) and two cardiac muscle (MLC_{1C}, MLC_{2C}) isoforms. Two isoforms of MLC_{1S}(a and b)

Table 1. Estimated Number of Structural Genes for Myosin Light Chains (MLC)[a,b]

	Rabbit	Other mammals	Chicken
Skeletal fast muscle	1F (2,4)	1F (6)	1F (4)
	2F (2)	2F (6)	2F (4)
	3F (2,4)	3F (6)	3F (4)
Slow skeletal muscle	1aS (5)	1aS (5)	1S (7)
	1bS (5)	1bS (5)	ND[c]
	2S (5)	2S (5)	2S (7)
Embryonic skeletal muscle	1 emb (14)	1 emb (8,10,14)	1 emb (14)
Cardiac atrial muscle	ND	1A (11,13)	1C (8,12[d])
	ND		
Cardiac ventricular muscle	1V (9)	1V (11,13)	1C (8,12[d])
	2V (9)	2V (11,13)	2C (8,12[d])
Smooth muscle	ND	1Sm (1)	1Sm (3)
	ND	2Sm (1)	2Sm (3)
Nonmuscle	ND	ND	1Sm (3)
	ND	ND	2Sm (3)
			Brain (3)
Minimal number of genes	9–14	13–14	11

[a] Abbreviations for methods used in studies cited below: comp., amino acid composition; elect., electrophoretic mobility; imm., immunochemical or immunohistochemical analysis; pept., peptide analysis; seq., amino acid sequence analysis.

[b] Numbers in parentheses refer to the following studies: (1) Leger and Focant (1973), elect., comp.; (2) Frank and Weeds (1974), seq.; (3) Burridge and Bray (1975), elect.; (4) Holt and Lowey (1975), comp., imm.; (5) Weeds (1976) seq.; (6) Syrory and Gutman (1977), elect.; (7) Hoh (1978), elect.; (8) Dalla Libera *et al.* (1979), elect.; (9) Sartore *et al.* (1979), imm.; (10) Whalen *et al.* (1979), elect.; (11) Cummins *et al.* (1980), elect.; (12) Maita *et al.* (1980), elect., seq.; (13) Price *et al.* (1980), elect.; (14) Whalen (1980), elect.

[c] ND, no data available.

[d] No difference is found between ventricular and atrial MLC pattern.

have further been distinguished in mammalian slow skeletal muscle (Weeds, 1976), while atrial and ventricular MLC_1 and MLC_2 consist of two distinct sets of polypeptides in mammals (Wikman-Coffelt and Srivastava, 1979; Cummins *et al.*, 1980).

In rat and chicken, all five adult isoforms of MLC are initially synthesized by presumptive fast and presumptive slow "primitive muscle cells" in primary cell culture (Whalen *et al.*, 1978, 1979, 1981; Keller and Emerson, 1980; Stockdale *et al.*, 1981a) as well as in early development *in vivo* (in fast muscle: Stockdale *et al.*, 1981b; in slow muscle: Rubinstein *et al.*, 1977; Whalen *et al.*, 1978; Pette *et al.*, 1979). In fetal and neunatal tissue, MLC_{3F} is however found in much reduced amounts (Dow and Stracher, 1971; Chi *et al.*, 1975; Rubinstein *et al.*, 1977; Whalen *et al.*, 1978; Roy *et al.*, 1979b). Whalen *et al.* (1978, 1979) have furthermore shown that a distinct embryonic isoform of MLC ($MLC_{1\ emb}$) is produced in fetal and neonatal rat muscle in addition to the five adult isoforms. The sequence of changes in isoform pattern of myosin subunits which occur as development and growth proceeds indicate that beyond the decision of producing large quantities of myosin, a modification of the pattern of gene expression occurs that ultimately determines the fiber type. In rat for instance, $MLC_{1\ emb}$ is gradually replaced by the adult MLC_{1F} in fast muscle (Whalen *et al.*, 1981), and $MLC_{1\ emb}$ and the three fast isoforms of MLC are similarly replaced by the two slow isoforms of MLC upon the commitment of the muscle to become the slow type (Whalen *et al.*, 1978). Similar developmental MLC isoform changes are observed in primary cell cultures (Whalen *et al.*, 1978, 1979, 1981); however, none of the skeletal muscle specific MLC, including $MLC_{1\ emb}$, are detected prior to cell fusion *in vitro* (Whalen *et al.*, 1978, 1979, 1981; Keller and Emerson, 1980; Stockdale *et al.*, 1981a).

As shown in Table 2, at least nine distinct forms of MHC have also been identified in vertebrates: although not all have been found in any one species. At least eight MHC have been found in chicken, six in rabbit and six in rat. Of those described, two or three can be found in fast and slow skeletal muscle, two in cardiac muscle and two in embryonic and neonatal muscle. In rat, the embryonic MHC isoform has been identified in fetal tissue as well as in primary muscle cell cultures. The neonatal MHC isoform has only been found during the first two weeks postpartum (Brevet and Whalen, 1978; Whalen *et al.*, 1979). Whalen *et al.* (1981) observed that this neonatal form is gradually replaced by the adult form which eventually predominates. In chicken, an embryonic form of MHC has also been found in embryonic pectoralis fast muscle. This embryonic MHC was shown to have some similarities with the adult fast MHC isoform (Rushbrook and Stracher, 1979).

One or two isoforms of MHC which react with antibodies prepared against either adult fast or slow MHC have also been observed in primary cell cultures of avian muscle (Rubinstein and Holtzer, 1979; Havaranis and Heywood, 1981; Thibault *et al.*, submitted). Rubinstein and Holtzer (1979) reported that muscle cell cultures, derived from presumptive slow as well as fast muscle, synthesized only the fast MHC isoform in premature as well as in fully mature myotubes. In contrast, we have observed that small amounts of

Table 2. Estimated Number of Structural Genes for Myosin Heavy Chain[a,b]

Rabbit	Rat	Chicken
Fast skeletal muscle[c] (2,4,5,9)	Fast skeletal muscle (8,9,10)	Fast skeletal muscle (3,5,7)
Slow skeletal muscle (4,5,9)	Slow skeletal muscle (8,9,10)	Slow skeletal muscle (5,7)
Embryonic skeletal muscle (1)	Embryonic skeletal muscle (8,10)	Embryonic skeletal muscle (7)
ND[d]	Neonatal skeletal muscle (10)	ND
Cardiac atrial muscle (9)	Cardiac atrial muscle (9)	Cardiac atrial muscle (6)
Cardiac ventricular muscle (9)	Cardiac ventricular muscle (9)	Cardiac ventricular muscle (6)
ND	ND	Smooth muscle (3)
ND	ND	Cytoplasmic platelet (3)
ND	ND	Cytoplasmic brain (3)
Minimal number of genes: 6–9	Minimal number of genes: 6–9	Minimal number of genes: 8–9

[a] Numbers in parentheses refer to the following studies: (1) Huszar (1972), seq.; (2) Starr and Offer (1973), seq.; (3) Burridge and Bray (1975), pept.; (4) Brevet and Whalen (1978), pept.; (5) d'Albis *et al.* (1979), elect.; (6) Dalla Libera *et al.* (1979), pept.; (7) Rushbrook and Stracher (1979), pept.; (8) Whalen *et al.* (1979), imm.; (9) Schwartz *et al.* (1980), imm.; (10) Whalen *et al.* (1981), imm.

[b] Abbreviations are as in Table 1.

[c] Two have been distinguished.

[d] ND, no data available.

slow MHC are also produced in early myotubes from cultured breast muscle cells. As differentiation proceeds, the production of the slow isoform increases so that in fully mature myotubes both fast and slow MHC are synthesized at similar rates. If the culture media is prepared with embryo extract depleted of brain and spinal cord tissues however, the appearance of the slow type of MHC is much reduced, and accordingly delayed. Interestingly, we observed that the addition of a sciatin-containing nerve extract (Markelonis and Oh, 1979) to depleted embryo extract reverses this effect, thereby increasing the rate of slow MHC synthesis until slow MHC is produced in similar amounts as fast MHC in the newly formed myotubes (Thibault *et al.*, submitted). These findings suggest that certain factors present in nerve tissue may well influence fiber type formation and that the mechanisms involved may be studied *in vitro*. Utilization of these kinds of systems make it possible to eventually analyze the mechanisms by which nerve influences the expression of the genes which determine the myofibrillar protein isoforms of muscle.

It is well established that cross-innervation of mammalian fast twitch and slow twitch muscle alters the isoform pattern of myosin with an induction of the reciprocal phenotypes for native myosin (Hoh, 1975; Hoh *et al.*, 1980), MLC (Sreter *et al.*, 1974; Weeds *et al.*, 1974) and MHC (Sreter *et al.*, 1975; Weeds and Burridge, 1975). Although avian muscle appears to be less susceptible than mammalian muscle with regard to interconversion of fiber types in response to neuronal modification (Hnik *et al.*, 1967; Close and Hoh, 1968), transplantation of a minced muscle of a different type, or cross-innervation performed early in development, can also produce some changes in avian muscle fiber types (Jirmanova *et al.*, 1971; Gordon and Vrbova, 1975; Gordon *et al.*, 1977b).

Chronic stimulation of fast muscle, using a firing pattern typical to that received by slow muscle, can also produce a conversion of fast myosin to slow myosin. This conversion involves both MHC and MLC (Sreter *et al.*, 1975; Pette and Schnez, 1977; Rubinstein *et al.*, 1978). Similarly, electrostimulation of myotubes in culture derived from presumptive fast muscle has been shown to increase the cellular level of the slow types of MLC (Srihari and Pette, 1981). Both denervation (Rubinstein and Kelly, 1978) and cordotomy (Buller *et al.*, 1960; Hoh and Dunlop, 1974; Hoh *et al.*, 1980; Margreth *et al.*, 1980) however produce opposite effects, thus preventing or reversing the development of slow twitch muscle and producing an isozyme shift similar to that observed in cross innervated slow muscle.

It can be seen from this discussion that a large number of myosin isoforms exist during the development of muscle in any organism. In addition, the sequential appearance and disappearance of these isoforms can only mean that the signals responsible for proper muscle differentiation are complex. Superimposed upon the events by which embryonic tissue differentiates into muscle are other selection mechanisms which ultimately yield a particular fiber type containing the characteristic myosin isoform. The mechanisms involved for such gene expression of the myosin isoforms are not as yet known.

3.2. Actin

Actin, first characterized as a muscle-specific protein, is now known to be an ubiquitous protein in eukaryotic cells. Actin has a well-defined role in the contractile apparatus and now its involvement in various other cellular activities is known (Pollard and Weihing, 1974; Poste *et al.*, 1975; Sanger, 1975; McIntosh *et al.*, 1976; Brown *et al.*, 1977). Both the amino acid sequence and structure of actin have been highly conserved during eukaryotic evolution, and show little species specificity (Vandekerckhove and Weber, 1978a, 1978d; Zechel and Weber, 1978; Barker and Dayhoff, 1980). Lower eukaryotes have one form of actin which is strikingly similar to mammalian cytoplasmic actins (Gordon *et al.*, 1977a; Schadat *et al.*, 1977; Vandekerckhove and Weber, 1978b, 1978d, 1980).

On the basis of isoelectric mobility and tryptic peptide mapping, three distinct actin species have been found in a variety of mammalian cell lines and tissues (Barrels and Gibson, 1976). Using a similar analysis, Zechel and Weber (1978) more recently observed four isoforms of actin in different vertebrate cells and tissues. One form (α-actin) has been found in both differentiated skeletal and cardiac muscle (Barrels and Gibson, 1976; Zechel and Weber, 1978) while two other forms (β-actin and γ-actin) are present in variable proportions in nonmuscle cells and tissues, as well as in undifferentiated muscle cells (Whalen *et al.*, 1976). These forms of actin focus at pI values slightly higher than α-actin. A fourth type similar to γ-actin was obtained from chicken gizzard smooth muscle (Zechel and Weber, 1978).

Table 3. Estimated Number of Structural Genes for Actin

	Distinct types of actin (seq.)[b]	References[a]
Striated muscle		
Skeletal	α-skel	(1–5)
Cardiac	α-card	(5)
Nonmuscle cytoplasmic		
	β-Nm	(3)
	γ-Nm	(3)
Smooth muscle		
Aorta	α-like Sm	(5)
Gizzard	γ-like Sm	(4)

[a] (1) Collins and Elzinga (1975), rabbit skeletal muscle; (2) Lu and Elzinga (1977), rabbit skeletal muscle; (3) Vandekerckhove and Weber (1978a), rabbit skeletal muscle, calf thymus, bovine brain, and SV40-transformed mouse 3T3 cells; (4) Vandekerckhove and Weber (1979a), chicken breast and gizzard muscle; (5) Vandekerckhove and Weber (1979b), bovine aorta, bovine heart, bovine fast skeletal, muscle, and rabbit slow skeletal muscle.

[b] seq., amino acid sequencing analysis.

As shown in Table 3, amino acid sequencing has indicated that 4–6 actin isoforms are expressed in higher vertebrates in a tissue-specific manner. In addition to the two nonmuscle cytoplasmic forms (β-actin and γ-actin), two striated muscle α-actins and two smooth muscle forms (α-like and γ-like) have been distinguished.

The complete amino acid sequences have been determined for rabbit skeletal muscle actin (Collins and Elzinga, 1975) and more recently for chicken skeletal muscle actin (Vandekerckhove and Weber, 1979a). This latter study showed that α-actin, found in skeletal muscle, was identical in both species. No differences have been found in either rabbit fast or slow muscle actins (Vandekerckhove and Weber, 1979b). Although cardiac and skeletal muscles apparently contain one species of actin (α-actin) as determined by isoelectric focusing, amino acid sequence analysis indicates that in fact cardiac and skeletal muscles actins are two distinct polypeptides which differ by four amino acids. The similar overall net charge between these actins reflects conservative replacements of two charged amino acids (Vandekerckhove and Weber, 1979b).

In addition to the γ-like actin species found in chicken gizzard muscle, smooth muscle is also known to contain a second type of actin (α-like) typical of bovine aorta tissue. Those two smooth muscle actins differ from each other by only three amino acid substitutions at the amino-terminal end of the molecule (Vandekerckhove and Weber, 1979b). The γ-like smooth muscle actin (gizzard) has been shown to be closely related to skeletal muscle α-actin, differing by only 6 amino acid substitutions (Vandekerckhove and Weber, 1979a). Four of the γ-like actin amino acid substitutions are shared by the α-like smooth muscle actin (aorta) (Vandekerckhove and Weber, 1979b). Fingerprint analysis of actin from chicken and bovine tissues suggest strongly that those two smooth muscle actin forms are not species specific but rather tissue specific (Vandekerckhove and Weber, 1978b, 1979b).

The two cytoplasmic isoforms of actin are also distinct from each other as shown by amino acid sequence analyses. They differ at four positions in their amino acid sequence and γ-actin is shorter than β-actin by one residue. Additional differences are, however, found when one compares the nonmuscle forms of actin to skeletal muscle actin (Vandekerckhove and Weber, 1978a, 1978c).

As seen for other muscle proteins, change in the isoform pattern of actins also occurs during development. Smooth gizzard muscle, for example, produces almost exclusively β-actin in early chick embryo. However, as development proceeds, γ-actin gradually increases and β-actin decreases. Parallel changes in the quantities of the corresponding mRNAs suggest that there is a differential expression of the actin structural genes during development (Saborio *et al.*, 1979).

A similar phenomenon appears to occur in skeletal muscle tissue. Although α-actin is the major form detected in differentiated cultural muscle cells, β- and γ-actins have been shown to be principal actin constitutents in myoblasts (Garrels and Gibson, 1976). Following myoblast fusion in culture, the relative amount of skeletal muscle specific α-actin increases until it changes from a minor actin component to the predominant actin species (Whalen *et al.*, 1976; Rubenstein and Spudich, 1977). The appearance of α-actin upon fusion of myoblasts corresponds to the accumulation of muscle specific contractile proteins during this stage of development (Allen *et al.*, 1979). The induction of α-actin and repression of β- and γ-actin synthesis during myogenesis again illustrates the phenomenon of differential expression of contractile protein structural genes during development (Schwartz *et al.*, 1980; Schwartz and Rothblum, 1980). An understanding of the mechanisms by which the various actin genes are activated and deactivated and the corresponding utilization of their mRNAs will be required before the differentiation process of muscle types can be understood.

3.3. Tropomyosin

Tropomyosin (TM) is a component of the contractile apparatus in skeletal muscle, smooth muscle and cardiac tissue. In association with the troponin complex, TM is intimately involved in the calcium regulatory system controlling the actin and myosin interaction (Phillips *et al.*, 1980; Seymour and O'Brien, 1980). Rod-shaped and highly α-helical, TMs form dimers assembled in a head-to-tail fashion. The resulting polar filaments lay in each of the two grooves formed by the double-stranded actin helix (Phillips *et al.*, 1980; Seymour and O'Brien, 1980).

TM is present in all types of muscle (see Table 4). Based upon the electrophoretic mobility on denaturing gels, the two subunits of TM are thought to have different apparent molecular weight (M_r). They are distinguished on this basis as α-TM (smaller M_r) and β-TM (larger M_r). The subunits also

Table 4. Estimated Number of Structural Genes for Tropomyosin[a]

	Rabbit		Chicken	
	α-TM	β-TM	α-TM	β-TM
Fast skeletal muscle	Sk (3,4,7) Sk (2,7)		Sk-S (8)	Sk (8)
Slow skeletal muscle		Sk (4,7) Sk (7)	Sk-F (8)	Sk (8)
Cardiac muscle	Sk (4,6)	None	C (8)	None
Smooth muscle	Sm (1,5)	Sm (1,5)	Sm (5,8)	Sm (5,8)
Nonmuscle	Nm (5)	Nm (5)	ND	ND
Minimal number of genes	4	4	4–5	2–3

[a] Numbers in parentheses refer to the following studies: (1) Cummins and Perry (1974), comp.; (2) Sodek *et al.* (1978), seq.; (3) Stone and Smillie (1978), seq.; (4) Mak *et al.* (1979), seq.; (5) Dabrowska *et al.* (1980b), elect.; (6) Lewis and Smillie (1980), seq.; (7) Mak *et al.* (1980), seq.; (8) Montarras *et al.* (1981), elect.
[b] Abbreviations are as in Table 1.

appear to migrate differently when isolated from different tissues. For instance α-TM and β-TM migrate a bit slower in smooth muscle (M_r: 37,000–39,000) than in skeletal muscle (M_r: 34,000–36,000), whereas in nonmuscle tissue the subunit(s) migrate faster (in brain, M_r: 29,000–30,000) (Bretscher and Weber, 1978; Dabrowska *et al.*, 1980a, 1980b).

Complete amino acid sequence analyses of rabbit skeletal muscle α-TM and β-TM reveal that both have 284 residues per chain despite the different apparent molecular weight (Mak *et al.*, 1980). A substitution at 39 positions is observed in β-TM when compared to α-TM. Most changes, however, involve conservative amino acid replacements. The small difference observed in mobility between the two chains can be attributed to two amino acids which give a higher negative charge to β-TM (Mak *et al.*, 1980).

More than two forms of TM have been shown to exist in rabbit skeletal muscle. Minor variants of the major α-TM and β-TM forms, differing at only a few amino acid positions have been detected (Sodek *et al.*, 1978; Mak *et al.*, 1980). The minor form of β-TM (β′), shows substitutions at 11 positions in the amino acid sequence and is estimated to exist in one tenth the amount of the corresponding major form (Mak *et al.*, 1980).

α-TM and β-TM are produced by muscle cells in culture and accumulate when fusion approaches a maximum (Allen *et al.*, 1978, 1979; Allen, 1980). Prior to fusion, another isoform of TM (smooth TM) is synthesized only to be replaced by α-TM and β-TM as myotubes form (Carmon *et al.*, 1978; Garrels, 1979). Two types of α-TM, characteristic of fast and slow contracting muscles, have been distinguished in avian adult (Roy *et al.*, 1979c) and embryonic (Montarras *et al.*, 1981) muscles. Montarras and co-workers showed that in culture the fast isoform of α-TM is synthesized along with β-TM while slow α-TM is not detected.

The ratio of α-TM:β-TM seems to vary according to the type of skeletal muscle studied, with α-TM being greater than β-TM in white muscle and both being approximately equal in red muscle (Cummins and Perry, 1974). It is believed that the proportion of α-TM and β-TM could be related to the speed of muscular contractions. In support of this, Dhoot and Perry (1979) have showed immunohistochemically that α-TM appears most restricted to fast-twitch fibers, and β-TM mostly to slow-twitch fibers in different mammals.

The ratio of α-TM:β-TM is also known to change during development in both avian and mammalian skeletal muscle (Roy et al., 1978, 1980). β-TM, apparently characteristic of slow muscles, is found in higher proportions in both embryonic muscle and rat embryonic cell line whereas α-TM:β-TM reaches about 0.3 (Carmon et al., 1978). α-TM gradually replaces β-TM as development proceeds, according to the muscle type involved (adult chicken pectoralis, 100% α-TM; Adult rabbit fast muscle, 80% α-TM; adult chicken and rabbit slow muscle, about 50% α-TM). The adult fast chicken muscle posterior lattissimus dorsi however does not appear to have high proportion of α-TM compared to their mammalian fast muscle counterparts Roy et al., 1980).

Amphlett et al. (1975) observed that the α-TM:β-TM ratio does not change when rabbit soleus muscle is cross-innervated with a nerve that normally innervates a fast muscle. Denervation does not appear to affect the TM pattern either (Roy et al., 1980). The subunit ratio of TM appears, however, to be affected by the firing frequency of electrical stimulation. Low frequency chronic stimulation of rabbit fast muscle has been shown to switch the α-TM:β-TM ratio towards a typical slow muscle TM pattern (80–55%) (Roy et al., 1979a).

A relationship between the TM composition of muscle and the speed of contraction is more easily observed in cardiac tissue. β-TM, which is absent in small fast beating hearts (e.g., chicken, rabbit), is however found in the slower cardiac muscles of large mammals (15–20% of total TM) (Leger et al., 1976). In rabbit, cardiac TM consists of a single polypeptide (homodimer) which comigrates with skeletal α-TM. In this species, amino acid sequence analysis indicates that cardiac TM is identical to the skeletal α-TM form (Mak et al., 1979; Lewis and Smillie, 1980).

In chicken, cardiac α-TM has been shown to co-migrate with one of the two forms of chicken skeletal α-TM. Montarras et al. (1981) however suggest that cardiac α-TM may be distinct from skeletal muscle α-TMs, implying that three different α-TM genes for cardiac and skeletal muscle exist in chicken while only one α-TM gene exists in rabbit.

Two forms of TM, differing in apparent molecular weight are present in smooth muscle. TM from smooth muscle has not been as well studied as skeletal muscle TM, and its biological function remains unclear since this muscle does not contain a well-defined troponin complex (see: Dabrowska et al., 1979; Adelstein and Eisenberg, 1980). Smooth and skeletal muscle TMs however have similar ability to bind actin and troponin as does skeletal TM, and can substitute for skeletal TM in a reconstituted skeletal regulatory system (Dabrowska et al., 1980b). Smooth muscle TMs are, however, distinct from skeletal muscle TMs as shown by their higher apparent molecular weight, their peptide maps and their amino acid composition (Cummins and Perry, 1974; Fine and Blitz, 1975).

In a similar manner to the myosin heavy and light subunits as well as to the actins, a battery or genes exists for the tropomyosin subunits. These genes are differentially expressed during development and the resulting proteins exist in precise isoforms in specific muscle fiber types. The amount of the individual tropomyosin subunits produced in one fiber type as compared to other fiber types appears to be a characteristic feature of differentiation as well. Therefore, the quantitative aspects of tropomyosin gene expression must be understood in addition to the qualitative aspects in order to understand the differentiation of muscle fiber types.

3.4. Troponin

Localized in the thin filaments of striated muscles along with actin and tropomyosin, troponin is implicated in the regulation of contractions. Troponin consists of three components: troponin C (TnC) which binds Ca^{2+}, troponin T (TnT) which binds tropomyosin, and troponin I (TnI) which is involved in the inhibition of the myosin and actin interaction. As shown in Table 5, the troponin components differ from one muscle type to the other, presumably reflecting variations in the regulation of contraction.

TnC is a member of a superfamily of related proteins which includes myosin light chains, calmodulin, parvalbumin and other calcium-binding proteins (Collins, 1974, 1976; Weeds and McLachlan, 1974; Means and Dedman, 1980; Barker and Dayhoff, 1980). TnC and calmodulin have the most conserved structures and the greatest ability to bind Ca^{2+}, while other more adapted proteins, like the alkali light chains, no longer retain the Ca^{2+}-binding function (see Barker and Dayhoff, 1980). The main form of TnC found in mammalian fast skeletal muscle differs immunohistochemically from two very similar forms found in cardiac and slow skeletal muscle (Dhoot *et al.*, 1979). In mixed skeletal muscles preparations, two isoforms of TnC are found and can be demonstrated immunohistochemically (Dhoot *et al.*, 1979). Antibodies raised against the main form of fast muscle TnC were shown to interact with type II fast twitch fibers and antibodies against the main form of slow muscle TnC were shown to interact with type I slow twitch fiber in all mammalian skeletal muscles tested so far. The existence of two isoforms of TnC in different striated muscle types has been confirmed recently in Wilkinson's laboratory (Wilkinson, 1980). The amino acid sequence of TnC from rabbit slow skeletal muscle was subsequently shown to be the same as that of rabbit cardiac muscle TnC, and therefore both proteins appear to be the products of a single gene. Cardiac and slow TnC were, however, clearly distinct from fast TnC. Fast skeletal TnC consists of four homologous regions each containing one calcium-binding site (Collins *et al.*, 1973, 1977), while cardiac and slow TnC appear to have lost one of these binding sites (Van Eerd and Takahashi, 1975; Wilkinson, 1980). Unlike mammals, avians seem to have only one form of muscle TnC. Thus far, only the fast type of TnC has

Table 5. Estimated Number of Structural Genes for Troponin[a,b]

	Rabbit	Other mammals	Chicken
Fast skeletal muscle	TnC-F (1,12,13)	TnC-F (4,12,13)	TnC-F (6)
Slow skeletal muscle	TnC-C (12,13,14)	TnC-C (12,13)	ND
Cardiac muscle	TnC-C (12,13,14)	TnC-C (5,12,13)	ND
Minimal number of genes	2	2	1 (to 2)
Fast skeletal muscle	TnI-F (2,9,10,13)	TnI-F (8,9,13)	TnI-F (11)
Slow skeletal muscle	TnI-S (7,9,10,13)	TnI-S (8,9,13)	ND
Cardiac muscle	TnI-C (3,9,10)	TnI-C (8,9)	ND
Minimal number of genes	3	3	1 (to 3)
Fast skeletal muscle	TnT-F (12,13)	TnT-F (12,13)	ND
Slow skeletal muscle	TnT-S (12,13)	TnT-S (12,13)	ND
Cardiac muscle	TnT-C (12)	TnT-C (12)	ND
Minimal number of genes	3	3	1–3

[a] Numbers in parentheses refer to the following studies: (1) Collins *et al.* (1973) seq.; (2) Wilkinson and Grand (1975), seq.; (3) Grand and Wilkinson (1976), seq.; (4) Romero-Herrera *et al.* (1976), seq.; (5) Van Eerd and Takahashi (1976), seq.; (6) Wilkinson (1976), seq.; (7) Grand and Wilkinson (1977), seq.; (8) Cummins and Perry (1978), imm.; (9) Dhoot *et al.* (1978), imm.; (10) Wilkinson and Grand (1978a), seq.; (11) Wilkinson and Grand (1978b), seq.; (12) Dhoot *et al.* (1979), imm.; (13) Dhoot and Perry (1979), imm.; (14) Wilkinson (1980), seq.
[b] Abbreviations are as in Table 1.

been observed in chicken breast and leg muscles (Hirabayshi and Perry, 1974).

Three different forms of TnI have been found in striated muscle. Immunohistochemical analysis demonstrated that in mammals fast TnI was normally localized exclusively in type II fibers (fast twitch) and slow TnI in type I fibers (slow twitch) (Dhoot *et al.*, 1978; Dhoot and Perry, 1979). The proportion of fast and slow TnI was shown to vary in human mixed muscles in accordance with the proportions of fiber type present (Cummins and Perry, 1978). The amino acid sequences of the three TnI isoforms in rabbit and the single chicken skeletal (fast) form have been established (Grand and Wilkinson, 1976, 1977; Wilkinson and Grand, 1975, 1978a,b). The isoforms of TnI are of different lengths (182, 179, 184, and 206 residues for chicken fast, and rabbit fast, slow, and cardiac TnIs respectively). Wilkinson and Grand (1978a) compared the four sequences, ignoring the extra terminal residues but including the deletions as substituted amino acids. They found that rabbit fast, slow, and cardiac TnIs differ from each other in amino acid substitutions and deletions by 40%.

TnT has not been as extensively studied as TnC and TnI. Rabbit cardiac and fast skeletal TnT have been shown to have different apparent molecular weights and amino acid compositions (Greaser and Gergely, 1973; Brekke and Greaser, 1976). More recent data suggests that TnT also probably exists in three different forms in mammalian striated muscles. Immunochemical studies show that the major forms of TnT present in mammalian cardiac and fast and slow skeletal muscles are different from each other (Dhoot *et al.*, 1979; Dhoot and Perry, 1979). Immunohistochemically fast TnT antibodies apparently react exclusively with type II fibers (fast twitch) and slow skeletal TnT with type I fibers (slow twitch), showing for the first time that fast and slow skeletal TnT forms are different in mammals.

Two forms of TnT, identified as leg and breast types, have been found in chicken muscle (Perry and Cole, 1974; Wilkinson, 1978). Their isoform pattern has been shown to change during development (Matsuda *et al.*, 1981). Leg type TnT is first isolated from embryonic breast muscle, where upon further development, the breast type of TnT is also synthesized and both types of TnT can then be observed in breast muscle. The production of the breast type of TnT continues to increase and that of the leg type TnT to decrease until it becomes undetectable. This appears to be coordinated with the TM isoform changes. A similar phenomenon is not observed in leg muscles which produce the leg type of TnT throughout development. In culture, myogenic cells of either origin (breast and leg) produce the leg type of TnT exclusively. Muscular dystrophy has been shown to perturb the isoform pattern of development in chicken breast muscle in which both types are found in late stages of development (Obinata *et al.*, 1980).

Similar developmental changes have been observed for the troponin components in mammalian skeletal muscles in which both fast and slow forms of troponin are observed in the same cells early in development (Dhoot and Perry, 1979, 1980). On the basis of immunohistochemical analyses, Dhoot and Perry (1980) demonstrated that the fast types of the troponin complex are found in every muscle cell in newborn mammals while only the presumptive type I (slow twitch) cells also contain the slow types of TnC, TnI and TnT. As development progresses, the amount of fast troponin decreases while that of slow troponin increases in those presumptive type I cells. These cells will eventually contain only the slow isoforms. Dhoot and Perry (1980) found similar perturbations of the isoform patterns in human myopathies as did Obinata *et al.*, (1980) in chicken.

Other changes in the normal program of expression in the troponin structural genes have also been reported. Indeed, Amphlett *et al.* (1975) have shown that cross-innervation of rabbit slow soleus muscle with a nerve normally innervating a fast contracting muscle shifted the proportion of fast and slow TnI toward that seen more typically in fast muscle.

Again it is seen that the troponin subunits also vary with regard to the developmental program which leads to the differentiation of specific fiber types. The mechanisms involved in the specific expression of the genes coding for the proper isoform contractile proteins is unknown. It will be important to determine the chromosomal organization of these various genes and the signals which determine their activation. For example, it is possible that all the specific genes for the fast contractile protein isoforms are clustered. Alternatively, all the isoforms of any particular protein may be clustered separately from the isoforms of the other contractile proteins. Finally, it is also possible that gene organization may have little effect in this overall expression of differentiated fiber types but rather the mechanisms may only require specific receptors for gene activation and posttranscriptional controls.

4. Conclusions

The central question surrounding myogenesis is what cellular mechanisms are involved in the qualitative and quantitative aspects of gene expression during the formation of differentiated muscle fiber types. Eukaryotes obviously have an extensive repertory of regulatory mechanisms by which the ultimate phenotype of a cell is determined. The task then is to identify which of these mechanisms the developing muscle utilize and what their relationships are to each other. These control mechanisms must allow the expression of a multitude of genes involved in general cell metabolism ("housekeeping genes") while allowing the expression of possibly 50–300 additional genes required for muscular contractibility and ultimately orgasmal movement. These cellular functions are themselves further specialized in various muscle fiber types through additional differential gene expression which ultimately form a variety of muscle types capable of specialized functions.

The first step in muscle differentiation may well be the committment of mesodermal cells to become muscle without regard to the particular type of muscle an individual line of cells eventually becomes. Subsequent signals from the cellular environment may be responsible for the terminal differentiation of a specific fiber type. This process can be accomplished by either the synthesis of embryonic contractile protein isoforms which are subsequently replaced by specialized adult isoforms or by the synthesis of all contractile protein isoforms with the eventual repression of those genes not required by the individual muscle type. Alternatively both of these sequential processes may exist simultaneously in differentiating muscle cells.

The mechanisms involved in these processes must initially rely on transcriptional controls because all macromolecules (RNA and proteins—whether structural or regulatory) arise from the transcription of DNA. In addition the involvement of posttranscriptional controls in myogenesis appears necessary; however, the precise mechanisms and relative involvement of these controls is not known. In order to precisely define gene expression during myogenesis, an understanding of the interrelationships of these various controls will be required.

It is most likely that a great deal of information will be amassed using highly specific DNA probes. These probes must, however be of sufficient size and known sequence to unequivocally identify any particular muscle specific gene. In addition, insight into the mechanisms by which super-coiled DNA is specifically altered to allow the expression of the genes required for muscle differentiation will be essential for completing our understanding of muscle development. This most likely will require new experimental approaches and technical advances.

References

Adelstein, R. S., and Eisenberg, E., 1980, Regulation and kinetics of the actinomyosin-ATP interaction, *Ann. Rev. Biochem.* **49:**921.

Affara, N. A., Daubas, P., Weydert, A., and Gros, F., 1980a, Changes in gene expression during myogenic differentiation. II. Identification of the proteins encoded by myotube-specific complementary DNA sequences, *J. Mol. Biol.* **140**:459.

Affara, N. A., Robert, B., Jacquet, M., Buckingham, M. E., and Gros, F., 1980b, Changes in gene expression during myogenic differentiation. I. Regulation of messenger RNA sequences expressed during myotube formation, *J. Mol. Biol.* **140**:441.

Allen, R. E., 1980, Disproportionate accumulation of myosin and tropomyosin in cultured muscle cells, *Eur. J. Cell Biol.* **21**:247.

Allen, R. E., Stromer, M. H., Goll, D. E., and Robson, R. M., 1979, Accumulation of myosin, actin, tropomyosin, and α-actinin in cultured muscle cells, *Dev. Biol.* **69**:655.

Allen, R. E., Stromer, M. H., Goll, D. E., and Robson, R. M., 1979, Accumulation of myosin, actin, tropomyosin, and α-actinin in cultured muscle cells. *Dev. Biol.* **69**:655.

Amphlett, G. W., Perry, S. V., Syska, H., Brown, M. D., and Vrbova, G., 1975, Cross innervation and the regulatory protein system of rabbit soleus muscle, *Nature* **257**:602.

Bag, J., and Sarkar, S., 1976, Studies on a nonpolysomal ribonucleoprotein coding for myosin heavy chains from chick embryonic muscles, *J. Biol. Chem.* **251**:7600.

Bag, J., and Sells, B. H., 1979, Heterogeneity of the nonpolysomal cytoplasmic (free) mRNA-protein complexes of embryonic chicken muscle, *Eur. J. Biochem.* **99**:507.

Barker, W. C., and Dayhoff, M. O., 1980, Evolutionary and functional relationships of homologous physiological mechanisms, *Bioscience* **30**:593.

Benoff, S., and Nadal-Ginard, B., 1979a, Cell-free translation of mammalian myosin heavy-chain messenger ribonucleic acid from growing and fused-L6E9 myoblasts, *Biochemistry* **18**:494.

Benoff, S., and Nadal-Ginard, B., 1979b, Most myosin heavy chain mRNA in L6E9 rat myotubes has a short poly (A) tail, *Proc. Natl. Acad. Sci. USA* **76**:1853.

Benoff, S., and Nadal, Ginard, B., 1980, Transient induction of poly (A)-short myosin heavy chain messenger RNA during terminal differentiation of L6E9 myoblasts, *J. Mol. Biol.* **140**:283.

Bester, A. J., Dunheim, G., Kennedy, D. S., and Heywood, S. M., 1980, Isolation of myosin messenger ribonucleoprotein particles which contain a protein fraction affecting myosin synthesis, *Biochem. Biophys. Res. Comm.* **92**:524.

Bester, A. J., Kennedy, D. S., and Heywood, S. M., 1975, Two classes of translational control RNA: Their role in the regulation of protein synthesis, *Proc. Natl. Acad. Sci. USA* **72**:1523.

Bogenhagen, D. F., Sakonju, S., and Brown, D. D., 1980, A control region in the center of the 5S RNA gene directs specific initiation of transcription: II. The 3' border of the region, *Cell* **19**:27.

Bowman, L. H., and Emerson, C. P., 1977, Post-transcriptional regulation of ribosome accumulation during myoblast differentiation, *Cell* **10**:587.

Bowman, L. H., and Emerson, C. P., 1980, Formation and stability of cytoplasmic mRNAs during myoblast differentiation: pulse-chase and density labeling analyses, *Dev. Biol.* **80**:146.

Bragg, P. W., Dym, H. P., and Heywood, S. M., 1980, Embryonic chick myosin heavy chain mRNA is poly(A)+, *FEBS Lett.* **113**:177.

Brawerman, G., 1981, The role of Poly(A) sequence in mammalian mRNA, *CRC Crit. Rev. Biochem.* **10**:1.

Brekke, C. J., and Greaser, M. L., 1976, Separation and characterisation of the troponin components from bovine cardiac muscle, *J. Biol. Chem.* **251**:866.

Bretscher, A., and Weber, K., 1978, Tropomyosin from bovine brain contains two polypeptide chains of slightly different molecular weights, *FEBS Lett.* **85**:145.

Brevet, A., and Whalen, R. G., 1978, Comparative structural analysis of myosin after limited tryptic hydrolysis by use of two-dimensional gel electrophoresis, *Biochimie* **60**:459.

Brown, S., Levinson, W., and Spudich, J. A., 1977, Cytoskeletal elements of chick embryo fibroblasts revealed by detergent extraction, *J. Supramol. Struct.* **5**:119.

Buckingham, M. E., 1977, Muscle protein synthesis and its control during the differentiation of skeletal muscle cells in vitro, in: *International Review of Biochemistry. Biochemistry of Cell Differentiation II*, Vol. 15 (J. Paul, ed.), pp. 269–332, University Park Press, Baltimore.

Buckingham, M. E., Cohen, A., and Gros, F., 1976, Cytoplasmic distribution of pulse-labelled poly(A)-containing RNA, particularly 26S RNA, during myoblast growth and differentiation, *J. Mol. Biol.* **103**:611.

Buckingham, M. E., and Gros, F., 1975, The use of metrizamide to separate cytoplasmic ribonucleoprotein particles in muscle cell cultures: A method for the isolation of messenger RNA, independent of its poly A content, *FEBS Lett.* **53**:355.

Buller, A. J., Eccles, J. C., and Eccles, R. M., 1960, Differentiation of fast and slow muscles in cat hind limb, *J. Physiol. (Lond.)* **150**:399.

Burridge, K., and Bray, D., 1975, Purification and structural analysis of myosins from brain and other non-muscle tissues, *J. Mol. Biol.* **99**:1.

Carmon, Y., Newman, S., and Yaffe, D., 1978, Synthesis of tropomyosin in myogenic cultures and in RNA-directed cell-free systems: qualitative changes in the polypeptides, *Cell* **14**:393.

Chi, J. C. H., Rubinstein, N., Strahs, N., Holtzer, H., 1975, Synthesis of myosin heavy and light chains in muscle cultures, *J. Cell Biol.* **67**:523.

Civelli, O., Vincent, A., Maundrell, K., Buri, J. F., and Schener, K., 1980, The translational repression of globin mRNA in free cytoplasmic ribonucleoprotein complexes, *Eur. J. Biochem.* **107**:577.

Close, R. I., and Hoh, J. F., 1968, Effects of nerve cross-union of fast-twitch and slow-graded muscle fibres in the toad, *J. Physiol. (Lond.)* **198**:103.

Coleman, J. R., and Coleman, A. W., 1968, Muscle differentiation and macromolecular synthesis, *J. Cell Physiol. (Suppl. 1)* **72**:19.

Collins, J. H., 1974, Homology of myosin light chains, troponin-C and parvalbumins deduced from comparison of their amino acid sequences, *Biochem. Biophys. Res. Comm.* **58**:301.

Collins, J. H., 1976, Homology of myosin DTNG light chain with alkali light chains, troponin C and parvalbumin, *Nature* **259**:699.

Collins, J. H., and Elzinga, M., 1975, The primary structure of actin from rabbit skeletal muscle. Completion and analysis of the amino acid sequence, *J. Biol. Chem.* **250**:5915.

Collins, J. H., Greaser, M. L., Potter, J. D., Horn, M. J., 1977, Determination of the amino acid sequence of troponin C from rabbit skeletal muscle, *J. Biol. Chem.* **252**:6356.

Collins, J. H., Potter, J. D., Horn, M. J., Wilshire, G., and Jackman, N., 1973, The amino acid sequence of rabbit skeletal muscle troponin C: gene replication and homology with calcium-binding proteins from carp and hake muscle, *FEBS Lett.* **36**:268.

Cummins, P., and Perry, S. V., 1974, Chemical and immunochemical characteristics of tropomyosins from striated and smooth muscle, *Biochem. J.* **141**:43.

Cummins, P., and Perry, S. V., 1978, Troponin I from human skeletal and cardiac muscles, *Biochem. J.* **171**:251.

Cummins, P., Price, K. M., and Littler, W. A., 1980, Foetal myosin light chain in human ventricle, *J. Muscle Res. and Cell Motil.* **1**:357.

Dabrowska, R., Nowak, E., and Drabikowski, W., 1980a, Comparative studies of chicken gizzard and rabbit skeletal tropomyosin, *Comp. Biochem. Physiol.* **65B**:75.

Dabrowska, R., Sherry, J. M. F., and Hartshorne, D. J., 1979, Phosphorylation of myosin: A possible regulatory mechanism in smooth muscle, in: *Motility in Cell Function* (F. A. Pepe, J. W. Sanger, and V. T. Nachmias, eds.), pp. 147–160, Academic Press, New York.

Dabrowska, R., Sosinski, and Drabikowski, W., 1980b, Comparative studies of various kinds of tropomyosin, in: *Plasticity of Muscle* (D. Pette, ed.), pp. 225–239, Walter de Gruyter, New York.

d'Albis, A., Pantaloni, C., Bechet, J. J., 1979, An electrophoretic study of native myosin isozymes and of their subunit content, *Eur. J. Biochem.* **99**:261.

Dalla Libera, L., Sartore, S., and Schiaffino, S., 1979, Comparative analysis of chicken atrial and ventricular myosins, *Biochim. Biophys. Acta* **581**:283.

DePhilip, R. M., Rudert, W. A., and Lieberman, S., 1980, Preferential stimulation of ribosomal protein synthesis by insulin and in the absence of ribosomal and messenger ribonucleic acid formation, *Biochemistry* **19**:1662.

Devlin, R. B., and Emerson, C. P., 1979, Coordinate regulation of contractile protein synthesis during myoblast differentiation, *Cell* **13**:599.

Devlin, R. B., and Emerson, C. P., 1980, Coordinate accumulation of contractile protein mRNAs during myoblast differentiation, *Dev. Biol.* **69:**202.

Dhoot, G. K., Frearson, N., and Perry, S. V., 1979, Polymorphic forms of troponin T and troponin C and their localization in striated muscle cell types, *Expt. Cell Res.* **122:**339.

Dhoot, G. K., Gell, P. G. H., and Perry, S. V., 1978, The localization of the different forms of troponin I in skeletal and cardiac muscle cells, *Expt. Cell Res.* **117:**357.

Dhoot, G. K., and Perry, S. V., 1979, Distribution of polymorphic forms of troponin components and tropomyosin in skeletal muscle, *Nature* **278:**714.

Dhoot, G. K., and Perry, S. V., 1980, The components of the troponin complex and development in skeletal muscle, *Expt. Cell Res.* **127:**75.

Doetschman, T. C., Dym, H. P., Siegel, E. J., and Heywood, S. M., 1980, Myoblast stored myosin heavy chain transcripts are precursors to the myotube polysomal myosin heavy chain mRNAs, *Differentiation* **16:**149.

Dow, J., and Stracher, A., 1971, Identification of the essential light chains of myosin, *Proc. Natl. Acad. Sci. USA* **68:**1107.

Dym, H. P., Kennedy, D. S., and Heywood, S. M., 1979, Sub-cellular distribution of the cytoplasmic myosin heavy chain mRNA during myogenesis, *Differentiation* **12:**145.

Early, P., Rogers, J., Davis, M., Calame, K., Bond, M., Wall, R., and Hood, L., 1980, Two mRNAs can be produced from a single immunoglobulin μ gene by alternative RNA processing pathways, *Cell* **20:**313.

Emerson, C. P., and Beckner, S. K., 1975, Activation of myosin synthesis in fusing and mononucleated myoblasts, *J. Mol. Biol.* **93:**431.

Engelke, D. R., Ng, S. Y., Shastry, B. S., and Roeder, R. G., 1980, Specific interaction of a purified transcription factor with an internal control region of 5S RNA genes, *Cell* **19:**717.

Fine, R. E., and Blitz, A. L., 1975, A chemical comparison of tropomyosins from muscle and non-muscle tissues, *J. Mol. Biol.* **95:**447.

Flavell, R. A., 1980, The transcription of eukaryotic genes, *Nature* **285:**356.

Ford, P. J., 1980, Polymerase III control region defined, *Nature* **287:**109.

Frank, G., and Weeds, A. G., 1974, The amino-acid sequence of the alkali light chains of rabbit skeletal-muscle myosin, *Eur. J. Biochem.* **44:**317.

Garrels, J. I., 1979, Changes in protein synthesis during myogenesis in a clonal cell line, *Dev. Biol.* **73:**134.

Garrels, J. I., and Gibson, W., 1976, Identification and characterization of multiple forms of actin, *Cell* **9:**793.

Geoghegan, T., Cereghini, S., and Brawerman, G., 1979, Inactive mRNA-protein complexes from mouse sarcoma-180 ascites cells, *Proc. Natl. Acad. Sci. USA* **76:**5587.

Gette, W. R., and Heywood, S. M., 1979, Translation of myosin heavy chain messenger ribonucleic acid in an eukaryotic initiation factor 3- and messenger-dependent muscle cell-free system, *J. Biol. Chem.* **254:**9879.

Gordon, D. J., Boyer, J. L., and Korn, E. D., 1977a, Comparative biochemistry of non-muscle actins, *J. Biol. Chem.* **252:**8300.

Gordon, T., Perry, R., Srihari, T., and Vrbova, G., 1977b, Differentiation of slow and fast muscles in chickens, *Cell Tiss. Res.* **180:**211.

Gordon, T., and Vrbova, G., 1975, The influence of innervation the differentiation of contractile peeps of developing chick muscles, *Pflugers Arch.* **360:**199.

Grand, R. J. A., and Wilkinson, J. M., 1976, The amino acid sequence of rabbit cardiac troponin I, *Biochem. J.* **159:**633.

Grand, R. J. A., and Wilkinson, J. M., 1977, The amino acid sequence of rabbit slow-muscle troponin I, *Biochem. J.* **167:**183.

Greaser, M. L., and Gergeley, J., 1973, Purification and properties of the components from troponin, *J. Biol. Chem.* **248:**2125.

Greenberg, J. R., 1975, Messenger RNA metabolism of animal cells, *J. Cell Biol.* **64:**269.

Grosschedl, R., and Birnstiel, M. L., 1980, Identification of regulatory sequences in the preclude sequences of an H2A histone gene by the study of specific deletion mutants in vivo, *Proc. Natl. Acad. Sci. USA* **77:**1432.

Guyette, W. A., Matusik, R. J., Rosen, J. M., 1979, Prolactin-mediated transcriptional and post-transcriptional control of casein gene expression, *Cell* **17**:1013.

Havaranis, A. S., and Heywood, S. M., 1981, Cytoplasmic utilization of liposome-encapsulated myosin heavy chain messenger ribonucleoprotein particles during muscle cell differentiation, *Proc. Natl. Acad. Sci. USA* **78**:6898.

Herrmann, H., Heywood, S. M., and Marchok, A. C., 1970, Reconstruction of muscle development as a sequence of macromolecular synthesis, in: *Current Topics in Developmental Biology*, Vol. 5 (A. A. Moscona and A. Monroy, eds.), pp. 181–234, Academic Press, New York.

Heywood, S. M., Dowben, R. M., and Rich, A., 1967, The identification of polyribosomes synthesizing myosin, *Proc. Natl. Acad. Sci. USA* **57**:1002.

Heywood, S. M., and Kennedy, D. S., 1976, Purification of myosin translational control RNA and its interaction with myosin messenger RNA, *Biochemistry* **15**:3314.

Heywood, S. M., and Kennedy, D. S., 1979, Messenger RNA affinity column fractionation of eucaryotic initiation factor 3 and associated proteins and the translation of myosin messenger RNA, *Arch. Biochem. Biophys.* **192**:270.

Heywood, S. M., Kennedy, D. S., and Bester, A. J., 1975, Stored myosin messenger RNA in embryonic chick muscle, *FEBS Lett.* **53**:69.

Hill, R. E., Lee, K. L., and Kenney, F. T., 1981, Effects of insulin on messenger RNA activities in rat liver, *J. Biol. Chem.* **256**:1510.

Hirabayshi, T., and Perry, S. V., 1974, An immunochemical study of the calcium ion-binding protein (troponin-C) and the inhibitory protein (troponin-I) of the troponin complex and their interaction, *Biochim. Biophys. Acta* **351**:273.

Hnik, P., Jirmanova, I., Vejklickey, L., and Zelena, J., 1967, Fast and slow muscles of the chick after nerve cross-union, *J. Physiol. (Lond.)* **193**:309.

Hoh, J. F. Y., 1975, Neural regionation of mammalian fast and slow muscle myosins: an electrophoretic analysis, *Biochemistry* **14**:742.

Hoh, J. F. Y., 1978, Light chain distribution of chicken skeletal muscle myosin isoenzymes, *FEBS Lett.* **90**:297.

Hoh, J. F. Y., and Dunlop, C., 1974, Transformation of slow twitch muscle to fast twitch muscle in peraplegic rat, in: *III International Congress on Muscle Diseases. Congress Series No. 334* (W. G. Bradley, ed.), pp. 50–51, Excerpta Medica, Amsterdam.

Hoh, J. F. Y., Kwan, B. T. S., Dunlop, C., and Kim, B. H., 1980, Effects of nerve cross-union and cordotomy on myosin isoenzymes in fast-twitch and slow-twitch muscles of the rat, in: *Plasticity of Muscle* (D. Pette, ed.), pp. 339–352, Walter de Gruyter, New York.

Holt, J., and Lowey, S., 1975, An immunological approach to the role of the low molecular weight subunits in myosin. I. Physical-chemical and immunological characterization of the light chains, *Biochemistry* **14**:4600.

Hu, S. L., and Manley, J. L., 1981, DNA sequence required for initiation of transcription in vitro from the major late promoter of adenovirus 2, *Proc. Natl. Acad. Sci. USA* **78**:820.

Huszar, G., 1972, Developmental changes of the primary structure and histidine methylation in rabbit skeletal muscle myosin, *Nature (New Biol.)* **240**:260.

Iatrou, K., and Dixon, J. H., 1978, Protamine messenger RNA: its life history during spermatogenesis in rainbow trout, *Fed. Proc.* **37**:2526.

Ilan, J., and Ilan, J., 1976, Requirement for homologous rabbit reticulocyte initiation factor 3 for initiation of α and β globin mRNA translation in a crude protozoal cell-free system, *J. Biol. Chem.* **251**:5718.

Jagus, R., and Kay, J. E., 1979, Distribution of lymphocyte messenger RNA during stimulation by phytohaemagglutinin, *Eur. J. Biochem.* **100**:503.

Jain, S. K., and Sarkar, S., 1979, Poly(riboadenylate)-containing messenger ribonucleoprotein particles of chick embryonic muscles, *Biochemistry* **18**:745.

Jenkins, W. A., Kaumeyer, J. F., Young, E. M., and Raff, R. A., 1978, A test for masked message: the template activity of messenger ribonucleoprotein particles isolated from sea urchin eggs, *Dev. Biol.* **63**:279.

Jirmanova, I., Hnik, P., and Zelena, J., 1971, Implantation of "fast" nerve into slow muscle in young chickens, *Physiol. Bohemoslov.* **20**:199.

Kabat, D., and Chappell, M. R., 1977, Competition between globin messenger ribonucleic acids for a discriminating initiation factor, *J. Biol. Chem.* **252**:2684.

Keller, L. R., and Emerson, C. P., 1980, Synthesis of adult myosin light chains by embryonic muscle cultures, *Proc. Natl. Acad. Sci. USA* **77**:1020.

Kennedy, D. S., Siegel, A., and Heywood, S. M., 1978, Purification of myosin mRNP translational control RNA and its inhibition of myosin and globin messenger translation, *FEBS Lett.* **90**:209.

Kistler, M. K., Ostrowski, M. C., and Kistler, W. S., 1981, Developmental regulation of secretory protein synthesis in rat seminal vesicle, *Proc. Natl. Acad. Sci. USA* **78**:737.

Konkel, D. A., Tilghman, S. M., and Leder, P., 1978, The sequence of the chromosomal mouse β-globin major gene: homologies in capping, splicing, and poly(A) sites, *Cell* **15**:1125.

Korn, L. J., and Brown, D. D., 1978, Nucleotide sequence of *Xenopus borealis* oocyte 5S DNA: Comparison of sequences that flank several related eucaryotic genes, *Cell* **15**:1145.

Leger, V., Bouvert, P., Schwartz, K., and Swynghedauw, B., 1976, A comparative study of skeletal and cardiac tropomyosins: subunits, thiol group content and biological activities, *Pflügers Arch.* **362**:271.

Leger, J. J., and Focant, B., 1973, Low molecular weight components of cow smooth muscle myosins: characterisation and comparison with those of striated muscle, *Biochim. Biophys. Acta* **328**:166.

Lewis, W. G., and Smillie, L. B., 1980, The amino acid sequence of rabbit cardiac tropomyosin, *J. Biol. Chem.* **255**:6854.

Lodish, H. F., 1974, Model for the regulation of mRNA translation applied to hemoglobin synthesis, *Nature* **251**:385.

Lu, R. C., and Elzinga, M., 1977, Partial amino acid sequence of brain actin and its homology with muscle actin, *Biochemistry* **16**:5801.

Luse, D. S., and Roeder, R. G., 1980, Accurate transcription initiation on a purified mouse β-globin DNA fragment in a cell-free system, *Cell* **20**:691.

Maita, T., Umegane, T., Kato, Y., and Matsuda, G., 1980, Amino acid sequence of the L-1 light chain of chicken cardiac muscle myosin, *Eur. J. Biochem.* **107**:565.

Mak, A. S., Lewis, W. G., and Smillie, L. B., 1979, Amino acid sequences of rabbit skeletal β- and cardiac tropomyosins, *FEBS Lett.* **105**:232.

Mak, A. S., Smillie, L. B., and Stewart, G. R., 1980, A comparison of the amino acid sequences of rabbit skeletal muscle α- and β-tropomyosins, *J. Biol. Chem.* **255**:3647.

Margreth, A., Dalla Libera, L., Salviati, G., and Ischia, N., 1980, Spinal transection and the post natal differentiation of slow myosin isoenzymes, *Muscle and Nerve* **3**:483.

Markelonis, G., and Oh, T. H., 1979, A sciatic nerve protein has a trophic effect on development and maintenance of skeletal muscle cells in culture, *Proc. Natl. Acad. Sci. USA* **76**:2470.

Matsuda, R., Obinata, T., and Shimada, Y., 1981, Types of troponin components during development of chicken skeletal muscle, *Dev. Biol.* **82**:11.

McIntosh, J. R., Cande, W. Z., Lazarides, E., McDonald, K., and Snyder, J., 1976, in: *Cell Mobility* (R. Goldman, T. Pollard and U. Rosenbaum, eds.), pp. 12–61, Cold Spring Harbor Laboratory, Cold Spring Harbor.

Means, A. R., and Dedman, J. R., 1980, Calmodulin—an intracellular calcium receptor, *Nature* **285**:73.

Medford, R. M., Wydro, R. M., Nguyen, H. T., and Nadal-Ginard, B., 1980, Cytoplasmic processing of myosin heavy chain messenger RNA: evidence provided by using a recombinant DNA plasmid, *Proc. Natl. Acad. Sci. USA* **77**:5749.

Montarras, D., Fiszman, M. Y., and Gros, F., 1981, Characterization of the tropomyosin present in various chick embryo muscle types and in muscle cells differentiated *in vitro*, *J. Bio. Chem.* **256**:4081.

Northemann, W., Schmelzer, E., and Heinrich, P. C., 1980, Characterization of 20-S and 40-S non-polysomal cytoplasmic messenger ribonucleoprotein particles from rat liver, *Eur. J. Biochem.* **112**:451.

Obinata, T., Takano-Ohmuro, H., and Matsura, R., 1980, Changes in troponin-T and myosin isozymes during development of normal and dystrophic chicken muscles, *FEBS Lett.* **120**:195.

O'Loughlin, J., Heywood, S. M., and Dym, H. P., 1980, Differential binding of myoblast and myotube eIF3 preparations to myosin heavy chain mRNA, *J. Cell Biol.* **87**:272a.

O'Loughlin, J., Lehr, L., Havaranis, A., and Heywood, S. M., 1981, Encapsulation of "core" eIF3, regulatory components of eIF3 and mRNA into liposomes, and their subsequent uptake into myogenic cells in culture, *J. Cell Biol.* **90**:160.

Paterson, B., and Strohman, R. C., 1972, Myosin synthesis in cultures of differentiating chicken embryo skeletal muscle, *Dev. Biol.* **29**:113.

Pelham, R. B., and Brown, D. D., 1980, A specific transcription factor that can bind either the 5S RNA gene or 5S RNA, *Proc. Natl. Acad. Sci. USA* **77**:4170.

Perry, S. V., and Cole, H. A., 1974, Phosphorylation of troponin and the effects of interactions between the components of the complex, *Biochem. J.* **141**:733.

Pette, D., and Schnez, U., 1977, Myosin light chain patterns of individual fast and slow-twitch fibres of rabbit muscles, *Histochemistry* **54**:97.

Pette, D., Vrbova, G., and Whalen, R. C., 1979, Independent development of contractile properties and myosin light chains in embryonic chick fast and slow muscle, *Pflügers Arch.* **378**:251.

Phillips, G. N., Fillers, J. P., and Cohen, C., 1980, Motions of tropomyosin. Crystal as metaphor, *Biophys. J.* **32**:485.

Pollard, T. D., and Weihing, R. R., 1974, Actin and myosin and cell movement, *CRC Crit. Rev. Biochem.* **2**:1.

Poste, G., Papahadjopoulos, D., and Nicolson, G. L., 1975, Local anesthetics affect transmembrane cytoskeletal control of mobility and distribution of cell surface receptors, *Proc. Natl. Acad. Sci. USA* **72**:4430.

Price, K. M., Littler, W. A., and Cummins, P., 1980, Human atrial and ventricular myosin light-chain subunits in the adult and during development, *Biochem. J.* **191**:571.

Revel, M., 1977, Initiation of messenger RNA translation into protein and some aspects of its regulation, in: *Molecular Mechanisms of Protein Biosynthesis* (H. Weissbach and S. Pestka, eds.), pp. 245–321, Academic Press, New York.

Rogers, J., Early, P., Carter, C., Calame, K., Bond, M., Hood, L., and Wall, R., 1980, Two mRNAs with different 3' ends encode membrane-bound and secreted forms of immunoglobulin μ chain, *Cell* **20**:303.

Romero-Herrera, A. E., Castillo, O., and Lehmann, H., 1976, Human skeletal muscle proteins. The primary structure of troponin C, *J. Mol. Evol.* **8**:251.

Rosenthal, E. T., Hunt, T., and Ruderman, J. V., 1980, Selective translation of mRNA controls the pattern of protein synthesis during early development of the surf clam, Spisula solidissima, *Cell* **20**:487.

Roy, R. K., Mabuchi, K., Sarkar, S., Mis, C., and Sreter, F. A., 1979a, Changes in tropomyosin subunit pattern in chronic electrically stimulated rabbit fast muscles, *Biochem. Biophys. Res. Comm.* **89**:181.

Roy, R. K., Sreter, F. A., Pluskal, M. G., and Sarkar, S., 1980, Tropomyosin transitions in avian and mammalian skeletal muscles, in: *Plasticity of Muscle* (D. Petter, ed.), pp. 241–254, Walter de Gruyter & Co., New York.

Roy, R. K., Sreter, F. A., and Sarkar, S., 1978, Tropomyosin subunits and myosin light chains of avian and mammalian muscles: differential gene expression during development, in: *Aging: Aging in Muscle*, Vol. 6 (G. Kaldor and W. J. DiBattista, eds.), pp. 23–48, Raven Press, New York.

Roy, R. K., Sreter, F. A., and Sarkar, S., 1979b, Changes in tropomyosin subunits and myosin light chains during development of chicken and rabbit striated muscles, *Dev. Biol.* **69**:15.

Roy, R. K., Sreter, F. A., and Sarkar, S., 1979c, Evidence for an intrinsic developmental program in avian and mammalian skeletal muscles, in: *Motility in Cell Function* (F. A. Pepe, J. W. Sanger and V. T. Nachmias, eds.), pp. 371–375, Academic Press, New York.

Rubenstein, P. A., and Spudich, J. A., 1977, Actin microheterogeneity in chick embryo fibroblasts, *Proc. Natl. Acad. Sci. USA* **74**:120.

Rubinstein, N. A., and Holtzer, H., 1979, Fast and slow muscles in tissue culture synthesise only fast myosin, *Nature* **280**:323.

Rubinstein, N. A., and Kelly, A. M., 1978, Myogenic and neurogenic contributions to the development of fast and slow twitch muscles in rat, *Develop. Biol.* **62**:473.

Rubinstein, N., Mabuchi, K., Pepe, F., Salmons, S., Gergely, J., and Sreter, F., 1978, *J. Cell Biol.* **79**:252.

Rubinstein, N. A., Pepe, F. A., and Holtzer, H., 1977, Myosin types during the development of embryonic chicken fast and slow muscles, *Proc. Natl. Acad. Sci. USA* **74**:4524.

Rushbrook, J. I., and Stracher, A., 1979, Comparison of adult, embryonic, and dystrophic myosin heavy chains from chicken muscle by sodium dodecyl sulfate/polyacrylamide gel electrophoresis and peptide mapping, *Proc. Natl. Acad. Sci. USA* **76**:4331.

Saborio, J. L., Segura, M., Flores, M., Garcia, R., and Palmer, E., 1979, Differential expression of gizzard actin genes during chick embryogenesis, *J. Biol. Chem.* **254**:11119.

Sanger, J., 1975, Presence of actin during chromosomal movement, *Proc. Natl. Acad. Sci. USA* **72**:2451.

Sarkar, S., Mukherjee, A. K., and Guha, C., 1981, A ribonuclease-resistant cytoplasmic 10S ribonucleoprotein of chick embryonic muscle, *J. Biol. Chem.* **256**:5077.

Sarkar, S., Mukherjee, S. P., Sutton, A., Mondal, H., and Chen, V., 1973, Isolation of messenger ribonucleic acid for myosin heavy chain, *Prep. Biochem.* **3**:583.

Sartore, S., Dalla Libera, L., and Schiaffino, S., 1979, Fractionation of rabbit ventricular myosins by affinity chromatography with insolubilized antimyosin antibodies, *FEBS Lett.* **106**:197.

Savouret, D-F., Loosfelt, H., Atger, M., and Milgrm, E., 1980, Differential hormonal control of a messenger RNA in two tissues. Uteroglobin mRNA in the lung and the endometrium, *J. Biol. Chem.* **255**:4131.

Schadat, F. H., Harris, H. E., and Epstein, H. F., 1977, Actin from the nematode, *Caenorhabditis elegans*, is a single electrofocusing species, *Biochim. Biophys. Acta* **493**:304.

Schwartz, K., Lompré, A.-M., Bouveret, P., Wisnewsky, C., and Swynghedauw, B., 1980a, Use of antibodies against dodecylsulfate-denatured heavy meromyosins to probe structural differences between muscular myosin isozymes, *Eur. J. Biochem.* **104**:341.

Schwartz, R. J., Haron, J. A., Rothblum, K. N., and Dugiaczyk, A., 1980b, Regulation of muscle differentiation: cloning of sequences from α-actin messenger ribonucleic acid, *Biochemistry* **19**:5883.

Schwartz, R. J., and Rothblum, K., 1980, Regulation of muscle differentiation: isolation and purification of chick actin messenger ribonucleic acid and quantitation with complementary deoxyribonucleic acid probes, *Biochemistry* **19**:2506.

Seymour, J., and O'Brien, E. J., 1980, The position of tropomyosin in muscle thin filaments, *Nature* **283**:680.

Siegel, E., 1981, Characterization of translational control RNA (tcRNA) isolated from embryonic chick muscles, Ph.D. Thesis, Univ. of Connecticut, Storrs.

Siegel, R., Mroczkowski, B., McCarthy, T., and Heywood, S. M., 1982, Characterization of translational control RNA (tcRNA) isolated from embryonic chick muscle, Submitted.

Sodek, J., Hodges, R. S., and Smillie, L. B., 1978, Amino acid sequence of rabbit skeletal muscle α-tropomyosin. The COOH-terminal half, *J. Biol. Chem.* **253**:1129.

Sreter, F. A., Elzinga, M., and Mabuchi, K., 1975, The N'-methylhistidine content of myosin in stimulated and cross-reinnervated skeletal muscles of the rabbit, *FEBS Lett.* **57**:107.

Sreter, F. A., Gergely, J., and Luff, A. L., 1974, The effect of cross reinnervation on the synthesis of myosin light chains, *Biochem. Biophys. Res. Comm.* **56**:84.

Srihari, T., and Pette, D., 1981, Myosin light chains in normal and electrostimulated cultures of embryonic chicken breast muscle, *FEBS Lett.* **123**:312.

Starr, R., and Offer, G., 1973, Polarity of the myosin molecule, *J. Mol. Biol.* **81**:17.

Stockdale, F. E., Baden, H., and Raman, N., 1981a, Slow muscle myoblasts differentiating *in vitro* synthesize both slow and fast myosin light chains, *Dev. Biol.* **82**:168.

Stockdale, F. E., Raman, N., and Baden, H., 1981b, Myosin light chains and the developmental origin of fast muscle, *Proc. Natl. Acad. Sci. USA* **78**:931.

Stone, D., and Smillie, L. B., 1978, The amino acid sequence of rabbit skeletal α-tropomyosin. The NH₂-terminal half and complete sequence, *J. Biol. Chem.* **253**:1137.

Strohman, R. C., Moss, P. S., and Micou-Eastwood, J., 1980, Antiserum to myosin and its use in studying myosin synthesis and accumulation during myogenesis, in: *Current Topics in Developmental Biology* Vol. 14, part. II (A. A. Moscona and A. Monroy, eds.), pp. 297–319, Academic Press, New York.

Strohman, R. C., Moss, P. S., Micou-Eastwood, J., Spector, D., Przybyla, A., and Paterson, B., 1977, Messenger RNA for myosin polypeptides: Isolation from single myogenic cell cultures, *Cell* **10**:265.

Syrovy, I., and Gutmann, E., 1977, Differentiation of myosin in soleus and extensor digitorum longus muscle in different animal species during development, *Pflügers Arch.* **369**:85.

Tolstoshev, P., Berg, R. A., Rennard, S. I., Bradley, K. H., Trapnell, B. C., and Crystal, R. G., 1981, Procollagen production and procollagen messenger RNA levels and activity in human lung fibroblasts during periods of rapid and stationary growth, *J. Biol. Chem.* **256**:3135.

Tsai, S. Y., M.-J., and O'Malley, B. W., 1981, Specific 5' flanking sequences are required for faithful initiation of in vitro transcription of the ovalbumin gene, *Proc. Natl. Acad. Sci. USA* **78**:879.

Umeda, P. K., Zak, R., Rabinowitz, M., 1980, Purification of messenger ribonucleic acids for fast and slow myosin heavy chains by indirect immunoprecipitation of polysomes from embryonic chick skeletal muscle, *Biochemistry* **19**:1955.

Vandekerckhove, J., and Weber, K., 1978a, Actin amino-acid sequences. Comparison of actins from calf thymus, bovine brain, and SV40-transformed mouse 3 + 3 cells with rabbit skeletal muscle actin, *Eur. J. Biochem.* **90**:451.

Vandekerckhove, J., and Weber, K., 1978b, At least six different actins are expressed in a higher mammal: an analysis based on the amino acid sequence of the amino-terminal tryptic peptide, *J. Mol. Biol.* **126**:783.

Vandekerckhove, J., and Weber, K., 1978c, Mammalian cytoplasmic actins are the products of at least two genes and differ in primary structure in at least 25 identified positions from skeletal muscle actins, *Proc. Natl. Acad. Sci. USA* **75**:1106.

Vandekerckhove, J., and Weber, K., 1978d, The amino acid sequence of *Physarum* actin, *Nature* **276**:720.

Vandekerckhove, J., and Weber, K., 1979a, The amino acid sequence of actin from chicken skeletal muscle actin and chicken gizzard smooth muscle actin, *FEBS Lett.* **102**:219.

Vandekerckhove, J., and Weber, K., 1979b, The complete amino acid sequence of actins from bovine aorta, bovine heart, bovine fast skeletal muscle, and rabbit slow skeletal muscle, *Differentiation* **14**:123.

Vandekerckhove, J., and Weber, K., 1980, Vegetative *Dictyostelium* cells containing 17 actin genes express a single major actin, *Nature* **284**:475.

Van Eerd, J. P., and Takahashi, K., 1975, The amino acid sequence of bovine cardiac troponin-C. Comparison with rabbit skeletal troponin-C, *Biochem. Biophys. Res. Comm.* **64**:122.

Van Eerd, J. P., and Takahashi, K., 1976, Determination of the complete amino acid sequence of bovine cardiac troponin C, *Biochemistry* **15**:1171.

Vincent, A., Civelli, O., Maundrell, K., and Scherrer, K., 1980, Identification and characterization of the translationally repressed cytoplasmic globin messenger-ribonucleoprotein particles from duck erythroblasts, *Eur. J. Biochem.* **112**:617.

Wasylyk, B., Kedinger, C., Corden, J., Brison, O., and Chambon, P., 1980, Specific in vitro initiation of transcription on conalbumin and ovalbumin genes and comparison with adenovirus-2 early and late genes, *Nature* **285**:367.

Weeds, A. G., 1976, Light chains from slow-twitch muscle myosin, *Eur. J. Biochem.* **66**:157.

Weeds, A. G., and Burridge, K., 1975, Evidence for reciprocal transformation of heavy chains, *FEBS Lett.* **57**:203.

Weeds, A. G., and McLachlan, A. D., 1974, Structural homology of myosin alkyl light chains, troponin C and carp calcium binding protein, *Nature* **252**:646.

Weeds, A. G., Trentham, D. R., Kean, C. J. C., and Buller, A. J., 1974, Myosin from cross-reinnervated cat muscles, *Nature* **247**:135.

Whalen, R. G., 1980, Contractile protein isozymes in muscle development, in: *Plasticity of Muscle* (D. Pette, ed.), pp. 177–189, Walter de Gruyter, New York.

Whalen, R. G., Butler-Browne, G. S., and Gros, F., 1976, Protein synthesis and actin heterogeneity in calf muscle cells in culture, *Proc. Natl. Acad. Sci. USA* **73**:2018.

Whalen, R. G., Butler-Browne, G. S., and Gros, F., 1978, Identification of a novel form of myosin light chain present in embryonic skeletal muscle tissue and cultured muscle cells, *J. Mol. Biol.* **126**:415.

Whalen, R. G., Butler-Browne, G. S., Sell, S., and Gros, F., 1979, Transitions in contractile protein isozymes during muscle cell differentiation, *Biochimie* **61**:625.

Whalen, R. G., Butler-Browne, G. S., Sell, S. M., Gros, F., Schwartz, K., Bouveret, P., and Pinset-Härström, I., 1981, Transitions in myosin isozymes during rat muscle development, in: *Mechanism of Muscle Adaptation to Functional Requirements* (F. Guba, G. Maréchal, and Ö. Takács, eds.), *Adv. Physiol. Sci.*, Vol. 24, pp. 305–315, Pergamon Press, New York.

Wikman-Coffelt, J., and Srivastava, S., 1979, Differences in atrial and ventricular myosin light chain LC_1, *FEBS Lett.* **106**:207.

Wilkinson, J. M., 1976, The amino acid sequence of troponin C from chicken skeletal muscle, *FEBS Lett.* **70**:254.

Wilkinson, J. M., 1978, The components of troponin from chicken fast skeletal muscle, *Biochem. J.* **169**:229.

Wilkinson, J. M., 1980, Troponin C from rabbit slow skeletal and cardiac muscle is the product of a single gene, *Eur. J. Biochem.* **103**:179.

Wilkinson, J. M., and Grand, R. J. A., 1975, The amino acid sequence of troponin I from rabbit skeletal muscle, *Biochem. J.* **149**:493.

Wilkinson, J. M., and Grand, J. A., 1978a, Comparison of amino acid sequence of troponin I from different striated muscles, *Nature* **271**:31.

Wilkinson, J. M., and Grand, J. A., 1978b, The amino-acid sequence of chicken fast-skeletal-muscle troponin I, *Eur. J. Biochem.* **82**:493.

Zechel, K., and Weber, K., 1978, Actins from mammals, bird and slime mold characterized by isoelectric focusing in polyacrylamide gels, *Eur. J. Biochem.* **89**:105.

Zevin-Lonkin, D., and Yaffe, D., 1980, Accumulation of muscle-specific RNA sequences during myogenesis, *Dev. Biol.* **74**:326.

7

Cloning of Contractile Protein Genes

Robert J. Schwartz and Edwin M. Stone

1. Introduction

The way in which genes specify the ordered appearance of contractile and regulatory proteins in differentiating muscle cells is an important problem in developmental biology. The existence of easily detectable markers of cell differentiation at every stage of myogenesis, the ease with which myoblasts can be grown *in vitro,* and the fact that muscle is an electrically excitable tissue combine to make skeletal muscle a rich model system for studying many aspects of cell differentiation in higher organisms. The potential for muscle cell development exists in DNA molecules that contain the appropriate sequences. The actual genesis of a myotube, however, requires regulated transcription of the DNA sequences into mRNA, and controlled translation of the mRNA into proteins. Investigators who have begun to attack these interrelated processes experimentally have seen similarities in the behavior of many muscle genes. For example, the genes coding for actin, myosin heavy chains, myosin light chains, troponins, tropomyosin, creatine kinase, and aldolase, all have more than one nonidentical copy per haploid genome. That is, they exist as small gene families with each member of a family encoding a slight variant (isoform) of the family gene. In most cases, the isoforms are expressed differently during development, both in time and tissue distribution. For example, during skeletal myogenesis, α-actin, creatine kinase M, aldolase A_4, and adult type myosin heavy and light chains are "switched on" and actively expressed, while embryonic myosins, β- and γ-actin, creatine kinase B, and aldolase C are all "switched off" and no longer synthesized.

Although protein analysis of the embryonic and differentiated tissues can

Robert J. Schwartz and Edwin M. Stone • Department of Cell Biology, Baylor College of Medicine, Texas Medical Center, Houston, Texas 77030.

reveal evidence of such switching, the elucidation of the switching mechanism requires the ability to detect alterations in the DNA structure and mRNA concentrations accompanying the switch. Recent advances in the recombinant cloning of eukaryotic DNA in bacteria have allowed a number of the mRNAs of muscle-specific proteins to be completely characterized. Cloned DNA copies of these mRNAs can be used in hybridization experiments to analyze the metabolism of a number of mRNA species during muscle differentiation. Specifically, such experiments can detect the appearance and disappearance of isoformic mRNAs and thereby determine the role transcription plays in the regulation of these genes. Cloned DNA sequences are also being used to measure the number of members in a gene family, to determine the location of the sequences within the genome, and to assay their transcriptional potential.

This review summarizes the current information on the recombinant DNA cloning of contractile protein genes as well as the application of this technology to the study of muscle differentiation.

2. Model Systems of Muscle Development

2.1. Muscle Development in Culture

Most of the studies of the biochemical changes underlying skeletal muscle differentiation have been undertaken *in vitro* because of the increased developmental synchrony of cultured cells in comparison to cells *in vivo*. The majority of these studies have utilized primary avian or mammalian cultures, but established myoblast lines are now gaining popularity because such cells can be cloned and selected for mutations. Mutant lines resistant to fusion (Loomis *et al.*, 1973), α-amanitin (Somers *et al.*, 1975), and ouabain (Luzzati, 1974), as well as one that exhibits temperature-sensitive differentiation (Loomis *et al.*, 1973), have been produced.

Differentiation of embryonic skeletal muscle in culture occurs in a series of well-defined stages. *Prefusion* is the period of proliferation of presumptive muscle cells. *Fusion* is the formation of multinucleated myotubes from the mononucleated myoblasts. *Postfusion* is the phase in which the myotubes acquire an excitable membrane and functional myofibrillar sarcomeres.

2.2. Prefusion

Cell division and DNA synthesis have essential roles in the differentiation of muscle cells (Okazaki and Holtzer, 1966; Bonner and Hauschka, 1974; Dienstman *et al.*, 1974). For example, after embryonic cells have been dissociated and placed in culture, at least one cell cycle must take place before fusion can occur (Stockdale *et al.*, 1964). Also, the addition of DNA synthesis inhibitors like FudR and cytosine arabinoside to cultures immediately after plating interferes with the formation of multinucleated cells and the appearance of contractile proteins (Coleman and Coleman, 1968; Fischbach, 1972; Doering and Fischman, 1974). Inhibition of myogenesis by the thymidine analog 5-bromodeoxyuridine (BudR) has received special attention due to its apparent selective effect on the expression of differentiated functions (Coleman *et al.*, 1969; Bischoff and Holtzer, 1970). The fusion of proliferating myoblasts can be stopped by growth in a low concentration of BudR (3×10^{-6} M; Bischoff and Holtzer, 1970). This inhibition occurs after one round of DNA synthesis and appears to result from the substitution of BudR for thymidine within single strands of the newly synthesized DNA (O'Neil and Stockdale, 1974). However, it has also been suggested that BudR could act by inhibiting glycosylation of membrane receptor proteins, thereby rendering the cells insensitive to external differentiation signals (Rogers *et al.*, 1975). In any case, contractile proteins are not synthesized under these conditions, even though overall protein synthesis and cell proliferation are unaffected. The BudR effects are reversible by simply replacing the BudR-containing medium with fresh standard medium.

The observation that mitosis is linked to differentiation has caused some to postulate that the mitosis preceding cell fusion is different in some way from the proliferative mitoses that preceded it. This idea led Holtzer and his collaborators to study the *in vitro* development of clones derived from embryonic chick muscle (Abbott *et al.*, 1974; Dienstman *et al.*, 1974). As a result of this study, they proposed a rather elaborate model of myogenesis in which cells either proliferate, forming "presumptive myoblasts," or undergo a final "quantal mitosis" to form "postmitotic myoblasts," which are irreversibly committed to myotube formation without an intervening cell division. Konigsberg and his students, however, provided evidence which convincingly supports an alternate model (Buckley and Konigsberg, 1974; Konigsberg, 1975). These investigators measured the phases of the cell cycle in cultured quail myoblasts and then used time lapse cinematography to determine the developmental fate of the two daughter cells resulting from a given mitosis (Konigsberg, 1975; Konigsberg *et al.*, 1978). They found that the G_1 phase of the cell cycle of proliferating myoblasts begins to lengthen randomly before fusion, and that fusion itself takes place during a G_1 phase. In addition, the daughter cells resulting from a single mitosis were observed to behave independently with respect to subsequent fusion. Thus, Konigsberg feels that the random developmental increase in the length of G_1 increases the probability of an event that requires two cells in the G_1 phase, that is, fusion. The fact that one daughter cell of a mitosis can participate in a fusion while the other divides again is a strong argument against the quantal mitosis theory.

In agreement with Konigsberg, Nadal-Ginard (1978) demonstrated that fusion of L_6 rat myoblasts occurs in the G_1 phase and that mitotic daughter cells are independent with respect to fusion. In addition, these experiments suggest that a molecular exchange occurs between myoblasts which increases the probability of their fusion.

2.3. Fusion

A variety of *in vitro* investigations have demonstrated that a multinucle-ated muscle fiber originates from the fusion of diploid myoblasts. Cooper and Konigsberg (1961) used cinematography to show the fusion of myoblasts into myotubes. Yaffe and Feldman (1965) used mixed cell cultures from such diverse species as rat, rabbit, calf, and chick to form hybrid myotubes. Similar-ly, Mintz and Baker (1967) showed that two distinct, homozygous strains of mouse myoblasts can fuse into myotubes. In contrast, mixed cultures of cells from heart or kidney, with myoblasts, demonstrated that fusion is tissue spe-cific (Yaffe and Feldman, 1965). That is, nonskeletal muscle cells do not enter myotubes when added to a fusing muscle culture.

Although myoblast fusion *in vitro* can be experimentally hastened or inhibited by a variety of factors, it remains to be proven whether any of these factors play a direct role in *in vivo* muscle development. Nevertheless, such factors are often useful for synchronizing cultures in biochemical experiments.

Horse serum and chicken embryo extracts have been found empirically to contain factors necessary for growth of myoblasts in culture; in addition, chicken embryo extracts have been found by a number of investigators to promote muscle differentiation in culture (de la Haba *et al.*, 1975). In-terestingly, cultured myoblasts themselves secrete "conditioning factors" into the culture media, and, such cell-conditioned medium is capable of stimulat-ing myoblast fusion when added to younger cultures (Doering and Fischman, 1977).

In an early search for optimum muscle culture conditions, Hauschka and Konigsberg (1966) studied the effect of a collagen substrate on the attachment of myoblasts to culture vessels. They found that collagen facilitated cell attach-ment to the dishes and promoted the formation of myotubes in the culture. Insulin is another agent capable of promoting myoblast fusion (de la Haba *et al.*, 1966). This hormone, which is present normally in developing embryos and which is added to most culture media, triggers a variety of metabolic responses in cultured cells, ranging from increased amino acid transport to elevated cAMP phosphodiesterase levels (Mandel and Pearson, 1974). Myo-blasts normally secrete nerve growth factor, a protein similar to insulin, into the culture medium (Murphy *et al.*, 1977); thus, NGF may be one of the "conditioning factors" responsible for promoting myoblast fusion and differentiation.

Perhaps the most important modulator of myoblast fusion is the calcium ion concentration. Myoblasts grow well in 100 μM Ca^{2+}, but will not fuse until the concentration is brought to a physiological 2 mM (Shainberg *et al.*, 1971). When myoblasts are grown in a low calcium medium with dialysed serum, they form confluent monolayers of mononucleated cells. Addition of Ca^{2+} to such cultures produces a synchronized fusion of the cells. Inhibition of fusion and myoblast differentiation can also be effected by adding various mem-brane-active compounds to myoblast cultures. Lectins, such as concanavalin

A, wheat germ agglutinin, and abrin (Den *et al.*, 1975), and lipid modification by addition of exogenous fatty acids (Prives and Shinitzsky, 1977) or treatment with phospholipase C (Keller and Nameroff, 1974) have both been reported to inhibit the *in vitro* formation of myotubes. In addition, drugs such as colchicine (Fukuda *et al.*, 1976) and colcemid (Holtzer *et al.*, 1973), which destabilize microtubules, and cytochalasin B (Sanger, 1974), which disrupts microfilaments, all block the fusion process. Finally, fusion can also be blocked by mitogenic tumor promoters such as phorbol myristic acetate (Cohen *et al.*, 1977), and transforming agents such as the Rous sarcoma virus (Fiszman, 1978).

2.4. Postfusion

In both mammalian and avian muscle, a temporal relationship has been demonstrated between the onset of cell fusion and the accumulation of certain muscle-specific proteins, such as myosin, actin, and creatine kinase (Coleman and Coleman, 1968; Shainberg *et al.*, 1971; Turner *et al.*, 1974). When fusion is arrested by Ca^{2+} depletion, the synthetic rates of these proteins diminish, and when the calcium is replaced, the synthetic rates increase. This type of study has led to the concept that cell fusion not only precedes muscle differentiation but may in fact trigger specific gene expression.

Fusion cannot, however, be considered the sole trigger for differentiation in all muscle systems, because Holtzer *et al.* (1975) have shown that in chick somites mononucleated cells assume a cross-striated contractile state. Furthermore, fusion-blocked cultured myoblasts acquire the following characteristics of differentiated muscle fibers: (1) visible cross-striated myofibrils (Coleman and Coleman, 1968; Okazi and Holtzer, 1965); (2) acetylcholine receptors (Paterson and Prives, 1973); (3) acetylcholine esterase molecules (Wilson *et al.*, 1973); (4) reaction with antimyosin and antitropomyosin antibodies (Holtzer *et al.*, 1973; Masaki and Yoshizaki, 1972); and (5) high levels of glycogen phosphorylase, aldolase, and creatine kinase (Turner *et al.*, 1976). Along the same lines, Emerson and Beckner (1975) have shown that myosin can be detected in mononucleated quail muscle cells when the cells are inactive in DNA synthesis. Thus, they conclude that gene expression may be triggered by a repression of DNA synthesis rather than by the initiation of cell fusion. This is consistent with the observation of Vertel and Fischman (1976) that myoblasts in Ca^{2+}-depleted cultures are mitotically inhibited. The latter investigators suggest that muscle cell fusion should be considered a morphological manifestation of muscle differentiation rather than a regulatory event (Vertel and Fischman, 1976).

2.5. Influence of Innervation on Muscle Development

It is clear from the study of muscle development in culture that the early stages of myogenesis are independent of neural control. However, the development of more mature muscle fibers appears to depend upon the nervous system. In most organisms, there are several types of skeletal muscle fibers that differ in their structural, contractile, metabolic, and membrane characteristics. Several studies suggest that these differences result from differences in the motor neurons that innervate the fibers during development (Buller *et al.*, 1960; Henneman and Olson, 1965). One histochemical finding is that all the muscle fibers innervated by a single α motor neuron display the same enzymatic profile, even though these fibers are intermixed with other fiber types in the intact muscle. Other studies indicate that the fiber phenotype in mature muscles is plastic and can be altered by removal or modification of the nerve supply (Guth, 1968). In birds and mammals, adult muscle can be classified as fast or slow twitch by physiological, biochemical, and immunological criteria. For example, slow twitch fibers have a longer twitch duration, lower maximum isometric tension, and a more oxidative metabolism than do fast twitch fibers. In addition, the fiber types contain different myosin heavy and light chains (Sarkar *et al.*, 1971), tropomyosins, and troponins (Perry, 1973). If one removes the nerve supplying a fast muscle and replaces it by grafting a nerve from a slow muscle, one can change the physiological phenotype of the muscle from fast to slow (Buller *et al.*, 1960). Prolonged tonic stimulation of a nerve to a fast muscle can effect a similar change, including a switch in the myosin light chain profile (Salmons and Sreter, 1976).

Thus, it appears that in both avian and mammalian muscles, normal differentiation can be altered by manipulation of the nerve supply. It should be pointed out, however, that many of these experiments suffer from the necessity of *in vivo* design. That is, the effects of denervation cannot be directly assessed in the face of the whole organism's hormonal and inflammatory response to the insult. Similarly, without painstaking histological monitoring, one cannot be sure that a grafted nerve is the only supply for the denervated muscle, nor can one argue that the fibers themselves actually change phenotype without verifying that new fibers do not arise from satellite cells following the change in innervation.

2.6. Nematode and Drosophila Muscle Systems

Even though the study of muscle culture systems has greatly increased our knowledge of muscle structure, function, and development, many gaps in our understanding remain. For example, the mechanism by which protein components are assembled into such precise arrays of thick and thin filaments is not known. Also, the function of several proteins associated with the myofilaments is not understood, and, despite recent advances, little is known about how the differential expression of muscle protein genes is controlled during development.

Two invertebrate organisms, *C. elegans* and *Drosophila*, may be the tools to use to answer these questions. The nematode *C. elegans* has a simple anatomy that is composed largely of muscle cells of a few distinct types; thus, its various contractile proteins can be easily isolated for biochemical analyses. In addition, *C. elegans* has a short and well-defined developmental period, which facilitates developmental studies. The short generation time also allows a variety of mutants that exhibit specific muscle defects to be isolated (Brenner, 1974).

The well-defined genetics of *Drosophila* and the ability to culture its embryonic skeletal muscles (Donady *et al.*, 1975) make it an ideal system for studying muscle development with mutants. In addition, one can use *in situ* hybridization to preparations of its polytene chromosomes to identify the chromosomal locations of specific genes.

3. Introduction to Contractile Proteins and Multigene Families

The rate of synthesis and accumulation of a number of enzymes and structural proteins changes dramatically during the course of myogenesis. The differentiated state is characterized by proteins that are unique to muscle, but whose isoforms are found in nonmuscle cells. The recent discovery of actin and myosin in many nonmuscle eukaryotic cells, as well as the long recognized structural and physiological differences among the various muscle fiber types, have stimulated a search for heterogeneity in the contractile proteins from different tissues. As a result, convincing evidence has accumulated for tissue-specific differences in actin, tropomyosin, and myosin heavy and light chains. Each different isoform of a given contractile protein is thought to be encoded by a separate gene, and thus the genome of an organism is thought to contain a multigene family for each contractile protein. Such multigene families are demonstrated for a given protein by: (1) finding differences in the protein isoforms that cannot be explained by posttranslational modification; (2) characterizing mRNAs with very similar coding regions, but divergent noncoding regions; and ultimately (3) characterizing the different natural genes in the genomic DNA.

The changes in type and concentration of various contractile proteins in myogenesis is thought to result from selective activation of the members of multigene families, but at present the mechanisms behind such selective activation are not totally understood. The following sections briefly introduce the contractile proteins that will be discussed in this review and present the experimental evidence for their multigene families.

3.1. Actin

Actin is found in most, if not all, eukaryotic cells, and in muscle it is the main constituent of the thin filaments. In thin filaments, the monomers of molecular weight 42,000 (G-actin) are polymerized into polar strands (F-actin) which wind around each other in a double helical fashion. The thin actin filaments interact with thick myosin filaments to activate the Mg^{2+}-dependent myosin ATPase, which, in turn, liberates the energy used in contraction (Mannherz and Goody, 1976).

In contrast to its relatively circumscribed role in the contractile apparatus of muscle cells, the functions of actin in nonmuscle cells are diverse (Pollard and Weihing, 1974). Some of the nonmuscle contractile processes believed to involve actin are chromosome movement (Sanger, 1975), cytokinesis (Schroeder, 1973), exocytosis (Berl *et al.*, 1973), phagocytosis (Stossel and Hartwig, 1976), and whole cell motility (Huxley, 1973). In most nonmuscle cells, actin is present in the form of 6-nm microfilaments (Ishikawa *et al.*, 1969). These are often found just inside the plasma membrane where they constitute either a diffuse matrix (Malech and Lentz, 1974), or thick bundles many micrometers in length (Lazarides and Weber, 1974). It has become apparent that these diverse actin functions do not result from the properties of a single polypeptide, but rather from a family of closely related proteins. Such multiple forms of actin have been identified. Gruenstein and Rich (1975) compared the tryptic peptide fragments of brain and muscle actins from the chicken and found several differences. Furthermore, nonmuscle actins from 3T3 fibroblasts and HeLa cells were also found to differ from chicken muscle actin by several tryptic peptides (Grunstein *et al.*, 1975). Such interspecies comparisons are meaningful because no differences have been detected between the peptide maps of chicken and mammalian muscle actins (Carsten and Katz, 1964).

Whalen *et al.* (1976) found that actin in dividing myoblasts exists in three forms with similar biochemical properties and molecular weights (42,000), but with different isoelectric points. Two of the forms, β- and γ-actin, are found in prefusion dividing myoblasts and also in cultured kidney cells. The most acidic of the three forms, termed α-actin, is the only species detectable in mature muscle tissue, and eventually replaces the β and γ forms in muscle cultures after the cells have fused. The relationship of these three proteins to each other was shown by careful tryptic map comparisons. Garrels and Gibson (1976) showed that thirty of forty radiolabeled peptides were common to all three proteins, and that the patterns obtained from β- and γ-actin were nearly identical. These nonmuscle actins contained six peptides not present in α-actin, while α-actin was found to contain four peptides not present in β- or γ-actin. Hunter and Garrels (1977) then used translation assays to show that the β- and γ-actins do not differ from α-actin merely by different degrees of phosphorylation, acetylation, or histidine methylation. Finally, Vandekerckhove and Weber (1978) showed that the heterogeneity in isoelectric points is due to differences in the amino acid sequence of the various actins. Their elegant work demonstrated that the overall amino acid sequence of the different actin isoforms is highly conserved, but that six major isoforms of the protein can be detected by the examination of their amino terminal tryptic peptides. This implies that the actins identified in skeletal muscle (α), cardiac muscle, and fibroblasts (β and γ), are products of different genes, and this conclusion is supported by recent gene dosage experiments that show that actin polypeptides are encoded by a small multigene family in a number of organisms (Kindle and Firtel, 1978; Fryberg *et al.*, 1980; Schwartz and Rothblum, 1980; Engel *et al.*, 1981).

3.2. Myosin

Myosin is the major component of muscle "thick" filaments and has a molecular weight of about 460,000 (Taylor, 1972). When denatured, it yields two "heavy chains" of 200,000 molecular weight and four "light chains" of about 20,000 molecular weight each. The native complex consists of a rod segment, composed of the helically entwined C-terminal ends of the two heavy chains, and two globular heads, each composed of the N-terminus of a heavy chain and two light chains (Lowey *et al.*, 1969; Weeds and Lowey, 1971). The globular heads of the myosin complex are both at the same end of the rod segment and contain the actin-binding sites as well as the myosin-associated ATPase activity (Squire, 1975). The rod segments of many myosin complexes interact to form a 14-nm thick filament from which the globular heads protrude to form cross-bridges with the 6-nm actin thin filaments during contraction.

Chemical and immunological studies of myosin heavy chains isolated from various muscle and nonmuscle tissues have provided considerable evidence for multiple allelic isoforms. A comparison of a methylhistidine-containing sequence from adult rabbit skeletal myosin with the corresponding fragment from cardiac myosin found three of the thirteen amino acids to differ (Huszar, 1972). The same polypeptide from fetal myosin was found to have two amino acid substitutions with respect to the adult molecule. Similarly, Burridge and Bray (1975) used cyanylation fingerprinting (cleavage at cysteine residues) to identify different myosins in five tissues of the adult chicken, namely, skeletal, cardiac, and smooth muscle, brain, and platelets.

In another supporting study, Pollard *et al.* (1974) showed that an antibody against human platelet myosin failed to react with myosin isolated from skeletal muscle. Thus, there is a difference in the types of myosin present in muscle and nonmuscle cells.

The polymorphic nature of myosin is underscored by experiments that found immunological differences between the myosins of fast and slow twitch fibers (Rubinstein *et al.*, 1977; Gauthier *et al.*, 1978). The early studies of this type, however, yielded confusing results. That is, while the myosins of fast and slow twitch fibers were immunologically distinct, both types appeared to be present in embryonic mammalian fast twitch muscle. Also, the enzymatic properties of the myosins isolated from embryonic muscles were often at odds with the type of myosin that appeared, immunologically, to be present. This situation was clarified when investigators discovered that separate embryonic and neonatal myosins exist, which cross-react with the "specific" antibodies made against the adult myosin types. Lowey *et al.* (1981) used extremely specific monoclonal antibodies to study myosins in the developing chicken pectoralis muscle, and found evidence for at least three types of myosin in that tissue: embryonic, neonatal, and adult fast twitch. They also showed that embryonic myosin is replaced by the neonatal form shortly after hatching, and that the adult form does not appear until fourteen weeks later. In another study, Whalen *et al.* (1981) examined the chymotryptic digestion patterns of various rat myosins. They found a distinct embryonic-type myosin heavy chain in both cultured muscle cells and fetal muscle tissue. They also demonstrated that during the first two weeks after birth, a distinct neonatal form appears in developing rat muscle, and that this is followed, on about the nineteenth postnatal day, by the appearance of one of the adult myosin types. The effect that this change in isoforms has on contractility in developing muscle has not yet been determined.

Multiple myosin heavy chain types have also been described in the nematode system (Zengel and Epstein, 1980). The principle isoform is a molecule of 210,000 molecular weight which comprises 90% of the myosin in adult nematodes. The minor isoform is 206,000 molecular weight and has been immunocytochemically localized to the animal's pharyngeal muscle. The major form is located in the body wall musculature of the wild type nematode, but at least one mutant strain (UNC-54) has been found with an abnormal body wall which contains a 203,000 molecular weight myosin heavy chain.

3.3. Myosin Light Chains

Light chains are proteins of 15,000–29,000 molecular weight that associate with the globular heads of heavy chains in the native myosin complex, and that are thought to influence the ATPase activity of these globular heads (Gazith *et al.*, 1970; Weeds and Lowey, 1971). Three different light chains can be isolated from some muscles, while two are found in others. Light chains from a given source are numbered according to decreasing molecular weight. For example, adult chicken breast muscle, which is composed almost entirely of fast twitch fibers, contains three different myosin light chains: LC_1 (25,000 molecular weight), LC_2 (18,000 molecular weight), and LC_3 (16,000 molecular weight). Two light chains are found in the slow twitch latissimus dorsi of the adult chicken, and these migrate similarly to LC_1 and LC_2 from fast twitch muscle when electrophoresed alone on a polyacrylamide gel. However, coelectrophoresis of the light chains from both of these sources shows that all five light chains differ in molecular weight (Weeds, 1976). Consequently, light chains sometimes receive a letter designation, f or s, of the fiber type from which they were isolated.

The light chains found in adult rabbit muscles are similar to those in the chicken. Fast twitch muscles contain three light chains: LC_1 (25,500 molecular weight), LC_2 (17,400 molecular weight), and LC_3 (15,100 molecular weight); slow twitch muscles have two: LC_1 (27,000 molecular weight), and LC_2 (18,000 molecular weight); and cardiac muscle contains two light chains of 27,000 and 19,000 molecular weight (Sarkar *et al.*, 1971, Weeds, 1976).

The appearance of different light chain phenotypes in developing muscle fibers has been extensively studied. Chi *et al.* (1975) found that before fusion the replicating myoblasts express fibroblast-type light chains, while after fusion a switch to muscle type LC_1 and LC_2 occurs. LC_3 does not appear in developing fast twitch fibers until much later—well after hatching in the chicken—and thus the expression of LC_3 is not tightly coupled with that of LC_1 and LC_2. In a similar study, Whalen *et al.* (1978) found an embryonic myosin light chain of the LC_1 type in both fetal muscles and cultured muscle cells of the rat. Synthesis of this protein, named LC_1emb, was found to decrease relative to the adult form after cell fusion (Whalen *et al.*, 1979). These experiments indicate that there are at least six separately regulated sets of light chain genes in the vertebrate genome: (1) nonmuscle or fibroblast type light chains; (2) embryonic LC_1; (3) LC_1 and LC_2 of fast twitch fibers; (4) LC_1 and LC_2 of slow twitch fibers; (5) LC_1 and LC_2 of cardiac fibers; and (6) LC_3 of fast twitch fibers.

Gauthier *et al.* (1978), using immunocytochemical techniques, studied a similar switch in fast twitch fat diaphragm. These investigators found that each embryonic diaphragm fiber contains both fast and slow type light chains, whereas each adult fiber contains only one or the other type. The fast twitch phenotype of the bulk diaphragm results from a predominance of fast twitch fibers in the adult. Stockdale *et al.* (1981), using two-dimensional polyacrylamide gel electrophoresis of light chains from chicken breast muscle, found that early in development the muscle synthesizes both fast and slow type light chains and utilizes both of them in the construction of myofibrils. Later in development, however, synthesis of the slow type chains ceases.

The experiments discussed above demonstrate that light chains are encoded by multigene families in a variety of organisms. The studies that probe the selective expression of family member genes during development will be discussed in later sections.

3.4. Tropomyosin

Tropomyosin molecules lie in the two grooves of double-helical F-actin filaments, and are thought to regulate contraction by reversibly covering the myosin ATPase binding sites on the actin molecules (Ebashi and Nonomura, 1973; Vibert *et al.*, 1972). This binding-site protection is in turn regulated by Ca^{2+}-sensitive troponin molecules bound to each tropomyosin. High Ca^{2+} concentrations favor the binding of myosin to actin and hence contraction. Avian and mammalian muscles contain two nonidentical tropomyosins named α and β. When dissociated with 8 M urea and β-mercaptoethanol, their migration on SDS-polyacrylamide gels reflects a substantial difference in molecular weight, with α migrating at 33,500 and β at 37,000 (Hodges and Smillie, 1970; Cummins and Perry, 1973). Nonmuscular tropomyosins from calf platelet and pancreas, mouse fibroblast, and rabbit brain have lower molecular weights of around 30,000 (Fine and Blitz, 1975; Clarke and Spudich, 1977). All tropomyosins have an acetylated N-terminus; a C-terminal leucine; and a distinctive cyanogen bromide fragment of 17,000 molecular weight, which provides a clear marker for identifying the protein (Sodek *et al.*, 1972).

The developmental appearance of tropomyosin in muscle cultures was quantitatively measured by the isotope dilution technique. Allen *et al.* (1978), found that both α- and β-tropomyosin begin to increase in content as the cells begin to fuse. Whalen *et al.* (1976) also noted that α-tropomyosin appears *de novo* after fusion in primary muscle culture. The increased synthesis of tropomyosin during muscle cell fusion is consistent with increased synthesis of other muscle proteins, including myosin and actin, but there is still some doubt about the levels of tropomyosin in prefusion cells. Holtzer *et al.* (1973) could not detect tropomyosin in prefusion myoblasts or fibroblasts using indirect immunofluorescence with antitropomyosin IgG. Similarly, Devlin and Emerson (1978), using two-dimensional gel electrophoresis of ^{35}S-labeled cellular proteins, also failed to detect tropomyosin in prefusion muscle cells. The results of this study, however, were based solely upon the assumption that two-dimensional gel electrophoresis allows a quantitative detection of all cellular proteins present. This assumption is questionable because isoelectric focusing does not always allow quantitative entry of the proteins into the gel, and because some proteins may be lost during preparation of the isoelectric focusing gel for the second dimension (O'Farrel, 1975). In contrast to the findings cited above, Lazarides (1975) and Chamley-Campbell *et al.* (1977) detected tropomyosin-containing fibers in prefusion chick myoblasts and fibroblasts by immunofluorescent techniques. In addition, Carmon *et al.* (1978), using the rat myoblast cell line L84, showed qualitatively that α- and β-tropomyosin are both present in replicating myoblasts. These investigators were also able to resolve β-tropomyosin into two components, β_1 and β_2, by isoelectric focusing. The β_1 component was present in a small amount at all developmental stages, whereas the more basic β_2 component was found to be specific for differentiated cultures. The existence of α- and β-tropomyosin in prefusion muscle cells is also supported by experiments that use cell-free translation assays to document the presence of tropomyosin mRNA in these cells. The difference between the observations of Devlin, Emerson, and Holtzer, on the one hand, and Lazarides, Chamley-Campbell, Carmon, Newman, and Yaffe, on the other, may simply reflect an increased sensitivity of tropomyosin detection in the experiments of the latter investigators.

The accumulation of tropomyosin in differentiating skeletal muscles *in vivo* has also been described. Roy *et al.* (1976) found that embryonic chicken breast muscle contains both α- and β-tropomyosin, with β as the predominant species. They also found that after hatching, as the fast twitch phenotype appears in the breast muscle, the β form totally disappears and is replaced by α. In contrast, an equimolar ratio of α- and β-tropomyosin was found to be maintained in the slow twitch muscles of the leg. The factors responsible for the switching of tropomyosin isoforms are not known, but the frequency of nerve impulses and the trophic influence of the nerve muscle interaction are the likely candidates at this time.

3.5. Troponin

Troponin is a complex of 74,000–76,000 molecular weight that works with tropomyosin, as described above, to regulate contraction in vertebrate skeletal muscle (Ebashi, 1963). Troponin consists of three subunits: TnC, which contains the calcium-binding site; TnT, which binds the troponin complex to tropomyosin; and TnI, which contains the contraction-inhibiting activity (Perry, 1973). Muscles that utilize myosin-linked rather than actin-linked calcium regulation, contain little or no troponin (Kendrick-Jones *et al.*, 1976; Frederiksen, 1976). For example, in smooth muscle and some invertebrate muscles, troponin is absent, and contraction is regulated by the binding of Ca^{2+} to one of the regulatory myosin light chains. Troponin *per se* has not been found in nonmuscle cells, but the ubiquitous calcium-binding protein calmodulin is probably related evolutionarily to troponin C (Means and Dedman, 1980). Evidence for troponin polymorphism is at present limited to the TnI subunit, which has been found to exist in different forms in cardiac, slow twitch, and fast twitch muscles (Amphlett *et al.*, 1975). There is no information yet on troponin synthesis in cultured myoblasts. This is particularly disappointing because troponin appears to be exclusively expressed in muscle, and would therefore be an excellent marker in developmental studies for the onset of expression of the muscle phenotype.

4. Prerecombinant Study of Contractile Protein Gene Expression

The rapid increase of synthesis and accumulation of contractile proteins during myogenesis, as well as the various cases of isoform switching, have caused much attention to be focused on the mechanisms regulating the expression of these genes. The earliest investigations into the control of muscle protein synthesis were directed at mRNA metabolism. Specifically, the aim was to determine whether mRNA was newly synthesized during certain stages of muscle development, or whether it was stored in the cytoplasm of replicating myoblasts.

One early approach to this problem was to inhibit transcription in developing muscle cultures with agents like actimomycin D. Normally, enzymes such as creatine kinase, myokinase, and glycogen phosphorylase are present at low levels in fresh primary muscle cultures, and then abruptly increase in activity after the cells begin to fuse. Shainberg *et al.* (1971) and Yaffe and Dym (1973) studied the effect of actinomycin D on the accumulation of these enzymes. Enzyme measurements in 3-day cultures indicated that creatine kinase (CK), myokinase, and glycogen phosphorylase activities continued to increase at their normal rates for 8–12 h following inhibition of RNA synthesis. These results suggested that mRNA for an enzyme is stored in a "masked" form before fusion and is translated after fusion. Later, however, Morris *et al.* (1976) examined the effect of actinomycin D on the isozymic change of CK in culture, and found that although there was a significant increase in the BB form of CK, there was no increase in the MB or MM isozymes which normally appear in myotubes. Thus, if the isozymic form of the protein is either undetermined or inappropriate, one must use caution when interpreting enzymatic activity or synthesis data as evidence of new gene expression. This is increasingly true as the evidence for multigene families accumulates for many contractile proteins.

In addition to the lack of isoform identification, many other problems hamper the interpretation of these early actinomycin D experiments (Schwartz, 1973). Foremost is the notorious cytotoxicity of the drug, which causes cell death in prolonged *in vitro* incubations (Honig and Rabinovitz, 1965). This toxicity problem is unavoidable, because when lower and hence less toxic concentrations of the drug are used, mRNA synthesis is not completely blocked. That is, at low concentrations, actinomycin D preferentially blocks ribosomal RNA synthesis, with relative mRNA sparing (Penman *et al.*, 1968). Finally, actinomycin D appears to indirectly affect protein synthesis by reducing the levels of initiation and elongation factors (Singer and Penman, 1972; Schwartz, 1973).

A more direct approach to the study of mRNA metabolism during myogenesis is to assay the mRNA species themselves, by translating mRNAs isolated from staged myogenic cells in a cell-free reticulocyte or wheat germ lysate. This approach, however, is limited in that it only detects translatable messengers and cannot be controlled for differences in translational efficiency (Cereghini *et al.*, 1979). In addition, the quantitative result of an *in vitro* translation experiment can be altered by the presence or absence of message discrimination factors, posttranscriptional mRNA modifications, and inhibitory molecules such as translation-controlling RNA (Heywood and Kennedy, 1976; Kennedy *et al.*, 1978). Thus, measurements of translational activity may be misleading when they are assumed to indicate the physical presence or absence of a particular mRNA species.

The most direct approach to the quantitation of specific mRNA species has become possible with the advent of molecular hybridization techniques and the ability to synthesize DNA copies (cDNA) of specific mRNAs. Using such cDNAs as hybridization probes, it is possible to directly determine the cellular concentrations of specific RNA sequences during development. cDNAs have also been used to establish the number of copies of a specific gene in the haploid genome (Sullivan *et al.*, 1973), and, perhaps most importantly, to insert eukaryotic gene sequences into prokaryotic cells (Rabbitts, 1976). The conditions necessary for the synthesis of full length DNA copies of mRNAs have been well worked out (Monahan, 1976), but it should be mentioned that the purity of the mRNA used as a template in the reaction has an enormous effect on the purity of the resultant cDNA. This is important to keep in mind, because the specificity of one's measurement with hybridization is proportional to the sequence purity of the cDNA probe. Recombinant DNA technology has greatly facilitated the production of pure cDNAs and will be discussed at length later. The next sections review the prerecombinant approaches to the study of muscle gene expression, which not only provided the preliminary data on muscle gene regulation but also generated the pure and semipure cDNAs used later for cloning.

4.1. Actin

Actin is synthesized in appreciable quantities during myogenesis and constitutes about 10% of the myofibrillar proteins in adult muscle. The similar abundance of actin mRNA in this tissue allows the detection of actin translation activity in total cell RNA extracts. This property greatly facilitates the purification of any specific mRNA by allowing the desired molecules to be followed through the various purification steps. As shown in Fig. 1, actin can be detected in cell-free translation products by its ability (1) to form ac-

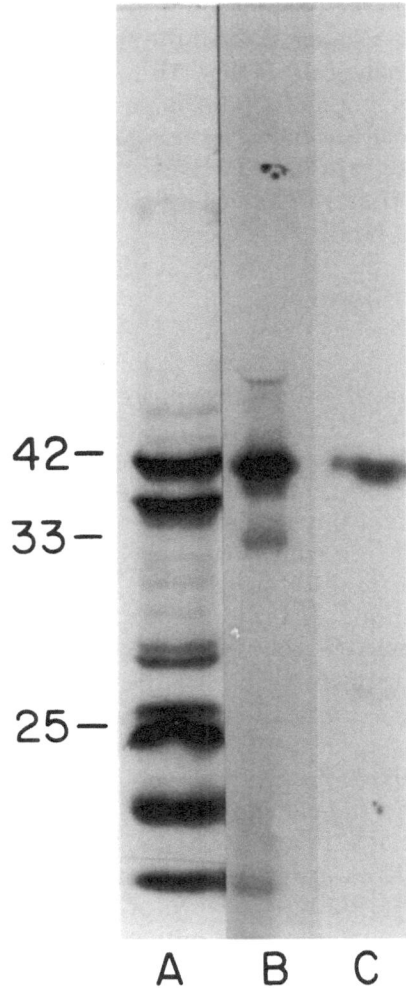

Figure 1. Translation product analysis of actin-containing mRNA. Total poly-A-enriched RNA (1 μg) from chicken breast muscle was translated in an mRNA-dependent reticulocyte lysate, and products were analyzed by: coprecipitation with chicken breast myosin (lane A), coprecipitation with F-actin (lane B), and binding to DNase-agarose (lane C). Molecular weight is expressed in thousands. (From Schwartz and Rothblum, 1980; courtesy of the editors.)

tomyosin (Paterson *et al.*, 1974), (2) to undergo globular to fibrillar polymerization (Bag and Sarkar, 1975), and (3) to bind DNase (Schwartz and Rothblum, 1980). The actomyosin complex is formed by the addition of purified chicken skeletal muscle myosin to the translation lysate in the presence of 0.4 M KCl. The complex is then precipitated by reducing the salt to 0.04 M NaCl, pelleted with centrifugation, and electrophoresed on an SDS-polyacrylamide gel. Since the cell-free translations are performed with a radioactive amino acid such as [^{35}S]methionine in the reaction mixture, the translation products can be identified by autoradiography of the dried gel. Densitometry of the autoradiogram in Fig. 1A shows that only 15% of the radioactivity in the actomyosin pellet is due to actin, with the remainder distributed in bands that comigrate with tropomyosin, the troponins, and the myosin light chains. Thus, the total radioactivity precipitated from a lysate with purified myosin is a relatively poor index of the degree of actin translation. In contrast, when the G-actin in the lysate is induced to form F-actin (with ATP and 0.1 M KCl) and then centrifuged, 80% of the radioactivity in the pellet is associated with actin on an analytical gel (Fig. 1B).

The best method for determining the amount of actin in a cell-free translation reaction, however, exploits one of the protein's most unusual properties: G-actin inhibits DNase activity by tightly binding to the nuclease in a 1:1 fashion (Lazarides and Lindberg, 1974; Mannherz *et al.*, 1975). The specificity of this binding allows an agarose–DNase conjugate to be used to quantitate the actin produced in translation assays. Figure 1C shows that the specificity of the DNase interaction is such that essentially all the radioactivity bound to the DNase–agarose comigrates with actin on the gel. When we used this method to quantitatively follow the actin mRNA purification (Schwartz and Rothblum, 1980), approximately 10% of the cell-free translation products of poly-A RNA from breast muscle were found to be actin, a finding that agrees well with the actin content of intact skeletal muscles. Furthermore, the translation product that was bound by the DNase-agarose was shown to be α-actin by cyanylation fingerprinting and two-dimensional gel electrophoresis.

Cell-free translation assays have been used by several investigators to detect actin mRNA activity in a number of cells and tissues. Storti and Rich (1976) found that cytoplasmic RNA from chick brain directs the synthesis of nonmuscle (β and γ) actin, while cytoplasmic RNA from embryonic chick thigh muscle directs the synthesis of muscle (α) actin. The overall actin mRNA content in myoblast cultures was shown by cell-free translation assays to increase 10- to 12-fold following myoblast fusion (Paterson *et al.*, 1974). Whalen *et al.* (1976) and Hunter and Garrels (1977) used primary calf myoblast cultures and the rat muscle cell line L6 to demonstrate that α-actin mRNA is undetectable by cell-free translation prior to fusion, but becomes the major actin mRNA activity after fusion. These were the first studies to specifically investigate the appearance of α-actin during myogenesis, but they did not directly quantitate the induction of α-actin mRNA or the contribution of nonmuscle actin mRNA species to the total actin mRNA content.

To directly quantitate the α-actin mRNA levels during myogenesis, we synthesized a cDNA hybridization probe using highly purified α-actin mRNA as a template (Schwartz and Rothblum, 1980). We began with the breast muscles of 3-week-old chickens because, in this tissue, actin mRNA comprises a large percentage of the total mRNA, and because α-actin is the only member of the actin multigene family to be expressed. The RNA was obtained by SDS–phenol extraction of total nucleic acids from frozen breast muscle, followed by the selective precipitation of the RNA with 3 M sodium acetate. Purification of the α-actin mRNA was then begun using cell-free translation and DNase–agarose binding to follow the actin mRNA activity through the various fractionations (Table I).

The total RNA was run over an oligo-dT-cellulose column that specifically binds RNA molecules containing poly-A tails. This resulted in a preparation of poly-A-containing mRNA that exhibited two prominent bands when electrophoresed in an agarose gel. One of the bands (band I) contained 135 times the α-actin translational activity of the total RNA. Further fractionation of the poly-A-enriched RNA on sucrose gradients removed RNAs greater than 18 S in size and increased the translational activity threefold. Sepharose 4B chromatography allowed the removal of most RNAs smaller than 15 S. When RNA from this stage was used in a translation assay, almost 95% of the products were either α-actin or its major contaminant, glyceraldehyde-3-phosphate dehydrogenase (GAPDH). Methylmercury hydroxide gel electrophoresis revealed that the molecular weights of actin mRNA (5.2×10^5) and GAPDH-mRNA (4.6×10^5) were too close to allow separation by column chromatography. Consequently, the partially purified RNA was electrophoresed on a disulfide-cross-linked, preparative polyacrylamide gel, which yielded two bands: band I, which contained a 1:1 ratio of actin to GAPDH-mRNA, and band II, which was pure GAPDH-mRNA by translation analysis. Full length cDNAs were made from each band and then hybridized to an excess of the RNA from which they were made. The hybridization of an excess of band II RNA to its cDNA yielded a $R_0t_{1/2}$ of 3×10^{-3}. This agrees with an mRNA sequence complexity of 1390 nucleotides, the length of GAPDH-mRNA, and thus confirms the homogeneity of band II. The same experiment using band I RNA and its cDNA suggested a complexity of 3190

Table 1. Purification of Actin mRNA Measured by Translational Activity

Stage of purification	Total [35S]-Met incorporation (cpm/ μg of RNA)	Radioactivity bound to DNase-agarose (cpm/ μg of RNA)	Percent incorporated into actin	x-fold purification
Total RNA	3,520	642	18	1
Poly-A RNA	218,140	22,230	10	35
Sucrose gradient	350,000	67,410	19	105
Sepharose 4B	395,000	185,650	47	289
Preparative polyacrylamide gels	360,780	184,130	51	210

nucleotides. This is consistent with two pure species of RNA: one of 1390 nucleotides (GAPDH-mRNA) and another of 1800 nucleotides (actin mRNA). Finally, the purified band II RNA was hybridized to band I cDNA and the mixture chromatographed on a hydroxylapatite column. The GAPDH-mRNA hybridized to its cDNA and was retained by the column, allowing highly purified actin cDNA to flow through.

This actin cDNA was then used to measure the concentration of actin mRNA in the starting RNA preparation. A hybridization experiment revealed that actin mRNA constituted only 0.09% of the starting RNA and had thus been purified 1400-fold by the steps described. Hybridization of the cDNA to muscle poly-A RNA, resulted in a $R_0t_{1/2}$ of 5×10^{-2}, indicating that actin mRNA represents 7.6% of the poly-A fraction. By assuming a content of 10^{-7} μg of poly-A RNA per diploid nucleus, actin mRNA was calculated to have 8800 copies for every diploid nucleus in mature chicken muscle.

Most of the work on actin gene expression during myogenesis was done after the actin cDNA was successfully inserted into bacteria and will be discussed in later sections of this chapter. One other interesting experiment, however, was performed with the cDNA made as described above. It was known that 20–30% of the highly abundant poly-A RNA sequences are transcribed from genes with more than five, but less than fifty, copies per haploid genome (Paterson and Bishop, 1977). It was also known that actin was one of the seven prominent species in the highly abundant class of muscle mRNA. These facts, coupled with the protein evidence of multiallelism, led us to design an experiment to measure the number of actin gene sequences in a haploid chicken genome. We used ovalbumin cDNA as a unique gene standard and measured its rate and extent of hybridization to genomic DNA. When these results were compared to those for actin cDNA, we found that there are ten actin sequences in the haploid chicken genome for every one ovalbumin sequence (Schwartz and Rothblum, 1980).

4.2. Myosin Heavy Chains

The levels of myosin heavy chains are low in prefusion myoblasts and early myotubes but increase markedly in postfusion to eventually account for 15% of the total protein synthesized by mature muscle cultures. Early investigators of myosin gene expression in muscle began by looking for myosin translation activity in RNA preparations. Heywood's group (Heywood and Nwagwa, 1968; Morris *et al.*, 1973) showed that in embryonic thigh muscles, a fraction of polyribosomes containing 50 to 60 ribosomes per message was capable of directing the synthesis of myosin heavy chains in a muscle cell lysate. A portion of the radioactive translation product was identified as myosin by (1) its coprecipitation with purified unradioactive myosin and (2) the comigration of the translated myosin with the pure carrier myosin when chromatographed on DEAE cellulose or electrophoresed on analytical polyacrylamide gels. A 26 S poly-A-containing RNA was then isolated from the heavy polysomes of embryonic thigh muscles and shown to promote translation in a reticulocyte lysate. Myosin heavy chains composed 10–40% of the products, and low resolution tryptic mapping failed to show any extra labeled peptides when compared to the digest of authentic myosin.

Heywood's finding was intriguing: an embryonic tissue with barely detectable levels of myosin contained a significant myosin translational activity (Heywood and Kennedy, 1976). This led to the suggestion that myosin heavy chain mRNA (MHC-mRNA) is transcribed prior to the accumulation of organized filaments and is stored in the cell cytoplasm in an inactive form. Strohman *et al.* (1977), however, obtained results that argue against such a storage of message. They isolated a poly-A-containing RNA fraction from muscle cultures that migrated at 26 S on formamide polyacrylamide gels and that was capable of directing the synthesis of myosin heavy chains in a reticulocyte lysate. The product was identified by precipitation with an antibody made to purified adult chicken skeletal muscle myosin. Using this translational assay, they found that the MHC-mRNA activity in cultured myoblasts increases 30-fold after fusion. In addition, the presence of a sequestered MHC-mRNA in prefusion myoblasts was ruled out because the RNA from both stages was isolated from the whole cell.

As we have already pointed out, however, there are a number of pitfalls involved in using cell-free translation assays to quantitate specific mRNAs. With myosin, a great difficulty lies in the large size of the mRNA, which causes it to translate poorly in comparison to smaller messages. Thus, a recurring problem in the early studies was the inability to obtain enough MHC-mRNA from muscle cultures to direct the translation of an unambiguous product. Even though denaturants such as methylmercury hydroxide can relax the secondary structure of large messages and thereby increase their translational activity (Payvar and Schimke, 1979), this approach has never been used to optimize myosin translations. Another problem is that a small species of RNA called translational control RNA (tcRNA) has been shown to inhibit the translation of myosin mRNA (Kennedy *et al.*, 1978). Variable levels of such tcRNA in different developmental stages could vastly affect translational assays, and this possibility precludes the use of such assays for any definitive study on myosin gene expression in myogenesis.

The more reliable method of determining MHC-mRNA levels by cDNA hybridization requires a pure mRNA fraction for the synthesis of the cDNA. In the earliest study, Buckingham *et al.* (1974), reported that a 26 S RNA fraction could be isolated from both pre- and postfusion calf muscle cultures. Even though this RNA species was uncharacterized with respect to the protein product for which it could code, it was taken to be MHC-mRNA. A short, 160-base cDNA was made to this 6000-base, 26 S RNA and then used to determine the content of MHC-mRNA in muscle cultures by RNA-cDNA hybridization. Since the cDNA probe was only 2.6% of the length of the MHC structural gene, and since back-hybridization to the 26 S RNA was not demonstrated, the specificity and complexity of the cDNA probe were completely unknown. Consequently, hybridizations using such a probe cannot be used to support any argument for the presence or absence of MHC-mRNA in muscle cultures.

Other investigators have also made cDNA probes to MHC-mRNA preparations of questionable purity. Robbins and Heywood (1976, 1978) synthesized a cDNA probe which back-hybridized to its template RNA with a $R_0t_{1/2}$ of 5.3. This indicates that MHC sequences make up only about 0.5% of the total cDNA preparation. Patrinou-Georgoulas and John (1977) utilized a similarly questionable 500-base cDNA probe to argue that replicating myoblasts contain no detectable MHC-mRNA, while large amounts are found in postmitotic myoblasts and fused myotubes. Although their findings agree in part with the translational data from Strohman's group, their cDNA probe was made to an impure MHC-mRNA preparation. This was apparent by a 22% hybridization of the probe to ribosomal RNA and by *in situ* hybridization of the probe to the nucleolus (John *et al.*, 1977). Furthermore, hybridization of their probe to both the MHC-mRNA preparation and to RNA from myogenic stages occurred over four log units of R_0t. Such inordinately broad hybridization curves are characteristic of either highly impure cDNA or extensively degraded RNA.

Even though a number of investigators were unable to purify MHC-mRNA to homogeneity, or to synthesize a sufficiently long cDNA with well-characterized properties, their work provided the semipure starting materials for the cloning experiments described in Section 6.6. In addition, some of MHC-mRNA's physical properties could be studied without achieving the purity required for cDNA synthesis. For example, the molecular weight of MHC-mRNA was determined by formamide polyacrylamide gel electrophoresis to be 2.2×10^6. This value is 10–20% higher than the minimum molecular weight of the mRNA predicted by the size of the protein and thus suggested the existence of a nontranslated segment at either the 3' or the 5' end. Mondal *et al.* (1974) then found a 170-nucleotide poly-A tail on the 3' end which is 2.9% of the total length. The length of poly A-tails on MHC-mRNA may be quite variable, however, because, as Benoff and Nadal-Ginard (1980) found, the majority of MHC-mRNA molecules from L_6E_9 myoblasts have short poly-A tails.

4.3. Myosin Light Chains

Myosin light chain mRNA has been found in both primary myoblast cultures and clonal myoblast cell lines, using cell-free translation assays (Devlin and Emerson, 1979; Yablonka and Yaffe, 1976). Interestingly, the light chain translation activity does not appear in the RNA of these cultures until after fusion. For example, two light chains, which comigrate with LC_1 and LC_2 from intact muscles on two-dimensional gels, can be translated from the RNA of postfusion rat muscle cultures, but not from prefusion cultures of the same cell line (Devlin and Emerson, 1979). LC_3 expression is more problematic. LC_3 is present in substantial amounts in adult rat thigh muscle, but is only synthesised to a minor extent in differentiated primary rat muscle cultures. Thus, the *in vitro* expression of a gene may not always reflect its behavior *in vivo* (Yablonka and Yaffe, 1976). However, in calf muscle, LC_3 is not a prominent species *in vivo* or *in vitro,* and, when RNA from postfusion calf muscle cultures is translated, the only light chains present in the products are LC_1 and LC_2 (Whalen *et al.,* 1977).

Arnold and Sidiqui (1979a) used embryonic chicken heart tissue to isolate and purify myosin light chain mRNA. Polysomes were first immunoadsorbed with a specific anti-light chain antibody, and then sized on a sucrose gradient. Finally, poly-A-containing sequences were selected by oligo-dT-cellulose chromatography, and it was found that this myosin light chain mRNA preparation makes up 2% of the total poly-A in embryonic chicken heart. When this RNA was electrophoresed on a denaturing polyacrylamide gel, it was resolved into two species of 1090 and 980 nucleotides in length. When translated *in vitro,* these RNAs directed the synthesis of light chains of 24,000 and 18,000 molecular weight respectively. These purified RNAs were later used for the recombinant DNA cloning of these sequences.

4.4. Tropomyosin

The controversy over the existence of tropomyosin in prefusion myoblasts, as well as the mounting evidence that much of muscle differentiation results from differential expression of multigene family members, prompted several investigators to characterize tropomyosin mRNA in developing muscle. Moss and Schwartz (1981) utilized a cell free translation assay to study α- and β-tropomyosin mRNA during myogenesis. They found that RNA from fibroblast cultures directed the synthesis of both α- and β-tropomyosin, which comigrated with tropomyosins from prefusion myoblasts and postfusion myotubes on both one- and two-dimensional polyacrylamide gels. In addition, the fibroblast tropomyosins could be precipitated with antibodies made against skeletal muscle tropomyosin. Thus, α- and β-tropomyosin appear to be synthesized in the same form throughout myogenesis. Moss and Schwartz (1981) went on to quantitate the tropomyosin mRNAs at various stages of muscle development. They found a $3:1$ ratio of $\alpha:\beta$ tropomyosin translational activity in the RNAs from all stages examined. Furthermore, they found that the tropomyosin translational activity per unit mass of poly-A RNA, increased only onefold from prefusion myoblasts to differentiated myotubes. Thus, the increase in tropomyosin mRNA per cell seen during development seems to simply parallel the increase in total poly-A during the same period, and therefore does not indicate a specific gene activation. As mentioned before, however, translational activity is a rather unreliable index of mRNA concentration, and thus the conclusion that α- and β-tropomyosins are constitutively expressed throughout myogenesis is tentative, pending definitive hybridization experiments.

Carmon *et al.* (1978) also studied tropomyosin mRNA by cell-free translation assays. They showed that RNA from both pre- and postfusion cultures of the rat muscle cell line L84 was capable of directing the synthesis of α- and β-tropomyosin. Moreover, they found that cell-free synthesized β-tropomyosin could be resolved into two species, β_1 and β_2. β_1 is the predominant form translated from RNA from prefusion cultures, while β_2 is the major form in late postfusion RNA.

These studies clearly demonstrate that tropomyosin mRNA is present in fibroblasts and prefusion myoblasts, and thereby support those experiments that reported the isolation of tropomyosins from such cultures. In addition, the apparent switch from β_1 to β_2 during myogenesis suggests that a gene switch is occurring in this multigene family, just as it is in the other families discussed above.

5. Major Recombinant DNA Techniques

Information about the structure and function of the eukaryotic genome has increased dramatically in the past few years. This progress is largely due to the advent of extremely powerful recombinant DNA technology that makes the "impossible" experiments of five years ago realizable today. Techniques have been developed for the isolation of single genes, which constitute barely a millionth part of an animal genome. Amplification of such isolated genes by growth in prokaryotes allows them to be structurally characterized with restriction enzymes (Arber, 1974; Nathans and Smith, 1975) and even to be completely and fairly rapidly sequenced (Maxam and Gilbert, 1977; Sanger *et al.*, 1977). The arrangement of a gene with respect to the rest of the genome can also be studied for developmental and evolutionary alterations in structure and position. One of the principal applications of the new technology is the production of extremely pure hybridization probes that can be used to accurately quantitate specific mRNAs and hence assess the activity of specific genes. Many of the fundamental but still unanswered questions about muscle development are being, and will be, solved using these methods.

5.1. Cloning of Recombinant DNA

A recombinant DNA molecule is one in which a DNA fragment of interest is joined to another DNA fragment, the cloning vector, which is capable of self-replication in a suitable host. Cloning vectors are usually either bacterial plasmids or bacteriophages which can efficiently replicate in bacteria. When a recombinant molecule is introduced into the proper bacterium, the origin of replication on the vector directs the replication of the entire molecule, and thus the DNA fragment of interest is copied along with the vector. The process of inserting recombinant molecules into bacteria is termed *transformation* when the vector is a plasmid, and *transfection* when the vector is a bacteriophage. Cloning of recombinant DNA is the process whereby the replicates of a given recombinant molecule are physically isolated from replicates of other recombinant molecules, so that a pure population of DNA results. Cloning with plasmid vectors is accomplished by streaking a nutrient plate with transformed bacteria so that colonies grow from single parent cells. Each colony can then be isolated and grown on a large scale and will only contain copies of the recombinant molecule present in the parent cell. With phage vectors, lawns of bacteria are transfected with such a low concentration of recombinant bacteriophage, that every plaque on the lawn represents a lytic cascade from a single parent phage. Such plaques can then be used to transfect large amounts of bacteria with identical recombinant molecules.

5.2. Plasmid Vectors

Plasmids are circular, self-replicating DNA molecules found naturally in many bacteria. One plasmid which has gained great popularity as a cloning vector is pBR322. This plasmid, which was derived from pBR313 by Bolivar *et al.* (1977), is "relaxed," which means that it can replicate in the absence of protein synthesis (Novick *et al.,* 1976). Chromosomal DNA in the *E. coli* RR1 host, however, requires protein synthesis for replication and can thus be selectively inhibited. As a result, chloramphenicol treatment of *E. coli* RR1 cultures containing pBR322 can cause up to 3000 copies of the plasmid to accumulate in each bacterium (Clewell, 1972). Several plasmids contain genes that confer antibiotic resistance upon the host, and pBR322 contains two such genes, one for resistance to ampicillin and the other to tetracycline. Five different restriction enzymes cut pBR322 only once: *Eco* RI, *Hind* III, *Bam* HI, *Pst* I, and *Sal* I (Sutcliffe, 1978). These enzymes linearize the circular plasmid without altering the sequences essential for self-replication. Cutting with *Pst* I, however, destroys the gene for ampicillin resistance. This property of the plasmid is extremely useful for detecting recombinant molecules.

Linearized plasmids can be joined to other DNA fragments in a variety of ways. Some restriction nucleases cut the DNA duplex in a staggered fashion, leaving a number of unpaired nucleotides at either end. These so called "sticky ends" are complementary to each other and to other molecules cut with the same enzyme, and, therefore, they will anneal with one another under the proper salt and temperature conditions. Such annealed fragments can then be covalently linked with DNA ligase (Mertz and Davis, 1972; Dugiaczyk *et al.,* 1975). Another linking method is to use terminal deoxynucleotidyl transferase to add a string of deoxycytosine nucleotides to the 3' ends of one fragment, and then to add a string of deoxyguanosine residues to the 3' ends of another fragment (Jackson *et al.,* 1972; Lobban and Kaiser, 1973). Such "tails" will anneal with complementary "tails" to form recombinant molecules and, unlike the "sticky end" method, where any end can anneal with any other, "dG/dC tailing" allows only heterologous fragments to be joined. Strings of dA and dT can also be used for "tailing." The "sticky end" method has one special advantage in that foreign sequences inserted into the vector can later be precisely removed using the restriction enzyme that originally generated the "sticky ends." dA/dT tails contain no specific restriction sites, but can be selectively melted apart with heat and salt, rendering them sensitive to a single-strand-specific nuclease (Hofstetter *et al.,* 1976). Perhaps the best overall method, however, is to use dG/dC tailing to ligate the foreign fragment to a *Pst* I-linearized plasmid (Dugaiczyk, 1975). Such tailing causes each half of the original *Pst* I site to regain its sensitivity to the restriction enzyme. Thus, the foreign DNA can be specifically removed from the vector with *Pst* I after amplification.

Once recombinant plasmids have been successfully formed, they are inserted into a bacterial host for replication. This transformation is usually accomplished by altering the bacterial membrane permeability with divalent calcium, thereby allowing the plasmids to enter the cells (Mandel and Higa, 1970). Once inside, the plasmids are amplified many times by self-replication and can be propagated indefinitely in bacterial cultures.

5.3. The Bacteriophage Vector

The bacteriophage lambda has been developed as a cloning vector for large fragments of DNA. One-third of the phage genome can be replaced with heterologous DNA without affecting the ability of the phage to grow lytically. Several groups of investigators have modified the wild type λ by eliminating certain restriction sites in the essential two-thirds of the genome. Two of these vectors, λ Charon (Blattner *et al.*, 1977) and λ gt WES (Leder *et al.*, 1977) are sensitive to certain restriction nucleases only in their dispensible third, and hence can have foreign DNA added there without altering the lytic ability of the phage. In fact, the formation of viable phage particles is dependent on the insertion of a DNA segment into the vector. Transfection is facilitated with calcium chloride in the same manner as transformation.

5.4. Genomic Libraries

Phage vectors such as λ Charon have been used to clone the entire genomes of certain eukaryotic organisms. A heterogeneous mixture of phage which collectively contains the genome of a given organism is called a gene bank or library for that organism. The construction of genomic libraries for *Xenopus* (Wahli and Dawid, 1980), chicken (Dodgson *et al.*, 1979), rat (Sargent *et al.*, 1979), mouse (Early *et al.*, 1979), and human (Lawn *et al.*, 1978) has followed the procedure that Maniatis *et al.* (1978) first devised for the construction of genomic libraries for *Drosophila*, silkmoth, and rabbit. Briefly, genomic DNA is fragmented by partial digestion with *Hae* III and *Alu* I restriction enzymes, and fragments of 14–22 kilobases (kb) in size are purified on a sucrose gradient. *Eco* R1-type sticky ends are added to the fragments, which are then ligated to linear bacteriophage λ molecules, which also possess sticky ends. The resulting concatameric DNA is mixed with phage packaging extracts and then added to bacterial cultures for transfection. The phage particles that result from the first transfection constitute the gene library, and, barring some preferential loss of certain sequences, contain the entire genome of the given organism.

If an investigator has a pure hybridization probe complementary to a portion of a gene, the procedure of Benton and Davis (1977) can be used to detect the phage particles that contain a genomic fragment bearing the gene. The technique is sensitive enough to detect one plaque containing the fragment of interest, among one million plaques containing other fragments. This method has been used to study the structure of genes in the normal genome, that is, to study the intervening and flanking sequences which are not present in cDNA made from mature mRNA.

5.5. Cloning of cDNA

In Section 4.3, we noted that impure cDNA preparations are not sufficiently specific for use as hybridization probes when quantitating individual mRNA species. This difficulty can be overcome by cloning the sequence; that is (1) inserting sequences from the impure preparation into a cloning vector, (2) inserting the vector into a host, and then (3) isolating the colony or plaque that contains the sequence of interest. Such an isolated clone can then be grown in bulk to provide a source of DNA that is totally free of the contaminating sequences present in the starting material. In this section, we will review the methods used for cloning cDNAs and the screening procedures that allow the desired clones to be identified.

The cDNA that is transcribed *in vitro* from an mRNA preparation with reverse transcriptase is predominantly single stranded. This is desirable if the cDNA is to be used directly as a hybridization probe, but undesirable if one wishes to insert the sequence into a double-stranded vector for cloning. In the latter case, a double-stranded cDNA must be synthesized. To do this, one can exploit an unusual property of reverse transcripts: the 3' hairpin loop. Efstratiadis *et al.* (1976) discovered a small double-stranded sequence (hairpin) at the 3' end of a rabbit globin cDNA which had been synthesized by reverse transcriptase. They also discovered that the small hairpin could serve as a primer sequence for *E. coli* DNA polymerase I, and that this enzyme could therefore synthesize the second DNA strand (once the template RNA was removed by treatment with alkali). Later, Monahan *et al.* (1976) demonstrated that when the mRNA and actinomycin D used in the synthesis of the first strand were removed, the reverse transcriptase itself was capable of synthesizing a complete second strand of the ovalbumin cDNA. A number of other workers (Rabbitts, 1976; Ulrich *et al.*, 1977) independently synthesized double-stranded cDNA's *in vitro*, using either reverse transcriptase, or DNA polymerase I, for the synthesis of the second strand.

Double-stranded cDNAs synthesized in this way contain a nucleotide loop at one end of the duplex. This can be removed with a nuclease that is specific for single strands. The resulting duplex molecule is then ligated to a cloning vector by one of the several methods described above. The cloning vector chosen for cloning cDNAs is almost invariably a plasmid. The bacteriophage vectors are used primarily for cloning large pieces of DNA, such as the 20-kb fragments in a genomic library. Chimaeric plasmids containing cDNA are used to transform bacteria which are subsequently plated at such a low density that each resultant colony is a clone of a single parent cell. One must then discover which colonies contain the cDNA of interest.

5.6. Isolation and Identification of Clones Bearing Specific DNA Fragments

Following transformation and cloning, the bacterial clones fall into four categories: (1) untransformed host bacteria, (2) bacteria transformed with the vector only, (3) bacteria transformed with recombinant plasmids containing unwanted sequences, and (4) bacteria transformed with recombinant plasmids carrying the desired sequence. By using a plasmid carrying antibiotic resistance genes, and a host that is normally sensitive to the antibiotics, clones in the first two categories can be easily separated from clones in the last two, in the following manner (Hamer and Thomas, 1976). Suppose the cDNAs are inserted into the *Pst* I site of pBR322, thereby destroying its ampicillin resistance gene, but leaving its tetracycline resistance gene intact. Also, suppose that *E. coli* RR1, which is sensitive to both antibiotics, is used as the host. Colonies of untransformed host bacteria will fail to grow on both ampicillin- and tetracycline-treated plates, while bacteria transformed with an intact native vector will grow on both. Clones carrying recombinant plasmids will grow on tetracycline but not on ampicillin plates. Thus, replica plating allows the identification of colonies that contain recombinant molecules, reducing the problem to identifying the clones that bear the foreign sequence of interest.

Grunstein and Hogness (1975) developed a method for rapidly screening a large number of recombinant clones for the presence of a given sequence. Replica plates are made from each plate of clones, and the colonies from one are physically transferred to a nitrocellulose filter. The bacteria are lysed on the filter under conditions that cause their DNA to tightly bind to the nitrocellulose. The purest hybridization probe available for a given sequence is then radioactively labeled and hybridized to the whole filter. When the filters are washed and apposed to X-ray film, those clones that contain the sequence of interest appear as dark spots on the autoradiogram. Viable bacteria from a positive clone are obtained from an undisturbed replica plate, and grown up in large quantities for further characterization. It is occasionally more expedient to first screen a set of nonclonal colonies, each made up of cells from a number of clones. When such a colony is found to be positive, the positive clone can be readily isolated by the Grunstein Hogness method.

If a reasonably pure cDNA probe does not exist, recombinant clones can still be screened for the presence of a certain gene with a variety of more laborious methods. These include hybridization-arrested translation (Paterson *et al.,* 1977), translation of clone-purified mRNA (Adams *et al.,* 1979), sequencing (Maxam and Gilbert, 1977; Sanger *et al.,* 1977), and solid phase radioimmunoassay of protein products synthesized in bacteria from cDNAs linked to a lac promoter (Villa-Komaroff *et al.,* 1978). The first three of these alternative screening methods are also used to independently test the identity of clones that are found to be positive by the Grunstein–Hogness procedure. They are therefore discussed in more detail below.

5.7. Characterization of Newly Cloned cDNAs: Restriction Enzyme Mapping

Type II restriction endonucleases cleave DNA at or near specific four-to-seven base pair sequences termed *restriction sites* (Roberts, 1976). A complete restriction digest of a homogenous population of DNA molecules yields a series of fragments that can be separated according to length by electrophoresis on polyacrylamide or agarose gels (Peacock and Dingman, 1968). When stained with ethidium bromide and viewed under UV light, these gels display a series of discrete bands that are characteristic of the digested molecule. By digesting a DNA species with several different enzymes, both singly and in combination, and by carefully determining the sizes of the resulting fragments with gel electrophoresis, it is possible to construct a detailed physical map of the restriction sites in the DNA species. Such a map is much easier to obtain than the nucleotide sequence of a DNA molecule, and, consequently, restriction mapping is often used in lieu of sequencing to characterize new cloned cDNAs.

One of the first steps in the study of a new recombinant plasmid is to determine the length of the foreign DNA "insert." A crude estimate of the insert size can be readily obtained by comparing the electrophoretic mobilities of the native vector and the recombinant plasmid. Since detailed restriction maps of the common vectors have been published (Bolivar *et al.*, 1977; Sutcliffe, 1978), a more accurate measurement of the insert can be made by digesting the plasmid with a restriction enzyme and comparing the resulting fragment sizes with the vector map. For example, suppose a cDNA which is free of *Hha* I restriction sites is inserted into the *Pst* I site of pBR322. The published restriction map of pBR322 indicates that the *Pst* I site is flanked by *Hha* I sites 360 bases apart (Sutcliffe, 1978). That is, *Hha* I digestion of native pBR322 yields many fragments including one that is 360 base pairs in length, containing the *Pst* I site. When the recombinant plasmid is digested with *Hha* I, the 360-bp fragment is no longer seen, because the insert is part of that fragment and causes it to migrate more slowly in a gel. The size of the "new" fragment is determined from its electrophoretic mobility, and the insert size is obtained by subtracting 360 base pairs from that value. If the cDNA had been inserted into the *Pst* I site with dG/dC tailing, so that new *Pst* I sites were generated on either side, the insert could be released from the vector with *Pst* I digestion and its size determined directly. However, if the insert also contained internal *Pst* I sites, more than one fragment would be released, and the total insert size would be determined by addition.

To construct a detailed restriction map of the insert, one begins by excising it from the vector and purifying it with preparative gel electrophoresis. To aid in the identification of fragments in the digestion steps, the insert is then labelled at its 5' termini with ^{32}P (Maxam and Gilbert, 1977). After labeling, the insert is exposed to enzymes, both singly and in combination, which cut it infrequently. The resulting fragment sizes, and the presence or absence of radioactivity, enable many of these restriction sites to be placed unambiguously on the map. Usually, individual fragments from the first round of digestions are redigested with other enzymes to increase the map's detail. When ambiguities remain in the map after the single and double digestions, they can often be resolved with partial digestion techniques.

Once a map of the cloned cDNA is constructed, it can be compared to restriction digest patterns of the original cDNA population to show that the cloned sequence originated from the intended pool of DNA, and also that the insert suffered no major rearrangements during the cloning procedure. A restriction map can also serve to select interesting portions of the insert for subcloning, and, perhaps most importantly, is a prerequisite for sequencing of the molecule.

5.8. DNA Sequencing

Many contractile proteins, including α-actin (Elzinga *et al.*, 1973; Vandekerckhove and Weber, 1978), α- and β-tropomyosin (Sodek *et al.*, 1972), troponin (Collins *et al.*, 1973), and myosin light chains (Frank and Weeds, 1974), have been totally or partially sequenced. For clones that contain a portion of the structural gene of such proteins, the ultimate proof of insert identity is the demonstration that the nucleotide sequence of the insert agrees with the amino acid sequence of the protein.

At present, there are two commonly used methods of DNA sequencing: Maxam and Gilbert's chemical modification and breakdown technique (Maxam and Gilbert, 1977) and Sanger's dideoxynucleotide technique (Sanger *et al.*, 1977). The latter technique can only be used to sequence single-stranded DNA fragments, which, until recently, could only be obtained by difficult strand separation of duplex molecules. As a result, the Maxam and Gilbert method has been the most widely used procedure for DNA sequencing. Interest in the Sanger technique, however, has been renewed by the development of a single-stranded cloning vector from bacteriophage M13 (Messing *et al.*, 1977).

M13mp7 is a single-stranded DNA phage that has a double-stranded replicative intermediate. Foreign DNA fragments can be ligated to linearized replicative intermediates in the same manner as to plasmid vectors. *E. coli* JM 103 cells are then transformed with these recombinant phage, and plated onto special nutrient agar plates. Plaques containing recombinant phage are selected by their lack of color, and then grown on fresh bacterial lawns. Later, single-stranded phage can be recovered from the culture supernatant, and sequenced directly by the Sanger method (Heidecker *et al.*, 1980).

The Maxam and Gilbert technique can be used to sequence both single- and double-stranded DNA molecules; all that is required is a homogenous population of DNA molecules, labeled at only one end with ^{32}P. Such a population is obtained by first labeling the 5' ends of a purified restriction fragment using polynucleotide kinase, and then either separating the two DNA strands on a gel or recutting the fragment with another restriction enzyme. As mentioned above, strand separation is rather difficult to achieve in practice; as a result, the latter method, which yields two sequenceable fragments, is most often used. The principle of this sequencing method is that a homogenous population of DNA molecules will migrate as a single band on a gel, but that if this population is treated with reagents that break the DNA molecules, a variety of radioactive fragments will be produced whose electrophoretic mobility is indicative of the distance between the cleavage site and the radiolabeled end of the molecule. Chemical reactions that cleave DNA at specific nucleotides, coupled with gel electrophoresis of such high resolution that single nucleotide differences in DNA molecules can be resolved, allow the sequence of an end-labeled fragment to be read right off the autoradiogram of a gel. That is, each lane on a gel is loaded with DNA that has been treated with reagents that cleave at specific nucleotides. The presence of a band at a given position on such a gel, indicates a cleavage by that lane's reagent at the corresponding nucleotide position in the DNA fragment. In this manner, as many as 300 nucleotides of a given fragment can be determined with a single gel.

Even though the nucleotide sequence of an insert is the ultimate identification, the effort involved in sequencing a fragment with the techniques described above renders them impractical for screening a large number of clones. When the available cDNA is so impure that the Grunstein–Hogness method yields a large number of false positives, the following two techniques can be used for rescreening these clones.

5.9. Hybridization-Arrested Translation (HART)

The HART technique was developed by Paterson *et al.* (1977) as a method of rapidly screening clones derived from impure cDNA. It is based upon the fact that when an mRNA is hybridized to a complementary DNA molecule, the message becomes incapable of directing the synthesis of a complete protein in a cell-free translation assay. Thus, clones are scored as positive for a given sequence when their DNA inhibits the translation of the protein encoded by that sequence. Only a fraction of a microgram of plasmid DNA is needed to inhibit the translation of an mRNA species, and thus this assay can be used to screen a large number of clones. The drawback to this method is that the significant results are negative, and hence must be regarded with the same skepticism with which one regards all proofs by negative result. An additional problem is that nearly full length cDNAs are necessary to completely inhibit translation, and smaller cDNAs can often yield questionable results. Nevertheless, hybridization-arrested translation successfully identified the clones containing the gene sequences of chicken albumin (Gordon *et al.*, 1978) and mouse heavy and light immunoglobulin chains (Schibler *et al.*, 1978). In addition, we used the HART procedure in the initial screening of our actin cDNA clones. As an example of this type of data, Fig. 2 illustrates

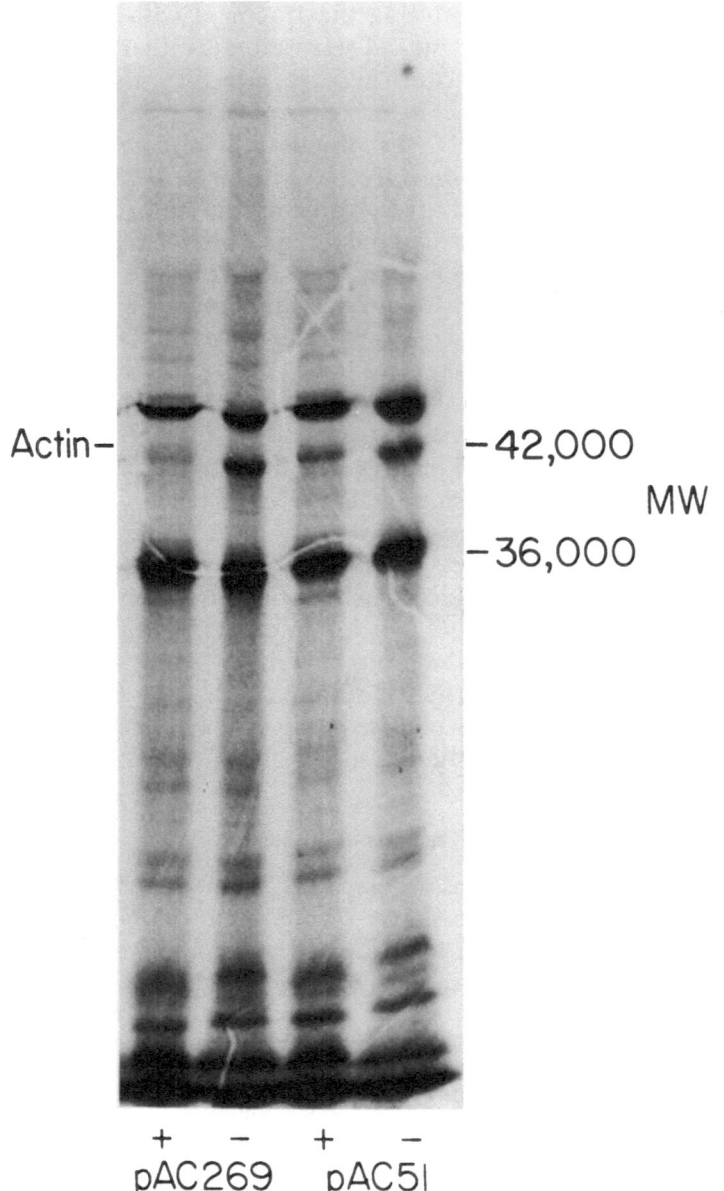

Figure 2.   Electrophoresis of the translation products of breast muscle RNA. Translations run in lanes marked (+) were inhibited by the addition of 0.45 μg of actin-cDNA-containing plasmid DNA, while those run in lanes marked (−) were controls without plasmid DNA. Note that the plasmid with the smaller insert (pAC51, 410-bp insert) inhibits the translation of actin (molecular weight 42,000) to a lesser degree than does the plasmid bearing a full-length cDNA (pAC269, 1350-bp insert).

the selective inhibition of actin translation by two cloned α-actin cDNAs, pAC269 and pAC51.

5.10. *Translation of Cloned-DNA-Purified mRNA*

Another widely used method for screening a large number of clones that are derived from an impure cDNA population is to use the plasmid DNA from a clone to purify its complementary message by affinity chromatography. Plasmid DNA from each clone is either covalently linked through a diazo linkage to finely divided cellulose (Noyes and Stark, 1975) or denatured and baked onto nitrocellulose fibers (Kafatos *et al.*, 1979). This DNA cellulose preparation is then incubated with poly-A RNA from a tissue in which the sequence of interest is actively expressed. The complementary mRNA species hybridizes to the recombinant DNA linked to the cellulose, and this preparation is then extensively washed to remove nonspecifically bound RNA. The specific mRNA is then eluted and translated in a cell-free lysate, yielding the protein product that is encoded by the DNA fragment borne on the cloned plasmid. This method has two major advantages over HART. First, a significant result is positive, that is, involves the synthesis of a specific protein; and second, the translated protein can be analyzed by a variety of techniques to verify its identity. One- and two-dimensional gel electrophoresis (O'Farrel, 1975), protein fingerprinting (Jacobson *et al.*, 1973), and precipitation with highly specific antibodies are commonly used for such protein identification.

We used this DNA cellulose assay to confirm the identity of a clone, pAC269, containing α-actin cDNA (Schwartz *et al.*, 1980). pAC269 DNA was covalently linked to diazobenzyloxymethyl cellulose by the method of Noyes and Stark (1975). Poly-A RNA from chick breast muscle was then hybridized to an excess of the pAc269-cellulose. This preparation was washed, and the bound RNA (5–10% of the input) was eluted and translated in a cell-free lysate. When the translation product was electrophoresed on an SDS–polyacrylamide gel, we found it to consist of a single species, with the same electrophoretic mobility as actin purified from chicken skeletal muscle (Fig. 3, lanes A and B). Further comparison of the translation product with purified actin was made by fragmentation of the proteins at their cysteine residues with the cyanylating reagent dithiodinitrocyanobenzoic acid (TNB–CN). The resulting fragments were electrophoresed on a 12% SDS–polyacrylamide gel, and when an autoradiogram of this gel was compared to a stained gel of a TNB–CN digest of purified actin, the patterns were found to be identical in fragment number, position, and relative density (Fig. 3, lanes C and D).

The translation product was also positively identified as actin by comigration with purified actin on the high resolution two-dimensional gel system of O'Farrell (Fig. 4). Finally, the translation product was found to bind to DNase I agarose, which implies that the actin synthesized *in vitro* is functional. These findings alone are unequivocal evidence that pAC269 DNA is complementary to actin mRNA. However, nucleic acid sequencing of the pAC269 insert is

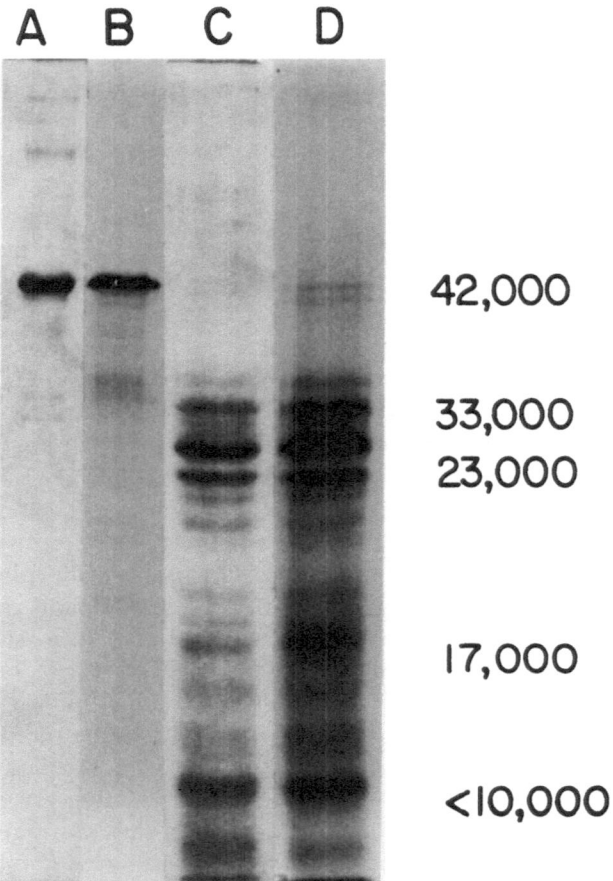

Figure 3.   Analysis of products translated from RNA selected by hybridization to pAC269 DNA. Total poly-A-containing RNA from muscle was hybridized to pAC269 cellulose, eluted, translated in a reticulocyte lysate in the presence of [^{35}S]methionine, and electrophoresed on a 12% SDS–polyacrylamide slab gel. Lane A contains authentic actin, purified from chicken skeletal muscle and stained with Coomassie blue. Lane B is an autoradiogram of the translation product electrophoresed on the same gel as lane A. Lane C contains a cyanylation digest of authentic actin stained with Coomassie blue. Lane D is an autoradiogram of the translation product which was cyanylation digested and run on the same gel as Lane C. (From Schwartz *et al.*, 1980; courtesy of the editors.)

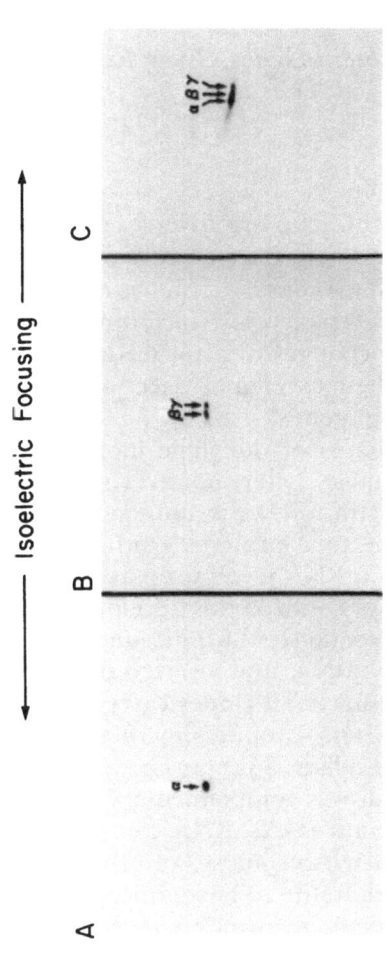

Figure 4. Two-dimensional gel electrophoresis of products translated from RNA selected by hybridization to pAC269 DNA-cellulose. Panels A–C are autoradiograms of the two dimensional gels. Panel A shows the single translation product of RNA selected by pAC269 from breast muscle poly-A RNA. Panel B shows the two translation products of RNA selected by pAC269 from brain poly-A RNA. Panel C shows the translation products of RNA selected by pAC269 from a mixture of muscle and brain poly-A RNA. (From Schwartz *et al.*, 1980; courtesy of the editors.)

now being done, and the preliminary results completely support our earlier conclusion.

It should be emphasized that such extensive analysis of a new clone and the product translated from its complementary mRNA is absolutely required before the cloned DNA is used as a hybridization probe to quantitate specific mRNA levels. A questionable probe renders such hybridization experiments meaningless.

6. Study of Contractile Protein Genes Using Recombinant DNA Technology

6.1. Invertebrate Actin

Although the functions of actin are diverse, ranging from muscle contraction to chromosome movement, the structure of actin is highly conserved within and between species. This conservation has made it possible to identify actin in a number of lower eukaryotes, including slime mold, yeast, sea urchin, and *Drosophila*. The changes in actin mRNA levels during development in these species make them excellent subjects for the study of the regulation and organization of actin genes.

In early germinating spores of the slime mold, *Dictyostelium discoidium*, actin mRNA is barely detectable. Later, however, in vegetative cells, actin is actively expressed; that is, actin mRNA is abundant, and actin protein comprises a large fraction of the total proteins synthesized. Finally, in the late stages of development, actin mRNA levels drop as the actin genes are deactivated. Firtel and his co-workers utilized this developmental system to pioneer the cloning of actin gene sequences. Kindle and Firtel (1978) randomly sheared *Dictyostelium* genomic DNA, and inserted the fragments into plasmid pMB9 using dA/dT tailing. Bacterial clones carrying recombinant plasmids were identified by the Grunstein–Hogness method, using ^{32}P pulse-labeled poly-A RNA from vegetative cells as a probe. One positive clone was found to contain a plasmid, M6, which was complementary to about 1% of the newly synthesized RNA from vegetative cells. RNA complementary to M6 was isolated by hybridization to M6DNA cellulose, and then translated *in vitro*. The two translation products were found to be identical, by both two-dimensional gel electrophoresis and tryptic fingerprinting, to the two major forms of actin which they had isolated from intact cells.

It should be pointed out that Vanderckhove and Weber (1980), who isolated and sequenced actin from two different slime molds, found only one actin species in both. It has been proposed that the heterogeneity found by Kindle and Firtel is due to incomplete acetylation of the major form (Rubenstein and Deuchler, 1979). Alternatively, some modification may have occurred during Kindle and Firtel's sample preparation that caused an artifactual heterogeneity, but it is also possible that the other investigators simply

Table 2. Current Contractile Protein Gene Clones

Gene	Species	Designation (cDNA/genomic)	Reference
Actin	Slime mold	M6 (genomic)	Kindle and Firtel, 1978
		pd2 actin 2 (genomic)	Kindle and Firtel, 1978
		pcDd actin A_1 (cDNA)	McKeown et al., 1978
		pcDd actin B_1 (cDNA)	McKeown et al., 1978
		pdDd actin 3, 5, 7 (genomics)	Firtel et al., 1979
Actin	Yeast	pYact I (genomic)	Ng and Abelson, 1980
		pYA 208 (genomic)	Gallwitz and Sures, 1980
Actin	*Drosophila*	K1 (genomic)	Tobin et al., 1980
		λ DmA_2 (genomic)	Fryberg et al., 1980
		λ $DmA_{1,3,4,5,6}$ (genomics)	Fryberg et al., 1981
Actin	Sea urchin	pSA38 (cDNA)	Merlino et al., 1980
		pSpG17 (genomic)	Durica et al., 1980
α-Actin	Chicken	pAC269 (cDNA)	Schwartz et al., 1980
		pAC51 (cDNA)	Schwartz et al., 1980
		pAC23 (cDNA)	Schwartz et al., 1980
		λAC5 (genomic)	Zimmer and Schwartz, 1981
		pα-actin 1 (cDNA)	Ordahl et al., 1980
		pα-actin 2 (cDNA)	Ordahl et al., 1980
		λGα-actin A1 (genomic)	Ordahl et al., 1980
β-Actin	Chicken	pA_1 (cDNA)	Cleveland et al., 1980
		pA_2 (cDNA)	Cleveland et al., 1980
γ-Actin	Chicken	pA_3 (cDNA)	Cleveland et al., 1980
α-Actin	Mouse	Plasmid 91 (cDNA)	Minty et al., 1980
		Unnamed λ clones (genomics)	Minty et al., 1981
α-Actin	Rat	p106 (cDNA)	Katcoff et al., 1980
		p254 (cDNA)	Katcoff et al., 1980
		p649 (cDNA)	Katcoff et al., 1980
		p749 (cDNA)	Katcoff et al., 1980
Actin	Human	λHRL21,23,24,25,34,35,45, 83,84 (genomics)	Engel et al., 1981
Myosin HC	Nematode	SG24 (genomic)	Karn, 1981
Myosin HC fast	Chicken		
		p251 (cDNA)	Umeda, et al., 1981
		p110 (cDNA)	Umeda et al., 1981
Myosin HC Embryonic	Rat		
		p82 (cDNA)	Nudel et al., 1980
		$CMH_{1,3,6}$ (genomics)	Nudel et al., 1980
		pMHC25 (cDNA)	Medford et al., 1980
		287 $A_{3,6}$ (genomics)	Wydro et al., 1981
Cardiac		287 A_1 (genomic)	Wydro et al., 1981
Fast		287 A_2 (genomic)	Wydro et al., 1981
Myosin LC Cardiac LC_2	Chicken	pM10 (cDNA)	Arnold and Siddiqui, 1979
Myosin LC_1	Quail	pc127 (cDNA)	Hastings and Emerson, 1981
Myosin LC	Rat		
LC_{2f}		103	Katcoff et al., 1980
$LC_{1,3}$		Unnamed (cDNAs)	Garfinkel and Nadal-Ginard, 1981
Myosin $LC_{1,3}$	Mouse	Unnamed (cDNAs)	Caravati et al., 1981
Tropomyosin	*Drosophila*	λDm85 (genomic)	Storti et al., 1981
Tropomyosin	Chicken	pTm 10 (cDNA)	MacLeod, 1981

failed to detect a protein heterogeneity that really exists in slime mold. The sequence data presented below suggests that this may be the case.

In any event, the clone-purified mRNA from *Dictyostelium* contained two distinct actin messages, 1.25 and 1.35 kb in length. This finding coupled with the protein heterogeneity, led Firtel to suggest the existence of at least two different functional genes in the slime mold. To support such an idea, two different experiments were performed to estimate the number of actin genes in the slime mold genome. In one, M6 DNA was hybridized to an excess of genomic DNA; while in the other, genomic DNA was cut with various restriction enzymes, electrophoresed on an agarose gel, and then transferred to nitrocellulose paper where it was hybridized to an M6 DNA probe. Both of these experiments indicated that the actin-coding region is repeated 17 times in each slime mold haploid genome (Kindle and Firtel, 1978; McKeown *et al.*, 1978). The latter type of experiment is often called a "Southern blot" because the method was developed by E. M. Southern (1975).

In 1979, Firtel *et al.* looked for heterogeneity in the actin multigene family directly, by sequencing the 5' ends of six actin genomic clones and one actin cDNA clone from *Dictyostelium*. They found that three of the genomic clones, and the cDNA clone, had subtle nucleotide differences but still coded for the same protein. Two other genomic clones, including M6, also differed slightly in nucleotide sequence, but, more importantly, encoded a protein that differed by two amino acids from the protein encoded by the first group of clones. The remaining genomic clone is probably a pseudogene, because it was found to have a large number of sequence differences in comparison to the other clones studied.

Recent experiments of McKeowen and Firtel (1981) have examined the developmental expression of five of the 17 *Dictyostelium* actin genes, and found that the different genes are expressed at different times during development. One gene, which encodes the 1.25-kb mRNA, begins to be expressed early in development and remains highly active throughout the multicellular stage. Another gene encodes a 1.35-kb mRNA, which is expressed in vegetative cells but not in earlier stages, while a third gene encodes a different 1.35-kb mRNA, which is expressed at low levels throughout development. The gene that specifies the actin containing two amino acid differences is also expressed at a low level throughout development; that is, its transcripts never make up more than 5% of the total actin mRNA population. Finally, the pseudogene described above never produces any detectable transcripts. Thus, just as in vertebrate muscle, the actin genes of *Dictyostelium* are independently expressed during development.

In another lower eukaryote, *Saccharamyces cervisiae* (baker's yeast), actin plays a structural role in the cytoskeleton, and a functional, contractile role in chromosomal condensation and budding. There is only one form of actin protein in yeast, and this is consistent with the presence of a single actin gene. The fact that actin is highly conserved allows a cloned actin sequence from one species to be used to screen clones from another. Therefore, Ng and

Abelson (1980) used one of Firtel's actin clones from *Dictyostelium* to screen a set of bacterial colonies containing pBR322 plasmids with yeast DNA inserts. One of the colonies was positive, but contained only a part of the yeast actin gene. Consequently, another yeast DNA clone bank was screened. This second bank consisted of 5000 bacterial colonies bearing pBR322 plasmids with yeast *Eco* R1 fragment inserts. One of the colonies in this collection harbored a plasmid which hybridized to the actin cDNA probe when the bank was screened by the Grunstein–Hogness method. This plasmid, pYact I, had a 3.8-kb yeast DNA insert that contained the entire yeast actin gene. The nucleotide sequence of this gene and its flanking regions was then determined. This revealed a 309 nucleotide intervening sequence within the fourth codon of the coding region, that is, near the 5′ end of the gene. In addition, a typical polymerase recognition site, a poly-A addition site, and standard splice sequences at the ends of the intervening sequence were found. Thus the yeast actin gene has all the major characteristics of a "typical" eukaryotic gene, and may serve as a simple system with which to study the mechanism of RNA splicing.

The sea urchin is another species which has been used to study actin genes and gene expression, and a dramatic increase in actin synthesis is seen during development in this organism. Actin is barely detectable in the morula, but is one of the most abundant proteins, only a short time later, in the blastula (Merlino *et al.*, 1980). A parallel increase in translatable actin mRNA is seen during the same period. To properly study the regulation of actin gene expression during sea urchin development, a pure, sea urchin actin hybridization probe was necessary. Merlino *et al.* (1980) made double-stranded cDNA from an actin mRNA-enriched population of sea urchin poly-A RNA. This cDNA was inserted into the *Bam* HI site of pBR322, which was then used to transform *E. coli*. After preliminary screening by antibiotic sensitivity and hybridization to the original (impure) cDNA population, clones were screened by using plasmid DNA to purify an mRNA that could then be translated for identification. In this way, one clone (pSA38) was found to selectively hybridize to an mRNA species that encoded a protein of 43,000 molecular weight. Peptide mapping and two-dimensional gel electrophoresis showed this protein to be actin. Restriction endonuclease and heteroduplex mapping of pSA38 showed the cDNA insert to be 1.5 kb long, that is, long enough that it would likely contain a large portion of the actin-coding region. The plasmid was then used as a hybridization probe to confirm that actin mRNA levels rise abruptly during sea urchin development, and are thus likely to be responsible for the increase in actin protein.

Durica *et al.* (1980) used sea urchin genomic DNA to construct two different actin clones that were identified by Grunstein–Hogness hybridization with a Drosophila actin probe. Both clones were shown to specifically hybridize to actin mRNA from a complex RNA population. DNA sequencing of one of the cloned inserts (pspG17) showed that it contains a single intervening sequence of at least 200 nucleotides, which begins immediately after the 121st codon of the coding sequence. In addition, Southern blot hybridization experiments and DNA-excess solution hybridizations suggest that there are 5–20 actin genes per haploid sea urchin genome.

Drosophila is another invertebrate organism used to study actin, and three forms of the protein, actins I, II and III, have been documented in this system (Storti *et al.*, 1978). Actin I is the most acidic form, and is found only in muscle tissues and differentiated muscle cultures. The presence of an actin multigene family with a muscle-specific member is very similar to the situation found in vertebrate organisms. A number of investigators have taken advantage of *Drosophila*'s small genome (Laird, 1971), well-characterized genetic map (Lindsley and Grell, 1968), and potential for *in situ* hybridization to its polytene chromosomes (Gall and Pardue, 1971) to study the organization and expression of actin genes.

Actin natural genes from *Drosophila* were isolated for study by using actin-specific hybridization probes from *Dictyostelium* and embryonic chicken muscle to screen *Drosophila* genomic libraries. Tobin *et al.* (1980) isolated a 7.2-kb chromosomal DNA fragment, K1, which contains nucleotide sequences complementary to the actin gene probes. *Drosophila* mRNA, selected by hybridization to immobilized K1 DNA, was translated *in vitro*, and yielded products which comigrated with actins I, II and III on two-dimensional electrophoretic gels. Similarly, Davidson and his co-workers (Fryberg *et al.*, 1980) isolated a 17.5-kb fragment of *Drosophila* DNA that bears actin sequences. This fragment was hybridized to total poly-A RNA, and the DNA/RNA hybrid molecules were examined with the electron microscope. The resulting "R-loop" structures indicated that the actin natural gene consists of a 70–170 nucleotide coding region at the 5' end, a 1.65-kb intervening sequence, and a 1.55-kb coding region at the 3' end. A portion of the actin gene from this genomic clone (λDMA_2) was then used as a hybridization probe to characterize the actin mRNA species in *Drosophila*. Poly-A RNA was fractionated according to size by electrophoresis on a methylmercury hydroxide agarose gel. Then, preserving the two-dimensional orientation of the RNA species in the gel, the RNA molecules were transferred and covalently linked to a sheet of diazobenzyl-oxymethyl-cellulose which was apposed to the gel. Hybridization of a radioactive *Drosophila* actin clone fragment to the paper, revealed three distinct actin mRNA species of 2.3, 1.95, and 1.65 kb in length. This agrees well with the presence of three actin protein species in *Drosophila*. This method of identifying RNA species by blot hybridization was developed by Alwine *et al.* (1977), and is often called "Northern blotting" because of its similarity to the Southern procedure described above.

Fryberg *et al.* (1980) also used the actin portion of λDMA_2 as a hybridization probe. Their Southern blots of restriction digested genomic DNA from *Drosophila* indicated six actin genes in the haploid genome. In a later study, the same group isolated and characterized these six closely related *Drosophila* actin genes (Fryberg *et al.*, 1981); and the structural comparison of these genes resulted in several unexpected findings. First, the position of the intervening sequence varied from gene to gene. For example, clone DMA_4 contains an actin gene with an intervening sequence in codon 13, while clone DMA$_6$ has its intervening sequence in codon 304. In addition, nucleic acid sequencing of selected portions of these genes shows that they all encode amino acid sequences that are similar to nonmuscle actin from vertebrates. Thus, although *Drosophila* contains three expressed isoforms, with one specific for its muscles, none of the *Drosophila* actin genes directs the synthesis of a protein similar to actin from vertebrate skeletal muscle.

As mentioned above, one of the advantages of the *Drosophila* system is the ability to use *in situ* hybridization to locate specific gene sequences on the individual polytene chromosomes. When the actin genes were examined in this manner, they were found to exist in six dispersed locations (Fryberg *et al.*, 1980). One of these regions consists of two tightly linked actin bands at position 88F of the polytene chromosomes (Tobin *et al.*, 1980). This dispersal of actin genes in the genome will have to be taken into account by any model which seeks to explain the mechanism of isoform switching.

6.2. Chicken Actin

Vertebrate organisms each contain several different actin isoforms, some of which are developmentally regulated and tissue specific. For example, embryonic chicken muscle contains two forms of actin, termed β and γ, which are totally replaced by the more acidic α-actin in mature muscle. To study this induction of α-actin in developing chicken muscle, we constructed an α-actin cDNA clone (Schwartz *et al.*, 1980). Double-stranded cDNAs were transcribed from an RNA population greatly enriched for actin mRNA. These were inserted into the *Pst* I restriction site of pBR322 and the resulting hybrids were used to transform *E. coli* RR1. The clones were screened by the Grunstein–Hogness method using an actin-enriched cDNA as a probe, and positive colonies were confirmed by translation of clone-purified RNA. The translation product which was directed by RNA complementary to one of the cloned plasmids, pAC269, was a protein of 42,000 molecular weight identified as α-actin by electrophoretic mobility, DNase I affinity, and cyanylation peptide mapping. In a similar fashion, Ordahl *et al.* (1980) identified two chicken actin cDNA clones. One, pα-actin 1, was found to contain sequences complementary to the mostly noncoding 3′ end of α-actin mRNA, while the other, pα-actin 2, contained a greater portion of the coding region.

We measured the insert of our pAC269 by restriction enzyme mapping and heteroduplex analysis and found it to be 1400 ± 50 nucleotides in length. Since α-actin contains 374 amino acids and thus requires a coding region of only 1122 nucleotides (Elzinga *et al.*, 1973), the insert of pAC269 is long enough to contain the entire coding region as well as some noncoding sequences.

The sequence homology between the various actin genes in the chicken multigene family is such that an α-actin-specific probe like pAC269 will cross-hybridize with the non-α-actin genes (Schwartz *et al.*, 1980). As a result, pAC269 could be used to measure the length of the actin mRNAs present in various chicken tissues by using the "Northern blot" method of Alwine *et al.* (1977). That is, unfractionated RNA from chick muscle, gizzard, and brain, was electrophoresed on a denaturing agarose gel, and then transferred and covalently linked to DBM paper. ^{32}P-labeled pAC269 DNA was subsequently hybridized to the RNA linked to the paper. Autoradiography of the blot revealed that actin mRNAs from the three tissues differ in molecular weight (Fig. 5). Skeletal muscle, which expresses only α-actin, contains a single major actin mRNA species of 1575 nucleotides. This agrees well with the size of the actin mRNA found by Hunter and Garrels (1977) in the rat L6 myoblast cell line. Furthermore, if one subtracts the posttranscriptionally appended 100-nucleotide poly-A tail from the total mRNA size of 1575, one can calculate that the insert of pAC269 contains 95% of the α-actin mRNA sequence. β- and γ-actin are the isoforms found in the cytoplasm of brain and most other nonmuscle cells of the chicken. The mRNA molecules that encode these isoforms are both 2200 nucleotides in length (Fig. 5, slot B), and thus contain about 1000 nucleotides of noncoding sequences, much more than α-actin mRNA. Gizzard contains two actin mRNA species (Fig. 5, slot C), the 2200-nucleotide β-actin mRNA and a 1240-nucleotide message that encodes smooth muscle actin.

The differences in noncoding regions among the actin mRNA species can be exploited for their individual quantitation in two ways. First, the size differences can be visualized with a Northern blot, as in Fig. 5, allowing one to qualitatively demonstrate the accumulation of certain actin mRNAs in a given tissue. Secondly, the 3' noncoding regions of the various mRNAs are highly divergent in nucleotide sequence, and, as a result, it is possible to construct hybridization probes from these 3' sequences that will only recognize a single member of the actin gene family. For example, Ordahl *et al.* (1980) showed that their 3' specific α-actin clone (pα-actin 1) does not efficiently hybridize to β- or γ-actin mRNA. Also, Cleveland *et al.* (1980) showed that a cloned 3' noncoding region of β-actin mRNA does not hybridize to γ-actin mRNA.

In addition to the differences in the noncoding regions of the actin mRNAs, some differences also exist in the coding regions, which give rise to the variation in amino acid sequence among the actin isoforms. Several investigators have shown that the number of nucleotide differences between two generally complementary nucleic acid molecules can be determined by observing the temperature at which the two molecules can be melted apart (Leder *et al.*, 1973; Ullman and McCarthy, 1973). The temperature at which two totally complementary molecules melt is used as a reference, and then the decrease in melting temperature caused by the incomplete base pairing between the two nonidentical molecules, called ΔTm, is determined by subtraction. ΔTm is directly related to the percentage of base mispairings between two hybridizing molecules. The proportionality constant has been estimated to be 1.6–3.4°C of ΔTm, for each 1% sequence divergence (Benz *et al.*, 1977).

We melted pAC269-DNA/βγ-actin-mRNA hybrids, and measured a drop in TM of 10–13°C compared to the melting temperature of the pure double-stranded pAC269 insert. This indicates a minimum of 2.9% and a maximum

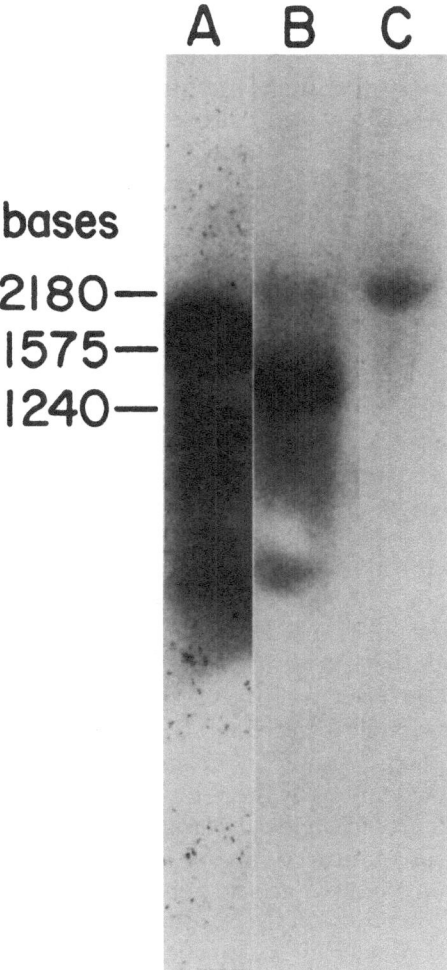

Figure 5. Detection of actin mRNA species with pAC269 DNA. Poly-A RNA samples from chicken breast muscle (5 µg, lane A), gizzard (20 µg, lane B) and embryonic brain (20 µg, lane C), were separated by electrophoresis under denaturing conditions on 2% agarose gels and transferred to DBM paper. Nick-translated ^{32}P-labeled pAC269 *Hha*I fragments containing the actin cDNA insert were hybridized to the RNA bound to the DBM paper, which was then autoradiographed with X-ray film. (From Schwartz and Rothblum, 1981; courtesy of the editors.)

of 8.1% sequence divergence between skeletal muscle and nonmuscle actin species. This divergence estimate is consistent with the 25 amino acid differences that Vandekerckhove and Weber (1978) found between the muscle and nonmuscle actins.

The ΔTm of greater than 10°C between muscle and nonmuscle actin sequences allowed us to develop stringent hybridization conditions in which pAC269 can anneal completely to α-actin mRNA, but does not hybridize at all to β- or γ-actin mRNA. These stringent hybridization conditions enabled us to separately quantitate the total actin mRNA content, and the α-actin-specific mRNA content, of cultured muscle cells at various developmental stages (Schwartz and Rothblum, 1981). As seen in Fig. 6, β- and γ-actin mRNA make up the vast majority of the actin mRNA present in prefusion myoblasts, about 2000 molecules per cell, while α-actin is present at the near-background level of 130 molecules per cell. When the myoblasts fuse however, the α-actin mRNA concentration per nucleus begins to increase dramatically, and within 24 hours reaches a level 270-fold greater than in prefusion cells. Conversely, β- and γ-actin mRNAs begin to disappear after fusion, and by the end of myotube formation, are undetectable by our techniques (≤ 50 molecules per nucleus). It should be stressed that this switch in actin mRNA types occurs in the absence of cell division and cell death, and thus represents a change in gene expression of healthy myotubes. Figure 6 also shows that α-actin mRNA reaches its peak concentration of 36,000 molecules per nucleus after the muscle cells have been in culture for about 100 h. Then, during the 100 h following the peak, the α-actin mRNA level falls to the maintenance concentration of 8000 molecules per nucleus, which is normally found in differentiated myotubes. This relative deinduction of α-actin mRNA during the second 100 h of culture is also occurring in the absence of cell death.

The induction of α-actin mRNA, and the disappearance of the β- and γ-species, could be effected by regulation at the transcriptional level. However, recent experiments in our laboratory (Zimmer and Schwartz, 1982) have provided evidence for the selective amplification of α-actin genes during chicken myogenesis. Nuclear DNA samples from adult tissues, as well as from embryonic and cultured myoblast stages, were digested with *Eco* RI and electrophoresed on agarose gels. The DNA fragments were transferred to nitrocellulose paper by the Southern procedure, and hybridized with a ^{32}P-labeled, cloned α-actin cDNA probe. Autoradiography of the blot revealed the same seven bands in every lane, plus two additional bands in lanes containing DNA from fusing cells. That is, prefusion cells and adult tissue appear to have at least seven actin genes in the genome, while fusing cells, both embryonic and cultured, appear to have at least nine. The *Eco* R1 fragments which bear the additional genes are 5.4 kb and 7.5 kb in length, and restriction mapping indicates that they contain α-actin genes. Moreover, the density of their corresponding bands on the Southern is much greater than that of the seven common bands. Densitometry indicates that the 5.4-kb species is 85-fold more concentrated than any of the seven common species in fusing muscle cells; and, that the 7.5-kb species is eight-fold more concentrated than the common species. Thus, the α-actin gene sequence appears to be amplified at least 90-fold during myoblast fusion. Interestingly, the amplified DNA species disappear from myotubes within 24 h of attaining their peak concentration. In addition, the gain and loss of the extra α-actin sequences closely parallels the rise and fall of the α-actin mRNA concentration. Thus, amplified DNA may be an intermediate in the developmental expression of α-actin.

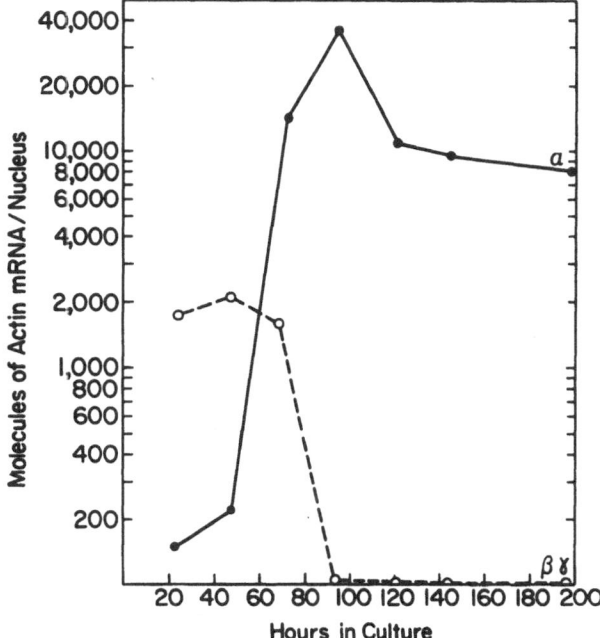

Figure 6. Switching of actin mRNA species during myogenesis. The content of α-actin mRNA (●) and β- and γ-actin mRNA (○) was derived from liquid hybridizations with pAC269 DNA under both highly stringent and nonstringent conditions. (From Schwartz and Rothblum, 1981; courtesy of the editors.)

The chromosomal arrangement of structural genes, and the characteristics of their intervening and flanking sequences can also be studied with recombinant DNA techniques. As mentioned before, recent technical advances using the λ bacteriophage have made it possible to construct and screen complete genomic libraries from higher eukaryotes (Maniatis *et al.*, 1978). Dodgson *et al.* (1979) used the λ Charon 4A vector to prepare a complete library of 14- to 22-kb DNA fragments from the chicken genome. Ordahl *et al.* (1980) screened this library and found 12 different clones containing actin sequences. Three of these were shown to contain α-actin sequences by their hybridization to the α-specific cDNA clone, pα-actin 1. Similarly, Schwartz and Rothblum (1981), isolated 26 clones containing actin sequences from the chicken library. The clones containing α-actin sequences were then identified by hybridization with ^{32}P-labeled α-actin mRNA, purified with pAC269 DNA-cellulose. Under stringent hybridization conditions, two of the 26 clones (λAC5, λAC25) hybridized to the α-specific probe, and thus contain α-actin natural gene sequences.

The initial restriction mapping of these α-actin natural gene clones shows the transcribed portion of the gene to be approximately 2700 nucleotides in length, including at least three small intervening sequences. We are now in the process of rescreening the 24 non-α-actin clones in search of the β-, γ-, cardiac, and smooth muscle actin natural genes.

6.3. Rodent Actin

Another vertebrate organism in which actin has been well studied is the rat. Yaffe and coworkers (Katcoff *et al.*, 1980) first prepared cDNAs to two RNA populations: one from proliferating rat myoblasts and another from differentiated myotubes. A comparison of the two cDNAs showed that differentiated myotubes contain some abundant messages that are absent or very rare in proliferating myoblasts. In cell-free translation assays, these abundant messages directed the synthesis of muscle-specific proteins, including actin. As a result, cDNA complementary to this RNA was used for cloning. Four of the resulting clones (named 106, 254, 649, and 749) hybridized strongly with cDNA enriched for actin sequences, but not with cDNA enriched for myosin light chain sequences. Positive translation assays and nucleic acid sequencing confirmed the presence of authentic actin sequences in these clones.

Plasmid p749 was found to specify a highly conserved region of the actin structural gene, codons 171 to 360, and thus it cross-hybridizes to RNA isolated from skeletal muscle, smooth muscle, cardiac muscle, and brain. Plasmid p106 on the other hand, specifies the last actin codon and the entire, divergent, 3' noncoding region of skeletal muscle actin mRNA. As a result, p106 is a specific probe for the rat α-actin sequences, while p749 is a general probe for the entire actin gene family (Shani *et al.*, 1981a).

Similarly, Minty *et al.* (1981a), constructed and isolated a mouse cDNA clone bearing actin sequences. Nucleic acid sequencing of a portion of their clone 91 showed it to contain α actin sequences. When removed from the plasmid and electrophoresed on an agarose gel, the insert was found to be 1350 nucleotides in length. Restriction mapping showed it to contain 90% of the α-actin coding region, plus 300 nucleotides of 3' noncoding sequences.

The rodent actin clones described above were then used as hybridization probes in "Northern blot" experiments to characterize the actin mRNA species present in various adult tissues and myogenic stages of rodents. These blots showed that RNA from adult skeletal muscle contains a single 1650 nucleotide actin mRNA (α), while nonmuscle RNA contains two actin messages (β and γ) that are 2100 nucleotides long (Minty *et al.*, 1981a; Katcoff *et al.*, 1980; Shani *et al.*, 1981b). The 500-nucleotide difference between muscle and nonmuscle actin messengers allowed them to be easily distinguished on the "Northerns." As a result, the relative concentrations of the mRNA species could be semiquantitatively determined by densitometry of the autoradiograms. In this manner, Shani *et al.* (1981b), found that poly-A-containing α-actin mRNA, increases 13-fold during rat myoblast differentiation, and that nonmuscle actin mRNAs decrease in concentration during the later stages of rat muscle development.

Southern blots of restriction digested genomic DNA have been used to estimate the number of actin genes in the haploid genomes of both mouse and rat. Minty *et al.* (1981a) used both 5' and 3'-specific actin gene probes to show that there are between 15 and 18 actin genes in the haploid genome of the mouse. In addition, when these blots were stringently washed, they indicated that the muscle specific actin isoform is encoded by a single gene. Likewise, Shani *et al.* (1981b) found evidence for 16 actin genes in the haploid genome of the rat.

Gene libraries of both rat and mouse have been successfully screened with actin cDNA probes. In the mouse, a series of actin natural gene clones were isolated, and one was identified as the α-actin natural gene by restriction mapping and R-loop experiments. This gene was found to have at least two, and probably more, small (< 150 bp) intervening sequences, similar to those observed in the chicken (Caravati *et al.*, 1981). In the rat, both α- and β-actin natural gene clones have been isolated. The former has at least six, and the latter at least five, small intervening sequences (Nudel *et al.*, 1981).

6.4. Human Actin

Engel *et al.* (1981) used cloned actin sequences from *Drosophila* and chicken to isolate 12 actin clones from a human DNA library. The presence of actin sequences in all of these clones was confirmed by translating the mRNA that hybridized to them, and then analyzing the translation product by one- and two-dimensional gel electrophoresis and peptide fingerprinting. Restriction mapping of these 12 clones revealed that each one contains a different actin gene. These same investigators then used the human actin clones to probe Southern blots of *Eco* RI digested human genomic DNA. They found that 25–30 of the *Eco* RI fragments contain actin sequences, and thus there are probably 25–30 separate actin genes in the human genome.

6.5. Actin Gene Organization

Most eukaryotic natural genes that have been studied have been found to contain regions of noncoding DNA interspersed in the coding regions (reviewed by Dawid and Wahli, 1979). The placement of these intervening sequences in globin genes (Maniatis *et al.*, 1980), vitellogenin genes (Wahli *et al.*, 1980), and ovalbumin genes (Royal *et al.*, 1979) is exactly the same for all members of the given gene family. With globin genes, this property extends across large evolutionary distances. For example, globin coding regions from *Xenopus* and humans are interrupted at exactly the same positions (Patient *et al.*, 1980), despite the fact that their ancestors diverged 350 millions years ago. The placement of intervening sequences in the actin gene family is much more divergent than the multigene families mentioned above. First, none of the several *Dictyostelium* actin genes sequenced by Firtel *et al.* (1979) appears to have an intervening sequence in its protein coding region. Second, the yeast actin gene is split at codon 4 (Gallwitz and Sures, 1980; Ng and Abelson, 1980), while one of the sea urchin actin genes is split between codons 121 and 122 (Durica *et al.*, 1980). Finally, in *Drosophila*, there is an intervening sequence in codon 13 of one actin gene, and one in codon 307 of two others (Fryberg *et al.*, 1981). Protein sequencing of the products from all these genes revealed a great similarity between these primitive actins and the β- and γ-actins of vertebrates, despite the differences in intervening sequence placement. As mentioned above, preliminary examination of the α-actin natural gene from the chicken discovered at least three small intervening sequences within the coding region. Even though the functional or evolutionary roles of intervening sequences are not understood at present, a great deal may be learned by comparing the α-actin natural gene to both cardiac and smooth muscle actin natural genes, since these three genes probably evolved more recently than their β and γ homologues.

6.6. Myosin Heavy Chains

Myosin heavy chains are the major contractile protein of mature muscle, comprising 20% of the total protein present (Nadal-Ginard, 1978), and, as a result, the developmental regulation of the MHC gene has been extensively studied. Early experiments provided circumstantial evidence that MHC expression is regulated at both the transcriptional and posttranscriptional level (John *et al.*, 1977; Buckingham *et al.*, 1974; Robbins and Heywood, 1976); however, as mentioned above, these studies often suffered from impure cDNA probes and inadequate identification of the MHC isoforms being measured. The latter problem has become especially important now that multiple MHC isoforms have been shown to appear and disappear during muscle development (Wahlen *et al.*, 1981; Lowey *et al.*, 1981).

Fortunately, our understanding of MHC gene structure and regulation has been advanced considerably with the recent construction of several cDNA and genomic (natural gene) clones bearing MHC sequences. Medford *et al.* (1980) made double-stranded cDNA against a high-molecular-weight polysome fraction from well-differentiated L_6E_9 rat myotubes. This cDNA was inserted into a plasmid vector and used to transform bacteria. One cloned plasmid, pMHC25, was shown to contain MHC sequences by (1) hybrid-arrested translation (HART), (2) translation of clone-purified MRNA and subsequent characterization of translation products, and (3) "Northern blot" analysis using pMHC25 as a probe to show that it hybridizes to an appropriately large (7100 nucleotide) mRNA population. Restriction mapping of pMHC25 showed that its MHC-cDNA insert is 680 base pairs long, or about 10% of the total MHC-mRNA length.

Similarly, Nudel *et al.* (1980) cloned and isolated a recombinant plasmid containing MHC-cDNA sequences which were transcribed from the mRNA of rat thigh muscle. The identity of this plasmid, p82, was confirmed in the same manner as pMHC25 above, except that the HART experiment was omitted. Umeda *et al.* (1981) constructed recombinant clones using cDNA made against the MHC-mRNA of 14-day embryonic chick skeletal muscles. Their positive clones were verified by comparing the nucleotide sequence of the inserts with the known C-terminal amino acid sequence of the myosin heavy chain. In this way, two different MHC clones were identified which differ in their 3' noncoding regions.

The construction of the MHC-cDNA clones described above has encouraged new, and more definitive studies of MHC gene expression. A good system for such a study is the cultured rat myoblast, L_6E_9, in which MHC synthesis is rapidly and transiently induced during development. That is, MHC protein synthesis increases in these cells until the concentration of heavy chains approaches that of differentiated muscle; then, MHC synthesis falls off to a basal maintenance level. To determine whether this induction of MHC protein synthesis is due to a similar induction of MHC-mRNA, Benoff and Nadal-Ginard (1980) used the pMHC25 sequence to quantitate the developmental changes in the MHC-mRNA concentration of L_6E_9 cells. They found that the rate of MHC synthesis at various stages of myogenesis is closely paralleled by the cytoplasmic concentrations of MHC-mRNA. In addition, they found that MHC-mRNA begins to accumulate at least 36 h before fusion, and that it increases from 200 molecules per cell in the replicating myoblast to 50,000 molecules per cell in six-day-old cultures. When these investigators measured the MHC-mRNA concentration in 10-day-old cultures, however, they found a concentration of only 3000 molecules per cell, which is consistent with the deinduction of MHC protein synthesis observed earlier. The induction and deinduction of MHC-mRNA in cultured rat muscle cells is very similar to the induction and deinduction of α-actin mRNA that has been observed in chicken muscle cultures (Schwartz and Rothblum, 1981). Importantly, these hybridization experiments clearly demonstrate that neither MHC or α-actin mRNA is stored as an untranslatable mRNP in prefusion myoblasts. In fact, recent evidence from Wydro *et al.* (1981) supports the idea that the developmental increase in MHC-mRNA concentration results from increased nuclear transcription.

The availability of MHC-cDNA clones has also facilitated the study of the haploid copy number and sequence organization of MHC natural genes. Wydro *et al.* (1981) used the pMHC25 clone as a probe in a "Northern" blot experiment to show that several MHC-mRNA species exist in the rat. When this cDNA was used to probe a Southern blot of genomic DNA from the rat, they were able to estimate the haploid MHC gene copy number to be at least 8. In reasonable agreement, Nudel *et al.* (1981) used their plasmid 82 in similar experiment and estimated the haploid copy number to be at least 5.

Wydro *et al.* (1981) used the pMHC25 sequence to isolate four MHC natural gene clones from a rat genomic library. Two of the clones, $287A_1$ and $287A_4$, were found to encode the adult cardiac muscle and adult skeletal muscle MHC isoforms, respectively; while the other two clones ($287A_3$, $287A_6$) appeared to specify embryonic forms of MHC. Electron microscopic R-loop analysis of embryonic MHC clone, $287A_3$, showed the coding sequence of this gene to be interrupted by at least 17 intervening sequences, ranging in length from less than 100 bp to more than 1000 bp. In addition, sequence analysis of $287A_3$ showed its coding regions to be 100% homologous to the cDNA clone pMHC25; thus, it appears that $287A_3$ contains the MHC gene that is expressed in the L_6E_9 cell line.

One of the most outstanding accomplishments in the myosin heavy chain field, and indeed, in molecular biology as a whole, is the recent work of

Jonathan Karn and Andrew McLachlan (Karn, 1981) in Sidney Brenner's laboratory. These investigators used the M13 single-stranded phage and Sanger's dideoxynucleotide sequencing technique to obtain the entire nucleic acid sequence of a myosin heavy chain gene. As a result, the entire 6000-amino-acid sequence of one MHC isoform is known, a feat that would have required years of work using standard peptide sequencing techniques. This amino acid sequence has allowed Karn and McLachlan to predict many of the structural features of the protein. For example, preliminary computer modeling suggests that surface charges can be optimally neutralized when two adjacent myosin heavy chain molecules are aligned in parallel but shifted by 98 residues. Such a shift would result in a 14.6-nm periodicity in a thick filament. Such detailed structural information about the MHC protein will be invaluable in studying many characteristics of myosin and myofilaments. For example, the molecular mechanism of myofilament assembly may be probed by comparing the MHC sequence of myofilament assembly mutants with this recently obtained wild-type sequence.

6.7. Myosin Light Chains

As discussed previously, the variety of light chain mRNA species that can be isolated from the tissues of a single organism, such as the chicken, demonstrate that the light chains are encoded by a multigene family. This gene family must be regulated by a very precise mechanism, because expression of the various isoforms is highly specific, both in time and tissue distribution. Moreover, this precise regulation must extend among, as well as within, gene families, because given sets of light chains are usually only expressed in conjunction with a specific heavy chain isoform. It is likely that such coordinate gene regulation is based upon some common gene element, such as a shared nucleotide sequence in the genes' flanking regions. The search for such common elements has been advanced considerably by the recent cloning of several myosin light chain genes.

In 1979, Arnold and Siddiqui developed a method of purifying light chain mRNA from embryonic chicken hearts. Their highly purified LC_2 mRNA fraction was then used as a template for the synthesis of a full-length double-stranded cDNA, which was subsequently inserted into pBR322 and used to transform bacteria. One of the resulting clones, pML10, was found by the HART procedure to partially inhibit the *in vitro* translation of LC_2. However, other methods of clone analysis have not been used to verify the identify of pML10, and, as a result, the presence of LC_2 sequences in this clone is questionable at this point.

Katcoff *et al.* (1980) reported the cloning of myosin LC_2 cDNA which had been made against a purified LC_2 mRNA from the rat. The resulting clone, 103, was verified by translating clone-purified mRNA and analyzing the translation product on a two-dimensional gel. In addition, "Northern" blots showed that the cloned sequence hybridizes to a single mRNA species, and that this species is only present in muscle tissue. This latter finding suggests that there is little sequence homology between LC_2 and the other light chains. Shani *et al.* (1981b) then used clone 103 as a hybridization probe to measure the LC_2 mRNA concentration at various stages of rat myogenesis. They found that prefusion myoblasts contain a very low level of LC_2 mRNA, but that after fusion, the concentration increases 50-fold. In addition, they found that the bulk of LC_2 mRNA contains poly-A sequences and is associated with active polysomes. This is another strong indication that the rate of synthesis of muscle-specific proteins closely follows the concentration of the given mRNAs, and that activation of stored inactive messengers is not a major mechanism in muscle gene regulation.

Very recently, the isolation of several other light chain cDNA clones has been reported. Hastings and Emerson (1981) described five clones that encode light chains from fast twitch (LC_1, LC_2 and LC_3) and slow twitch (LC_1 and LC_2) quail muscles. Similarly, cDNA clones specifying the fast twitch LC_1 and LC_3 of mouse (Caravatti *et al.*, 1981) and rat (Garfinkel and Nadal-Ginard, 1981) have been reported. With these cDNAs it will be possible to determine whether the various light chain mRNAs are transcribed by completely separate natural genes, or whether two or more mRNAs are transcribed from the same natural gene, as is the case for α-amylase mRNAs (Young *et al.*, 1981).

6.8. Tropomyosin

Two groups of investigators have reported the isolation of clones containing tropomyosin gene sequences. MacLeod (1981) constructed recombinant plasmids using cDNAs made against chicken skeletal muscle α-tropomyosin mRNA. Positive clones were initially selected by the Grunstein–Hogness method with the original cDNA as a probe. These were then retested by translating clone-purified mRNA *in vitro* and analyzing the products on two-dimensional polyacrylamide gels. One cloned plasmid, pTm10, hybridized to mRNA, which directed the synthesis of a protein that comigrates with α-tropomyosin. This plasmid contains a 1150-bp insert and hybridizes on "Northern" blots to a single mRNA species (slightly smaller than 18 S) that is only present in skeletal muscle. This indicates that the sequences of cardiac and β-tropomyosin genes must differ extensively from that of α-tropomyosin, despite the many physical and biological similarities among the proteins they encode. When the stringency of hybridization to the "Northern" blots was reduced, pTm10 hybridized weakly to an mRNA species from heart, which may prove to be cardiac tropomyosin mRNA. The great specificity of pTm10 for α-tropomyosin mRNA will be very useful in studying the appearance and accumulation of this message species during myogenesis. In particular, it will help determine whether skeletal α-tropomyosin mRNA is constitutively expressed in fibroblasts and replicating myoblasts of the chicken.

Another group, Storti *et al.* (1981), has recently cloned a segment of *Drosophila* genomic DNA that contains tropomyosin genes. They began by synthesizing a ^{32}P-labeled cDNA probe using mRNA from differentiated

Drosophila myotubes as a template. This probe was used to screen a *Drosophila* genomic library and 300 positive clones were isolated. These clones were then further characterized by translating the mRNAs that hybridized to their inserts and subsequently electrophoresing the translation products on two-dimensional gels. In this manner, one clone (λ Dm85) was found to contain gene sequences encoding a protein (or proteins) which comigrates with the tropomyosins. The translation product was also shown to contain tropomyosin by partial chymotryptic digestion. ^{3}H-labeled cRNA was synthesized from λ Dm85 and then hybridized *in situ* to a polytene chromosome preparation. In this experiment the probe hybridized to a single site at position 88F 2–5, which suggests that tropomyosin genes might be closely linked on a single chromosome in *Drosophila*. This notion was later confirmed when Storti found three tandem tropomyosin genes in the genomic insert of λ Dm85. "Northern" blots show that four different messengers (1.4, 1.5, 2.3, and 2.8 kb in length) are transcribed from these genes. The 2.3- and 2.8-kb mRNAs may be transcribed from the same gene, and a developmental study found that these messages are present during the first 5 h of myogenesis in culture, and thus probably encode nonmuscle tropomyosins. By early fusion, these mRNA species have disappeared and are replaced by the 1.5-kb message. The 1.4-kb message appears at late fusion. The study of the tandemly linked genes on λ Dm85 may give us great insight into the mechanism of selective gene expression within multigene families.

7. Directions of Future Research

Future research in muscle development will focus largely on the regulation of expression of multigene families. The rapid accumulation of contractile protein mRNAs during myogenesis, which we have reviewed in this chapter, strongly suggests that these genes are regulated at the level of transcription. However, it is certainly possible that posttranscriptional regulation may also play a role. This matter will be clarified in the future by pulse-chase experiments which will measure transcription rates and decay rates of specific messages.

One of the most challenging problems that remain in this field is to discover the molecular basis for transcriptional regulation. Evidence from several sources suggests that transcriptional regulation may be based upon conformational changes in chromatin. That is, pancreatic DNase I has been shown to preferentially digest a variety of transcriptionally active genes, including globin (Weintraub and Groudine, 1976), ovalbumin (Garel and Axel, 1976); heat-shock genes (Beissman *et al.*, 1977), ribosomal genes (Mathis and Gorovosky, 1977), and integrated viral genes (Panet and Cedar, 1977; Flint and Weintraub, 1977). This preferential digestion is thought to result from a conformational change in chromatin which is necessary, but not sufficient, for transcriptional activity, because genes that are expressed initially, but later inactivated, retain their DNase I sensitivity (Weintraub and Groudine, 1976;

Beismann *et al.*, 1977). Moreover, a gene's sensitivity to DNase I appears to be independent of the rate at which its transcribed (Garel *et al.*, 1977), and the sensitivity may extend well beyond the transcriptional borders of the gene, into the flanking regions (Benyajati and Worcel, 1976). Thus, while the chromatin changes detected by DNase I digestion may be a major factor in transcriptional regulation, at least one other regulatory element is required to explain the lack of expression of some DNase I-sensitive gene sequences.

A good candidate for such an additional regulatory mechanism is the methylation of cytosine residues in DNA. Riggs (1975) showed that the inactivated X chromosome of female mammals is hypermethylated. When restriction enzymes sensitive to methylation were discovered, several investigators showed that hypomethylation of the following genes is correlated with their expression: ribosomal genes (Bird, 1978), β-globin (Van der Ploeg and Flavell, 1980), ovalbumin and conalbumin (Mandel and Chambon, 1979), and viral genes (Vardiman *et al.*, 1980); Desroseirs *et al.*, 1979). Even though this is circumstantial evidence, one interesting experiment supports the idea that lack of methylation leads to gene expression. Jones and Taylor (1980) added two blockers of DNA methylation (5-aza-2'-deoxycytidine and 5-fluro-2'-deoxycytidine) to cultured mouse embryo cells, and found that the drugs induced terminal differentiation in the cells. Similarly, when these agents were added to cells of the C3H101/2 cell line, the cells developed contractile elements characteristic of striated muscle. Now that a variety of contractile protein gene sequences have been cloned, experiments can be performed to reveal the roles that chromatin conformation and DNA methylation play in the differential expression of members of multigene families.

Another regulator of transcription that will be extensively studied in the future is the nucleotide sequence itself. That is, certain sequences in the gene undoubtedly influence the binding and activity of the RNA polymerase and other associated proteins. The development of an *in vitro* transcription assay for RNA polymerase III genes (Wu, 1978; Bogenhagen *et al.*, 1980) has allowed much to be learned about the gene sequences necessary for transcription. Similarly, the recent progress in *in vitro* transcription of polymerase II genes (Luse and Roeder, 1980; Wasylyk *et al.*, 1980) will allow many aspects of sequence-mediated regulation of muscle genes to be determined in the future.

As methods are worked out for inserting eukaryotic genes into eukaryotic cells, the ultimate muscle gene regulation experiments will become possible. That is, cloned sequences will be altered and inserted into myoblasts where the effect of the alteration can be observed under otherwise normal conditions. In this way, sequences responsible for isoform switching may be found, and the function of various protein domains in sarcomere assembly could be investigated.

Finally, while transcriptional gene regulation is to be expected for many genes, one must not forget the possibility of gene control at the DNA level. The transposons of yeast (Cameron *et al.*, 1979; Calos and Miller, 1980), rearrangement of immunoglobulin genes (Tonegawa *et al.*, 1978), amplification of ribosomal (Brown and Dawid, 1968), chorion (Spradling and Ma-howald, 1980), and actin (Zimmer and Schwartz, 1982) genes are all re-minders of the diversity of regulatory mechanisms that may be involved in any developmental process, including myogenesis.

ACKNOWLEDGMENTS. We thank Drs. Mary Stone and Jeffrey Thurston for their help in preparing this manuscript. This work was supported by U.S. Public Health Service Grant NS-15050 and by a grant from the Muscular Dystrophy Association of America. R. J. S. is a recipient of a U.S. Public Health Service Research Career Development Award.

References

Abbott, J., Schiltz, J., Dienstman, S., and Holtzer, J., 1974, The phenotypic complexity of myo-genic clones, *Proc. Natl. Acad. Sci. USA* **71**:1506.

Adams, S. L., Alwine, J. C., deCrombrugghe, B., and Pastan, I., 1979, Use of recombinant plasmids to characterize collagen RNAs in normal and transformed chick embryo fibroblasts, *J. Biol. Chem.* **254**:4935.

Allen, R. E., Stromer, M. H., Goll, D. E., and Robson, R. M., 1978, Synthesis of tropomyosin in cultures of differentiating muscle cells, *J. Cell Biol.* **76**:98.

Alwine, J. C., Kemp, D. J., and Stark, G. R., 1977, Method for detection of specific RNAs in agarose gels by transfer to diazobenzyloxymethyl-paper and hybridization with DNA probes, *Proc. Natl. Acad. Sci. USA* **74**:5350.

Amphlett, G. W., Perry, S. V., Syska, H., Brown, M. D., and Urbura, G., 1975, Cross innervation and the regulatory protein system of rabbit soleus muscle, *Nature (London)* **257**:602.

Arber, W., 1974, DNA modification and restriction, *Progr. Nuc. Acid Res.* **14**:1.

Arnold, H. H., and Siddiqui, M. A. Q., 1979a, Control of embryonic development: isolation and purification of chick heart myosin light chain mRNA and quantitation with a cDNA probe, *Biochemistry* **18**:647.

Arnold, H. H., Siddiqui, M. A. Q., 1979b, Cloning of synthetic deoxyribonucleic acid that codes for embryonic cardiac myosin light-chain polypeptide, *Biochemistry* **18**:5641.

Bag, J., and Sarkar, S., 1975, Cytoplasmic nonpolysomal messenger ribonucleoprotein containing actin messenger RNA in chicken embryonic muscles, *Biochemistry* **14**:3800.

Beissmann, H., Wadsworth, S., Levy, B., and McCarthy, B. J., 1977, Correlation of structural changes in chromatin with transcription in the *Drosophila* heat-shock response, *Cold Spring Harbor Symp. Quant. Biol.* **42**:829.

Benoff, S., and Nadal-Ginard, B., 1980, Transient induction of poly(A) short myosin heavy chain messenger RNA during terminal differentiation of L_6E_9 myoblasts, *J. Mol. Biol.* **140**:283.

Benton, W. D., and Davis, R. W., 1977, Screening λgt recombinant clones by hybridizing to single plaques *in situ, Science* **196**:180.

Benyajati, C., and Worcel, A., 1976, Isolation, characterization, and structure of the folded interphase genome of Drosophila melanogaster, *Cell* **9**:393.

Benz, E. J., Giest, C. E., Steggels, A. W., Barker, J. E., and Nienhus, A. W., 1977, Hemoglobin switching in sheep and goats: preparation and characterization of complementary DNAs specific for the α, β, and γ-globin mRNAs of sheep, *J. Biol. Chem.* **252**:1908.

Berl, S., Puszkin, S., and Nicklas, W. J., 1973, Actomyosin-like protein in brain, *Science* **179**:441.

Bird, A. D., 1978, Use of restriction enzymes to study eukaryotic DNA methylation: II. The symmetry of methylated sites supports semi-conservative copying of the methylation pattern, *J. Mol. Biol.* **118**:49.

Bischoff, R., and Holtzer, H., 1970, Inhibition of myoblast fusion after one round of DNA synthesis in 5-bromodeoxyuridine, *J. Cell Biol.* **44**:134.

Blattner, F. R., Williams, B. G., Blechl, A. E., Denniston-Thompson, K., Faber, H. E., Furlong, L. A., Grunwald, D. J., Kiefer, D. O., Moore, D. D., Schumm, J. W., Sheldon, E. L., and Smithies, O., 1977, Charon phages: safer derivatives of bacteriophage lambda for DNA cloning, *Science* **196**:161.

Bogenhagen, D. F., Sakonju, S., and Brown, D., 1980, A control region in the center of the 5S RNA gene directs specific initiation of transcription: II. The 3' border of the region, *Cell* **19**:27.

Bolivar, F., Rodriquez, R. L., Betlach, M. C., and Boyer, H. W., 1977, Construction and characterization of new cloning vehicles I. Ampicillin-resistant derivatives of the plasmid pMB9, *Gene* **2**:75.

Bonner, P. H., and Hauschka, S. D., 1974, Clonal analysis of vertebrate myogenesis I. Early developmental events in the chick limb, *Devl. Biol.* **37**:317.

Brenner, S., 1974, The genetics of *caenorhabditis Elegans*, *Genetics* **77**:71.

Brown, D. D., and Dawid, I. B., 1968, Specific gene amplification in oocytes, *Science* **165**:349.

Buckingham, M. E., Caput, D., Cohen, A., Whalen, R. G., and Gros, F., 1974, The synthesis and stability of cytoplasmic messenger RNA during myoblast differentiation in culture, *Proc. Natl. Acad. Sci. USA* **71**:1466.

Buckley, P. A., and Konigsberg, I. R., 1974, The avoidance of stimulatory artifacts in cell cycle determinations *in vitro*, *Dev. Biol.* **37**:186.

Buller, A. J., Eccles, J. C., and Eccles, R. M., 1960, Interactions between motor neurons and muscles in respect of the characteristic speeds of their responses, *J. Physiol.* **150**:417.

Burridge, K., and Bray, D., 1975, Purification and structural analysis of myosins from brain and other non-muscle tissues, *J. Mol. Biol.* **99**:1.

Calos, M. P., and Miller, J. H., 1980, Transposable elements, *Cell* **20**:579.

Cameron, J. R., Loh, E. Y., and Davis, R. W., 1979, Evidence for transposition of dispersed repetitive DNA families in yeast, *Cell* **16**:739.

Caravati, M., Minty, R. A., Weydert, A., Alonso, S., Cohen, A., Daubas, P., and Buckingham, M. E., 1981, The messengers coding for myosins and actins: their expression during terminal differentiation of a mouse muscle cell line, in *Molecular and Cellular Control of Muscle Development*, p. 17, Sept. 8, Cold Spring Harbor, New York.

Carmon, Y., Newman, S., and Yaffe, D., 1978, Synthesis of tropomyosin in myogenic cultures and in RNA-directed cell-free systems: qualitative changes in the polypeptides, *Cell* **14**:393.

Carsten, M. E., and Katz, A. M., 1964, Actin: a comparative study, *Biochim. Biophys. Acta* **90**:534.

Cereghini, S., Geoghegan, T., Bergmann, I., and Brawerman, G., 1979, Studies on the efficiency of translation and on the stability of actin messenger ribonucleic acid in mouse sarcoma ascites cells, *Biochemistry* **18**:3153.

Chamley-Campbell, J., Campbell, G. E., Groschel-Stewart, V., and Burnstock, G., 1977, FITC-labeled antibody staining of tropomyosin, *Cell Tissue Res.* **183**:153.

Chi, J. C. H., Rubinstein, N., Strahs, K., and Holtzer, H., 1975, Synthesis of myosin heavy and light chains in muscle cultures, *J. Cell Biol.* **67**:523.

Clarke, M., and Spudich, J. A., 1977, Nonmuscle contractile proteins: the role of actin and myosin in cell motility and shape determination, *Annu. Rev. Biochem.* **46**:797.

Cleveland, D. W., Lopata, M. A., McDonald, R. J., Cowan, N. J., Rutter, W. J., and Kirschner, M. W., 1980, Number and evolutionary conservation of α and β tubulin and cytoplasmic β and λ actin genes using specific cloned cDNA probes, *Cell* **20**:95.

Clewell, D. B., 1972, Nature of Col E_1 plasmid replication in *Escherichia coli* in the presence of chloramphenicol, *J. Bacteriol.* **110**:667.

Cohen, R., Pacifici, M., Rubinstein, N., Biehl, J., and Holtzer, H., 1977, Effect of a tumour promoter on myogenesis, *Nature (London)* **266**:538.

Coleman, J. R., and Coleman, A. W., 1968, Muscle differentiation and macromolecular synthesis, *J. Cell. Physiol.* **72** (Suppl 1):19.

Coleman, J. R., Coleman, A. W., and Hartline, E. J. H., 1969, A clonal study of the reversible inhibition of muscle differentiation by the halogenated thymidine analog 5-Bromodeoxyuridine, *Dev. Biol.* **19**:527.

Collins, J. H., Potter, J. D., Horn, M. J., Wilshire, G., and Jackman, N., 1973, The amino acid sequence of rabbit skeletal muscle tropenin C, gene replication and homology with calcium-binding proteins from carp and hake muscle, *FEBS Lett.* **36**:268.

Cooper, W. G., and Konigsberg, I. R., 1961, Dynamics of myogenesis *in vitro, Anat. Rec.* **140**:195.

Cummins, P., and Perry, S. V., 1973, The subunits and biological activity of polymorphic forms of tropomyosin, *Biochem. J.* **133**:765.

Dawid, I. B., and Wahli, W., 1981, Application of recombinant DNA technology to questions of developmental biology: a review, *Dev. Biol.* **69**:305.

Den, H., Malinzak, D. A., Keating, H. J., and Rosenberg, A., 1975, Influence of concanavalin A, wheat germ, agglutinin, and soybean agglutinin on the fusion of myoblasts *in vitro, J. Cell Biol.* **67**:826.

Desrosiers, R. C., Mulder, C., and Fleckenstein, B., 1979, Methylation of *herpes virus saemiri* DNA in lymphoid tumor cell lines, *Proc. Natl. Acad. Sci. USA* **76**:3839.

Devlin, R. B., and Emerson, C. P., 1978, Coordinate regulation of contractile protein synthesis during myoblast differentiation, *Cell* **13**:599.

Devlin, R. B., and Emerson, C. P., 1979, Coordinate accumulation of contractile protein mRNAs during myoblast differentiation, *Dev. Biol.* **69**:202.

Dienstman, S. R., Biehl, J., Holtzer, S., and Holtzer, H., 1974, Myogenic and chondrogenic lineages in developing limb buds grown *in vitro, Dev. Biol.* **39**:83.

Dodgson, J. B., Strommer, J., and Engel, J. D., 1979, Isolation of the chicken β-globin gene and a linked embryonic β-like globin gene from a chicken DNA recombinant library, *Cell* **17**:879.

Doering, J. L., and Fischman, D. A., 1974, The *in vitro* cell fusion of embryonic chick muscle without DNA synthesis, *Dev. Biol.* **36**:225.

Doering, J. L., and Fischman, D. A., 1977, A fusion-promoting macromolecular factor in muscle coordinated medium, *Exp. Cell Res.* **105**:437.

Donady, J. J., Seecof, A. L., and Dewhurst, S., 1975, Actinomycin D-sensitive periods in the differentiation of *Drosophila* neurons and muscle cells *in vitro, Differentiation* **4**:9.

Dugaiczyk, A., Boyer, H. W., and Goodman, H. M., 1975, Ligation of *Eco*RI endonuclease generated DNA fragments into linear and circular structures, *J. Mol. Biol.* **96**:171.

Durica, D. S., Schloss, J. A., and Crain, W. E., Jr., 1980, Organization of actin gene sequences in the sea urchin: molecular cloning of an intron containing DNA sequence coding for a cytoplasmic actin, *Proc. Natl. Acad. Sci. USA* **77**:5683.

Early, P. W., Davis, M. M., Kaback, D. B., Davidson, N., and Hood, L., 1979, Immunoglobin heavy chain gene organization in mice: analysis of a myeloma genomic clone containing variable and α constant regions, *Proc. Natl. Acad. Sci. USA* **76**:857.

Ebashi, S., 1963, Third component participating in the superprecipitation of "natural actomyosin," *Nature* **200**:1010.

Ebashi, S., and Endo, M., 1980, Calcium ion and muscle contraction, *Prog. Biophys. Mol. Biol.* **18**:125.

Ebashi, S., and Nonomura, Y., 1973, Proteins of the myofibril, in: *The Structure and Function of Muscle* (G. H. Bourne, ed.), Vol. 3, *Physiology and Biochemistry*, Academic Press, New York.

Efstratiadis, A., Kafatos, F. C., Maxam, A. M., and Maniatis, T., 1976, Enzymatic *in vitro* synthesis of globin genes, *Cell* **7**:279.

Elzinga, M., Collins, J. H., Kuehl, W. M., and Adelstein, R. S., 1973, Complete amino acid sequence of actin of rabbit skeletal muscle, *Proc. Natl. Acad. Sci. USA* **70**:2687.

Emerson, C. P., and Beckner, S. K., 1975, Activation of myosin synthesis in fusing and mono-nucleated myoblasts, *J. Mol. Biol.* **93**:431.

Engel, J. N., Gunning, P. W., and Kedes, L., 1981, Isolation and characterization of human actin genes, *Proc. Natl. Acad. Sci. USA* **78**:4674.

Fine, R. E., and Blitz, A. L., 1975, A chemical comparison of tropomyosins from muscle and non-muscle tissues, *J. Mol. Biol.* **95**:447.

Firtel, R. A., Timm, R., Kimmel, A. R., and McKeown, M., 1979, Unusual nucleotide sequences at the 5′ end of actin genes in *Dictyostelium discoideum, Proc. Natl. Acad. Sci. USA* **76**:6206.

Fischbach, G. D., 1972, Synapse formation between dissociated nerve and muscle cells in low density cell cultures, *Dev. Biol.* **28**:407.

Fiszman, M. Y., 1978, Morphological and biochemical differentiation in RSV transformed chick embryo myoblasts, *Cell Differ.* **7**:89.

Flint, S. J., and Weintraub, H. M., 1977, An altered subunit configuration associated with the actively transcribed DNA of integrated adenovirus genes, *Cell* **12**:783.

Frank, G., Weeds, A. G., 1974, The amino-acid sequence of the alkali light chains of rabbit skeletal-muscle myosin, *Eur. J. Biochem.* **44**:317.

Frederiksen, D. W., 1976, Myosin-mediated Ca^{++} regulation of actomyosin-adenosinitriphosphatase from porcine aorta, *Proc. Natl. Acad. Sci. USA* **73**:2706.

Fryberg, E. A., Kindle, K. L., Davidson, N., and Sodja, A., 1980, The actin genes of *Drosophila:* a dispersed multigene family, *Cell* **19**:365.

Fryberg, E. A., Bond, B. J., Hershey, N. D., Mixter, K. S., and Davidson, N., 1981, The actin genes of *Drosophila:* protein coding regions are highly conserved but intron positions are not, *Cell* **24**:107.

Fukuda, J., Henkart, M. P., Fischback, G. D., and Smith, T. G., Jr., 1976, Physiological and structural properties of colchicine-treated chick skeletal muscle cells grown in tissue culture, *Dev. Biol.* **49**:395.

Gall, J., and Pardue, M., 1971, Nucleic acid hybridization in cytological preparations, in: *Methods in Enzymology* (L. Grossman and K. Moldave, eds.), pp. 470–480, Academic Press, New York.

Gallwitz, D., and Sures, I., 1980, Structure of a split yeast gene; complete nucleotide sequence of the actin gene in *Saccharomyces cerevisiae, Proc. Natl. Acad. Sci. USA* **77**:2546.

Garel, A., and Axel, R., 1976, Selective digestion of transcriptionally active ovalbumin genes from oviduct nuclei, *Proc. Natl. Acad. Sci. USA* **73**:3966.

Garel, A., Zolan, M., and Axel, R., 1977, Genes transcribed at diverse rates have a similar conformation in chromatin, *Proc. Natl. Acad. Sci. USA* **74**:4867.

Garfinkel, L. I., and Nadal-Ginard, B., 1981, Cloning and Characterization of Muscle mRNAs, in: *Molecular and Cellular Control of Muscle Development*, p. 37, Sept. 8, Cold Spring Harbor, New York.

Garrels, J. I., and Gibson, W., 1976, Identification and characterization of multiple forms of actin, *Cell* **9**:793.

Gauthier, G. F., Lowey, S., and Hobbs, A. W., 1978, Fast and slow myosin in developing muscle fibers, *Nature* **274**:25.

Gazith, J., Himmelfarb, S., and Harrington, W. F., 1970, Studies on the subunit structure of myosin, *J. Biol. Chem.* **245**:15.

Gordon, J. L., Burns, A. T. H., Christmann, J. L., and Deely, R. G., 1978, Cloning of a double-stranded cDNA that codes for a portion of chicken preproalbumin: a general method for isolating a specific DNA sequence from partially purified mRNA, *J. Biol. Chem.* **253**:8629.

Gruenstein, E., and Rich, A., 1975, Non-identity of muscle and non-muscle actins, *Biochem. Biophys. Res. Commun.* **64**:472.

Gruenstein, E., Rich, A., and Weihing, R. R., 1975, Actin associated with membranes from 3T3 mouse fibroblast and HeLa cells, *J. Cell Biol.* **64**:223.

Grunstein, M., and Hogness, D. S., 1975, Colony hybridization: a method for the isolation of cloned DNAs that contain a specific gene, *Proc. Natl. Acad. Sci. USA* **72**:3961.

Guth, L., 1968, "Trophic" influences of nerve on muscle, *Physiol. Rev.* **48**:645.

de la Haba, G., Cooper, G. W., and Elting, V. C., 1966, Hormonal requirements for myogenesis of striated muscle *in vitro:* insulin and somatotropin, *Proc. Natl. Acad. Sci. USA* **56**:1719.

de la Haba, G., Kamali, H. M., and Tiede, D. M., 1975, Myogenesis of avian striated muscle *in vitro:* role of collagen in myofiber formation, *Proc. Natl. Acad. Sci. USA* **72**:2729.

Hamer, D. H., and Thomas, C. A., Jr., 1976, Molecular cloning of DNA fragments produced by restriction endonucleases *Sal* I and *Bam* I, *Proc. Natl. Acad. Sci. USA* **73**:1537.

Harris, S. E., Rosen, J. M., Means, A. R., and O'Malley, B. W., 1975, Use of a specific probe for ovalbumin messenger RNA to quantitate estrogen induced gene transcripts, *Biochemistry* **14**:2072.

Hastings, K. E. M., and Emerson, C. P., Jr., 1981, cDNA clone analysis of mRNA regulation during myogenesis, in: *Molecular and Cellular Control of Muscle Development*, p. 22, Sept. 8, Cold Spring Harbor, New York.

Hauschka, S. D., and Konigsberg, I. R., 1966, The influences of collagen in the development of muscle clones, *Proc. Natl. Acad. Sci. USA* **55**:119.

Heidecker, G., Messing, J., and Gronenborn, B., 1980, A versatile primer for DNA sequencing in the M13mp2 cloning system, *Gene* **10**:69.

Henneman, E., and Olson, C. B., 1965, Relations between structure and function in the design of skeletal muscles, *J. Neurophysiol.* **28**:581.

Heywood, S. M., and Kennedy, D. S., 1976, Translational control in embryonic muscle, *Biochemistry* **15**:3314.

Heywood, S. M., Kennedy, D. S., and Bester, A. J., 1975, Stored myosin messenger in embryonic chick muscle, *FEBS Lett.* **53**:69.

Heywood, S. M., and Nwagwa, M., 1968, *De novo* synthesis of myosin in a cell-free system, *Proc. Natl. Acad. Sci. USA* **60**:229.

Hodges, R. S., and Smillie, L. B., 1970, Chemical evidence for chain heterogeneity in rabbit muscle tropomyosin, *Biochem. Biophys. Res. Commun.* **41**:987.

Hofstetter, A., Schambock, S., Van Den Berg, J., and Weissmann, C., 1976, Specific excision of the inserted DNA segment from hybrid plasmids constructed by the poly(dA)/poly(dT) method, *Biochim. Biophys. Acta* **454**:587.

Holtzer, H., Sanger, J. W., Ishikawa, H., and Strahs, K., 1973, Selected topics in skeletal myogenesis, *Cold Spring Harbor Symp. Quant. Biol.* **37**:549.

Holtzer, H., Croop, J., Dienstman, S., Ishikawa, H., and Somlyo, A. P., 1975, Effects of cytochalasin B and colcemid on myogenic cultures, *Proc. Natl. Acad. Sci. USA* **72**:513.

Honig, G. R., and Rabinovitz, M., 1965, Actinomycin D: Inhibition of protein synthesis unrelated to effect on template RNA synthesis, *Science* **149**:1504.

Hunter, T., and Garrels, J. I., 1977, Characterization of the mRNAs for α-, β-, and γ-actin, *Cell* **12**:767.

Huszar, G., 1972, Developmental changes of the primary structure and histidine methylation in rabbit skeletal muscle myosin, *Nature New Biol.* **240**:260.

Huxley, H. E., 1973, Muscular contraction and cell motility, *Nature* **243**:455.

Ishikawa, H., Bischoff, R., and Holtzer, H., 1969, Formation of arrowhead complexes with heavy meromyosin in a variety of cell types, *J. Cell Biol.* **43**:312.

Jackson, D., Symons, R., and Berg, P., 1972, Biochemical method for inserting new genetic information into DNA of Simian Virus 40: circular SV40 DNA molecules containing lambda phage genes and the galactose operon of *E. coli*, *Proc. Natl. Acad. Sci. USA* **69**:2904.

Jacobson, G. R., Schaffer, H. M., Stark, G. R., and Vanaman, T. C., 1973, Specific chemical cleavage in high yield at the amino peptide bonds of cysteine and cystine residues, *J. Biol. Chem.* **248**:6583.

John, H. A., Patrinou-Georgoulas, M., and Jones, K. W., 1977, Detection of myosin heavy chain mRNA during myogenesis in tissue culture by *in vitro* and *in situ* hybridization, *Cell* **12**:501.

Jones, P. A., and Taylor, S. M., 1980, Cellular differentiation, cytidine analogs, and DNA methylation, *Cell* **20**:85.

Kafatos, F. C., Jones, C. W., and Efstratiadis, A., 1979, Determination of nucleic acid sequence homologies and relative concentrations by a dot hybridization procedure, *Nuc. Acid Res.* **7**:1541.

Karn, J., 1981, Nucleotide sequence of a myosin heavy chain gene; genetical and protein structural implications, in: *Molecular and Cellular Control of Muscle Development*, p. 25, Sept. 8, Cold Spring Harbor, New York.

Katcoff, D., Nudel, U., Zevin-Sonkin, D., Carmon, Y., Shani, M., Lehrach, H., Frischauf, A. M., and Yaffe, D., 1980, Construction of recombinant plasmids containing rat muscle actin and myosin light chain DNA sequences, *Proc. Natl. Acad. Sci. USA* **77**:960.

Keller, J. M., and Nameroff, M., 1974, Induction of creatine phosphokinase in cultures of chick skeletal myoblasts with concomitant cell fusion, *Differentiation* **2**:19.

Kendrick-Jones, J., Szentkiralyi, E. M., and Szent-Gyorgyi, A. G., 1976, Regulatory light chains in myosins, *J. Mol. Biol.* **104**:747.

Kennedy, D. S., Siegel, E., and Heywood, S. M., 1978, Purification of myosin mRNA translational control RNA and its inhibition of myosin and globin messenger translation. *FEBS Lett.* **90**:209.

Kindle, K. L., and Firtel, R. A., 1978, Identification and analysis of *Dictyostelium* actin genes, a family of moderately repeated genes, *Cell* **15**:763.

Konigsberg, I. R., 1975, The culture environment and its control of myogenesis, in: *Regulation of Cell Proliferation and Differentiation*, (W. W. Nichols and D. G. Murphy, eds.), pp. 105–137, Plenum, New York.

Konigsberg, I. R., Sollmann, P. A., and Mixter, L. O., 1978, The duration of terminal G_1 of fusing myoblasts, *Dev. Biol.* **63**:11.

Laird, C. D., 1971, Chromatid structure: relationship between DNA content and nucleotide sequence diversity, *Chromosoma* **32**:378.

Lawn, R. M., Fritsch, E. F., Parker, R. C., Blake, G., and Maniatis, T., 1978, The isolation and characterization of linked δ- and β-globin genes from a cloned library of human DNA, *Cell* **15**:1157.

Lawson, G. M., Tsai, M. J., and O'Malley, B. W., 1980, Deoxyribonuclease I sensitivity of the nontranscribed sequences flanking the 5' and 3' ends of the ovomucoid gene and the ovalbumin and its related X and Y genes in hen oviduct nuclei, *Biochemistry* **19**:4403.

Lazarides, E., 1975, Tropomyosin antibody: the specific localization of tropomyosin in nonmuscle cells, *J. Cell Biol.* **65**:549.

Lazarides, E., and Lindberg, U., 1974, Actin *is* the naturally occurring inhibitor of deoxyribonuclease I, *Proc. Natl. Acad. Sci. USA* **71**:4742.

Lazarides, E., and Weber, K., 1974, Actin antibody: the specific visualization of actin filaments in non-muscle cells, *Proc. Natl. Acad. Sci. USA* **71**:2268.

Leder, P., Ross, J., Gielen, J., Packman, S., Ikawa, Y., Aviv, H., and Swan, D., 1973, Regulated expression of mammalian genes: globin and immunoglobin as model systems, *Cold Spring Harbor Symp. Quant. Biol.* **38**:753.

Leder, P., Tiemeier, D., Enquist, L., 1977, EK₂ derivatives of bacteriophage lambda useful in the cloning of DNA from higher organisms: the λgtWES system, *Science* **196**:175.

Lindsley, D. L., and Grell, E. H., 1968, *Carnegie Inst. Washington Publication* **6**:27.

Lobban, P. E., and Kaiser, A. D., 1973, Enzymatic end to end joining of DNA molecules, *J. Mol. Biol.* **78**:453.

Loomis, W. F., Jr., Wahrmann, J. P., and Luzzati, D., 1973, Temperature-sensitive variants of an established myoblast line, *Proc. Natl. Acad. Sci. USA* **70**:425.

Lowey, S., Slayter, H. S., Weeds, A. G., and Baker, H., 1969, Substructure of the myosin molecule I subfragments of myosin by enzymatic degradation, *J. Mol. Biol.* **42**:1.

Lowey, S., Benfield, P. A., Gauthier, G. F., LeBlanc, D. D., and Waller, G., 1981, Characterization of myosins from embryonic and developing chicken pectoralis muscle, in: *Molecular and Cellular Control of Muscle Development*, p. 3, Sept. 8, Cold Spring Harbor Laboratory.

Luse, D., and Roeder, R. G., 1980, Accurate transcription initiation on a purified mouse β-globin DNA fragment in a cell-free system, *Cell* **20**:691.

Luzzati, D., 1974, Characteristics of ouabain resistant clones isolated from a myoblast established line, *Biochimie* **56**:1567.

MacLeod, A. R., 1981, Construction of bacterial plasmids containing sequences complementary to chicken α tropomyosin mRNA, *Nuc. Acids Res.* **9**:2675.

Malech, H. L., and Lentz, T. L., 1974, Microfilaments in epidermal cancer cells, *J. Cell Biol.* **60**:473.

Mandel, J. L., and Chambon, P., 1979, DNA methylation: organ specific variations in the methylation pattern within and around ovalbumin and other chicken genes, *Nuc. Acids Res.* **7**:2081.

Mandel, M., and Higa, A., 1970, Calcium-dependent bacteriophage DNA infection, *J. Mol. Biol.* **53**:159.

Mandel, J. L., and Pearson, M. L., 1974, Insulin stimulates myogenesis in a rat myoblast line, *Nature (London)* **251**:618.

Maniatis, T., Hardison, R. C., Lacy, E., Lauer, J., O'Connell, C., Quon, D., Sim, G. K., and Efstratiadis, A., 1978, The isolation of structural genes from libraries of eukaryotic DNA, *Cell* **15**:687.

Maniatis, T., Fritsch, E. F., Lauer, J., and Lawn, R. M., 1980, The molecular genetics of human hemoglobins, *Ann. Rev. Genetics* **14**:145.

Mannherz, H. G., Barrington-Leigh, J., Leberman, R., and Pfrang, H., 1975, A specific 1:1 G actin: DNase I complex formed by the action of DNase I on F actin, *FEBS Lett.* **60**:34.

Mannherz, H. G., and Goody, R. S., 1976, Proteins of contractile systems, *Ann. Rev. Biochem.* **45**:428.

Masaki, T., and Yoshizaki, C., 1972, The onset of myofibrillar protein synthesis in chick embryo *in vivo, J. Biochem.* **71**:755.

Mathis, D. J., and Gorovsky, M. A., 1977, Structure of rDNA-containing chromatin of *tetrahymena pyriformis* analyzed by nuclease digestion, *Cold Spring Harbor Symp. Quant. Biol.* **42**:773.

Maxam, A. M., and Gilbert, W., 1977, A new method for sequencing DNA, *Proc. Natl. Acad. Sci. USA* **74**:560.

McKeown, M., and Firtel, R. A., 1981, Differential expression and 5′ end mapping of actin genes in *Dictyostelium, Cell* **24**:799.

McKeown, M., Taylor, W. C., Kindle, K. L., Firtel, R. A., Bender, W., and Davidson, N., 1978, Multiple heterogeneous actin genes in *Dictyostelium, Cell* **15**:789.

Means, A. R., and Dedman, J. R., 1980, Calmodulin—an intracellular calcium receptor, *Nature* **285**:73.

Medford, R. M., Wydro, R. M., Nguyen, H. T., and Nadal-Ginard, B., 1980, Cytoplasmic processing of myosin heavy chain messenger RNA: evidence provided by using a recombinant DNA plasmid, *Proc. Natl. Acad. Sci. USA* **77**:5749.

Merlino, G. T., Water, R. D., Chamberlain, J. P., Jackson, D. A., El-Gewely, M. R., and Kleinsmith, L. J., 1980, Cloning of sea urchin actin gene sequences for use in studying the regulation of actin gene transcription, *Proc. Natl. Acad. Sci. USA* **77**:765.

Mertz, J. E., and Davis, R. W., 1972, Cleavage of DNA by R$_I$ restriction endonuclease generates cohesive ends, *Proc. Natl. Acad. Sci. USA* **69**:3370.

Messing, J., Gronenborn, B., Muller-Hill, B., and Hofschneider, P. H., 1977, Filamentous coliphage M13 as a cloning vehicle: insertion of a *Hind*II fragment of the *lac* regulatory region in M13 replicative form *in vitro, Proc. Natl. Acad. Sci. USA* **74**:3642.

Minty, A. J., Caravatti, M., Benoit, R., Cohen, A., Daubas, P., Weydert, A., Gros, F., and Buckingham, M. E., 1981a, Mouse actin messenger RNAs: construction and characterization of a recombinant plasmid molecule containing a complementary DNA transcript of mouse α-actin mRNA, *J. Biol. Chem.* **256**:1008.

Minty, A. J., Alonso, S., Caravatti, M., Cohen, A., Daubas, P., and Buckingham, M., 1981b, The number and structure of mouse actin genes, in: *Molecular and Cellular Control of Muscle Development,* p. 27, Sept. 8, Cold Spring Harbor, New York.

Mintz, B., and Baker, W. W., 1967, Normal mammalian muscle differentiation and gene control of isocitiate dehydrogenase synthesis, *Proc. Natl. Acad. Sci. USA* **58**:592.

Monahan, J. M., Harris, S. E., Woo, S. L. C., Robberson, D. L., and O'Malley, B. W., 1976, The synthesis and properties of the complete complementary DNA transcript of ovalbumin mRNA, *Biochemistry* **15**:223.

Mondal, H., Sutton, A., Chen, V., and Sarkar, S., 1974, Highly purified mRNA for myosin heavy chain: size and polyadenylic acid content, *Biochem. Biophys. Res. Commun.* **56**:988.

Morris, G. E., Buzash, E. A., Rourke, A. W., Tepperman, K., Thompson, W. C., and Heywood, S. M., 1973, Myosin messenger RNA: studies on its purification, properties, and translation during myogenesis in culture, *Cold Spring Harbor Symp. Quant. Biol.* **37**:535.

Morris, G. E., Piper, M., and Cole, R., 1976, Do increases in enzyme activities during muscle differentiation reflect expression of new genes? *Nature* **263**:76.

Moss, M., and Schwartz, R., 1981, Regulation of tropomyosin gene expression during myogenesis, *Mol. and Cell Biol.* **1**:289.

Murphy, R. A., Singer, R. H., Saide, J. D., Pantazis, N. J., Blanchard, M. H., Byron, K. S., Arnason, B. G. W., and Young, M., 1977, Synthesis and secretion of a high molecular weight form of nerve growth factor by skeletal muscle cells in culture, *Proc. Natl. Acad. Sci. USA* **74**:4496.

Nadal-Ginard, B., 1978, Commitment, fusion, and biochemical differentiation of a myogenic cell line in the absence of DNA synthesis, *Cell* **15**:855.

Nathans, D., and Smith, H. O., 1975, Restriction endonucleases in the analysis and restructuring of DNA molecules, *Ann. Rev. Biochem.* **44**:273.

Ng, R., and Abelson, J., 1980, Isolation and sequence of the gene for actin in *Saccharomyces cerevisiae, Proc. Natl. Acad. Sci. USA* **77:**3912.

Novick, R. P., Clowes, R. C., Cohen, S. N., Curtiss, R., III, Datta, N., and Falkow, S., 1976, Uniform nomenclature for bacterial plasmids: a proposal, *Bacteriol. Rev.* **40:**168.

Noyes, B. E., and Stark, G. R., 1975, Nucleic acid hybridization using DNA covalently coupled to cellulose, *Cell* **5:**301.

Nudel, U., Katcoff, D., Carmon, Y., Zevin-Sonkin, D., Levi, Z., Shaul, Y., Shani, M., and Yaffe, D., 1980, Identification of recombinant phages containing sequences from different rat myosin heavy chain genes, *Nucl. Acids. Res.* **8:**2133.

Nudel, U., Shani, M., Zakut, R., Katcoff, D., Calvo, J., Carmon, Y., Finer, M., and Yaffe, D., 1981, Studies on the structure of genes coding for contractile proteins, in: *Molecular and Cellular Control of Muscle Development*, p. 14, Sept. 8, Cold Spring Harbor Laboratory, New York.

O'Farrell, P. H., 1975, High resolution two dimensional electrophoresis of proteins, *J. Biol. Chem.* **250:**4007.

Okazaki, K., and Holtzer, H., 1965, An analysis of myogenesis *in vitro* using fluorescein-labeled antimyosin, *J. Histochem. Cytochem.* **13:**726.

Okazaki, K., and Holtzer, H., 1966, Myogenesis: fusion, myosin synthesis, and the mitotic cycle, *Proc. Natl. Acad. Sci. USA* **56:**1484.

O'Neill, M. C., and Stockdale, F. E., 1974, 5-Bromodeoxyuridine inhibition of differentiation. Kinetics of inhibition and reversal in myoblasts, *Dev. Biol.* **37:**117.

Ordahl, C. P., Tilghman, S. M., Ovitt, C., Fornwald, J., and Largen, M. T., 1980, Structure and developmental expression of the chick α-actin gene. *Nucl. Acids Res.* **8:**4989.

Panet, A., and Cedar, H., 1977, Selective degradation of integrated murine leukemia proviral DNA by deoxyribonucleases, *Cell* **11:**933.

Paterson, B. M., and Bishop, J. O., 1977, Changes in the mRNA population of chick myoblasts during myogenesis *in vitro, Cell* **12:**751.

Paterson, B., and Prives, J., 1973, Appearance of actylcholine receptor in differentiating cultures of embryonic chick breast muscle, *J. Cell Biol.* **59:**241.

Paterson, B. M., Roberts, B. E., and Yaffe, D., 1974, Determination of actin messenter RNA in cultures of differenitating embryonic chick skeletal muscle, *Proc. Natl. Acad. Sci. USA* **71:**4467.

Paterson, B. M., Roberts, B. E., and Kuff, E. L., 1977, Structural gene identification and mapping by DNA·mRNA hybrid-arrested cell-free translation, *Proc. Natl. Acad. Sci. USA* **74:**4370.

Patient, R. K., Elington, J. A., Kay, R. M., and Williams, J. G., 1980, Internal organization of the major adult α- and β-globin genes of *X. laevis, Cell* **21:**565.

Patrinou-Georgoulos, M., and John, H. A., 1977, The genes and mRNA coding for the heavy chains of chick embryonic skeletal myosin, *Cell* **12:**491.

Payvar, F., and Schimke, R. T., 1979, Methylmercury hydroxide enhancement of translation and transcription of ovalbumin and conalbumin mRNAs, *J. Biol. Chem.* **254:**7636.

Peacock, A. C., and Dingman, C. W., 1968, Molecular weight estimation and separation of ribonucleic acid by electrophoresis in agarose-acrylamide composite gels, *Biochemistry* **7:**668.

Penman, S., Vesco, C., and Penman, M., 1968, Localization and kinetics of formation of nuclear heterodisperse RNA, cytoplasmic heterodisperse RNA, and polyribosome-associated messenger RNA in HeLa cells, *J. Mol. Biol.* **34:**49.

Perry, S. V., 1973, Variation in the contractile and regulatory proteins of the myofibril with muscle type, in: *Exploratory Concepts in Muscular Dystrophy II* (A. Milhorat, ed.), Excerpta Medica, Amsterdam, pp. 319–330, American Elsevier Publ. Co., New York.

Pollard, T., Thomas, S., and Niederman, R., 1974, Human platelet myosin 1. Purification by a rapid method applicable to other nonmuscle cells, *Anal. Biochem.* **60:**258.

Pollard, T. D., and Weihing, R. R., 1974, Actin and myosin and cell movement, in *Critical Reviews in Biochemistry, Vol. 2* (G. Fasman, ed.), pp. 1–65, CRC Press, Cleveland, Ohio.

Prives, J., and Shinitzky, M., 1977, Increased membrane fluidity precedes fusion of muscle cells, *Nature (London)* **268:**761.

Rabbitts, T. H., 1976, Bacterial cloning of plasmids carrying copies of rabbit globin messenger RNA, *Nature* **260:**221.

Razin, A., and Riggs, A. D., 1980, DNA methylation and gene function, *Science* **210:**604.

Riggs, A. D., 1975, X inactivation, differentiation, and DNA methylation, *Cytogenet. Cell Genet.* **14**:9.

Robbins, J., and Heywood, S. M., 1976, Preparation and characterization of myosin copy DNA, *Biochem. Biophys. Res. Commun.* **68**:918.

Robbins, J., and Heywood, S. M., 1978, Quantification of myosin heavy chain mRNA during myogenesis, *Eur. J. Biochem.* **82**:601.

Roberts, R. J., 1976, Restriction endonucleases, *CRC Crit. Rev. Biochem.* **4**:123.

Rogers, J., Ng, S. K. C., Coulter, M. B., and Sanwal, B. D., 1975, Inhibition of myogenesis in a rat myoblast line by 5-bromodeoxyuridine, *Nature (London)* **256**:438.

Roy, R. K., Potter, J. D., and Sarkar, S., 1976, Characterization of the Ca^{2+}-regulatory complex of chick embryonic muscles: polymorphism of tropomyosin in adult and embryonic fibers, *Biochem. Biophys. Res. Commun.* **70**:28.

Royal, A., Garapin, A., Cami, B., Perrin, F., Mandel, J. L., LeMeur, M., Bregegeyne, F., Gannon, F., LePennec, J. P., Chambon, P., and Kourilsky, P., 1979, The ovalbumin gene region: common features in the organization of three genes expressed in chicken oviduct under hormonal control, *Nature* **279**:125.

Rubenstein, P., and Deuchler, J., 1979, Acetylated and non-acetylated actins in *Dictyostelium discoideum*, *J. Biol. Chem.* **254**:11142.

Rubinstein, N. A., Pepe, F. A., and Holtzer, H., 1977, Myosin types during the development of embryonic chicken fast and slow muscles, *Proc. Natl. Acad. Sci. USA* **74**:4524.

Salmons, S., and Sreter, F. A., 1976, Significance of impulse activity in the transformation of skeletal muscle types, *Nature* **263**:30.

Sanger, F., Nicklen, S., and Coulson, A. R., 1977, DNA sequencing with chain terminating inhibitors, *Proc. Natl. Acad. Sci. USA* **74**:5463.

Sanger, J. W., 1974, The use of cytochalasin B to distinguish myblasts from fibroblasts in cultures of developing chick striated muscles, *Proc. Natl. Acad. Sci. USA* **71**:3621.

Sanger, J. W., 1975, Presence of actin during chromosomal movement, *Proc. Natl. Acad. Sci. USA* **72**:2451.

Sargent, T. D., Wu, J. R., Sala-Trepat, J. M., Wallace, R. B., Reyes, A. A., and Bonner, J., 1979, The rat serum albumin gene; analysis of cloned sequences, *Proc. Natl. Acad. Sci. USA* **76**:3256.

Sarkar, S., Sreter, F. A., Gergely, J., 1971, Light chains of myosins from white, red, and cardiac muscles, *Proc. Natl. Acad. Sci. USA* **68**:946.

Schibler, U., Marcu, K. B., and Perry, R. P., 1978, The synthesis and processing of the messenger RNAs specifying heavy and light chain immunoglobulins in MPC-11 cells, *Cell* **15**:1495.

Schroeder, T. E., 1973, Actin in dividing cells: contractile ring filaments bind heavy meromyosin, *Proc. Natl. Acad. Sci. USA* **70**:1688.

Schwartz, R. J., 1973, Control of glutamine synthetase synthesis in the embryonic chick neural retina: a caution in the use of actinomycin D, *J. Biol. Chem.* **248**:6426.

Schwartz, R. J., and Rothblum, K., 1980, Regulation of muscle differentiation: isolation and purification of chick actin messenger ribonucleic acid and quantitation with complementary deoxyribonucleic acid probes, *Biochemistry* **19**:2506.

Schwartz, R. J., and Rothblum, K., 1981, Gene switching in Myogenesis: differential expression of the chicken actin multigene family, *Biochemistry* **20**:4122.

Schwartz, R. J., Haron, J. A., Rothblum, K. N., and Dugaiczyk, A., 1980, Regulation of muscle differentiation: cloning of sequences from α-actin messenger ribonucleic acid, *Biochemistry* **19**:5883.

Shainberg, A., Yagil, G., and Yaffe, D., 1971, Alterations of enzymic activities during muscle differentiation *in vitro*, *Dev. Biol.* **25**:1.

Shani, M., Nudel, U., Zevin-Sonkin, D., Zakut, R., Givol, D., Katcoff, D., Carmon, Y., Reiter, J., Frischauf, A. M., and Yaffe, D., 1981a, Skeletal muscle actin mRNA. Characterization of the 3′ untranslated end, *Nucl. Acids Res.* **9**:579.

Shani, M., Zevin-Sonkin, D., Saxel, O., Carmon, Y., Katcoff, D., Nudel, U., and Yaffe, D., 1981b, The correlation between the synthesis of skeletal muscle actin, myosin heavy chain and myosin light chain and the accumulation of corresponding mRNA sequences during myogenesis, *Dev. Biol.* (in press).

Singer, R. H., and Kessler-Icekson, G., 1978, Stability of polyadenylated RNA in differentiating myogenic cells, *Eur. J. Biochem.* **88:**395.

Singer, R. H., and Penman, S., 1972, Stability of HeLa cell mRNA in actinomycin, *Nature* **240:**100.

Sodek, J., Hodges, R. S., Smillie, L. B., and Jurasek, L., 1972, Amino-acid sequences of rabbit skeletal tropomyosin and its coiled coil structure, *Proc. Natl. Acad. Sci. USA* **69:**3800.

Somers, D. G., Pearson, M. L., and Ingles, C. J., 1975, Isolation and characterization of an α-amanitin-resistant rat myoblast mutant cell line possessing α-amanitin-resistant RNA polymerase II, *J. Biol. Chem.* **250:**4825.

Southern, E. M., 1975, Detection of specific sequences among DNA fragments separated by gel electrophoresis, *J. Mol. Biol.* **98:**503.

Spradling, A. L., and Mahowald, A. F., 1980, Amplification of genes for chorion proteins during oogenesis in *Drosophila melangaster, Proc. Natl. Acad. Sci. USA* **77:**1096.

Spudich, J. A., and Watt, S., 1971, The regulation of rabbit skeletal muscle contraction 1. Biochemical studies of the interaction of the tropomyosin-troponin complex with actin and the proteolytic fragments of myosin. *J. Biol. Chem.* **246:**4866.

Squire, J. M., 1975, Muscle filament structure and muscle contraction, *Ann. Rev. Biophys. Bioeng.* **4:**137.

Stockdale, F., Okazaki, K., Nameroff, M., and Holtzer, H., 1964, 5-Bromodeoxyuridine: effect on myogenesis *in vitro, Science* **146:**533.

Stockdale, F. E., Raman, N., and Baden, H., 1981, Myosin light chains and the developmental origin of fast muscle, *Proc. Natl. Acad. Sci. USA* **78:**931.

Storti, R. V., Horovitch, S. J., Scott, M. P., Rich, A., and Pardue, M. L., 1978, Myogenesis in primary cell cultures from *Drosophila melangaster:* protein synthesis and actin heterogeneity during development, *Cell* **13:**589.

Storti, R. W., Blautch, V., Szwast, A., Mischke, D., and Pardue, M. L., 1981, Molecular cloning, characterization and organization of *Drosophila* muscle genes, in: *Molecular and Cellular Control of Muscle Development,* p. 21, Sept. 8, Cold Spring Harbor Laboratory, New York.

Stossel, T. P., and Hartwig, J. H., 1976, Interactions of actin, myosin, and a new actin-binding protein of rabbit pulmonary macrophages. II. Role in cytoplasmic movement and phagocytosis, *J. Cell Biol.* **68:**602.

Strohman, R. C., Moss, P. S., Micou-Eastwood, J., Spector, D., Przybyla, A., and Paterson, B., 1977, Messenger RNA for myosin polypeptides: isolation from single myogenic cell culture, *Cell* **10:**265.

Sullivan, D., Palacios, R., Stavnezer, J., Taylor, J. M., Faras, A. J., Kielly, M. L., Summers, N. M., Bishop, J. M., and Schimke, R. T., 1973, Synthesis of a deoxyribonucleic acid sequence complementary to ovalbumin messenger ribonucleic acid and quantification of ovalbumin genes, *J. Biol. Chem.* **248:**7530.

Sutcliffe, J. G., 1978, pBR322 restriction map derived from the DNA sequence: accurate DNA size markers up to 4361 nucleotide pairs long, *Nucl. Acids Res.* **5:**2721.

Taylor, E. W., 1972, Chemistry of muscle contraction, *Annu. Rev. Biochem.* **41:**577.

Tobin, S. L., Zulauf, E., Sanchez, F., Craig, E. A., and McCarthy, B. J., 1980, Multiple actin related sequences in the *Drosophila melangaster* genome, *Cell* **19:**121.

Tonegawa, S., Maxam, M., Tizard, R., Bernard, O., and Gilbert, W., 1978, Sequence of a mouse germ-like gene for a variable region of an immunoglobulin light chain, *Proc. Natl. Acad. Sci. USA* **75:**1485.

Turner, D. C., Maier, V., and Eppenberger, H. M., 1974, Creatine kinase and aldolase isoenzyme transitions in cultures of chick skeletal muscle cells, *Dev. Biol.* **37:**63.

Turner, D. C., Gmur, R., Siegrist, M., Burckhardt, E., and Eppenberger, H. M., 1976, Differentiation in cultures derived from embryonic chicken muscle, *Dev. Biol.* **48:**258.

Tilghman, S. M., Tiemeier, D. C., Polsky, F., Edgell, M. H., Seidman, J. F., Leder, A., Enquist, L. W., Norman, B., and Leder, P., 1977, Cloning specific segments of the mammalian genome:

bacteriophage λ containing mouse globin and surrounding gene sequences, *Proc. Natl. Acad. Sci. USA* **74**:4406.

Ullman, J. S., and McCarthy, B. J., 1973, The relationship between mismatched base pairs and the thermal stability of DNA duplexes I. Effects of depurination and chain scission, *Biochim. Biophys. Acta* **294**:405.

Ullrich, A., Shine, J., Chirgwin, R., Pictet, R., Tischer, E., Rutter, W. J., and Goodman, H. M., 1977, Rat insulin genes: construction of plasmids containing the coding sequences, *Science* **196**:1313.

Umeda, P. K., Kavinsky, C., Sinha, M., Hsu, H. J., Jakovcic, S., and Rabinowitz, M., 1981, Molecular cloning of myosin heavy chain cDNAs from chick embryo skeletal muscle, in: *Molecular and Cellular Control of Muscle Development*, p. 24, Sept. 8, Cold Spring Harbor Laboratory, New York.

Vandekerckhove, J., and Weber, K., 1978, Mammalian cytoplasmic actins are the products of at least two genes and differ in primary structure in at least 25 identified positions from skeletal muscle actins, *Proc. Natl. Acad. Sci. USA* **75**:1106.

Vandekerckhove, J., and Weber, K., 1979, The complete amino acid sequence of actins from bovine aorta, bovine heart, bovine fast skeletal muscle, and rabbit slow skeletal muscle, *Differentiation* **14**:123.

Vandekerckhove, J., and Weber, K., 1980, Vegetative *Dictyostelium* cells containing 17 actin genes express a single major actin, *Nature* **284**:475.

Van der Ploeg, L. H. T., and Flavell, R. A., 1980, DNA methylation in the human γδβ-globin locus in erythroid and nonerythroid tissues, *Cell* **19**:947.

Vardiman, L., Neuman, R., Kuhlmann, I., Sutter, D., and Doerfler, W., 1980, DNA methylation and viral gene expression in adenovirus-transformed and injected cells, *Nucleic Acid Res.* **8**:2461.

Vertel, B. M., and Fischman, D. A., 1976, Myosin accumulation in mononucleated cells of chick muscle cultures, *Dev. Biol.* **48**:438.

Vilbert, P. J., Haselgrove, J. C., Lowy, J., and Poulsen, F. R., 1972, Structural changes in actin-containing filaments of muscle, *J. Mol. Biol.* **71**:757.

Villa-Komaroff, L., Efstratiadis, A., Broome, S., Lomedico, P., Tizard, R., Naber, S. P., Chick, W. L., and Gilbert, W., 1978, A bacterial clone synthesizing proinsulin, *Proc. Natl. Acad. Sci. USA* **75**:3727.

Wahli, W., and Dawid, I. B., 1980, Isolation of two closely related vitellogenin genes, including their flanking regions, from a *Xenopus laevis* gene library, *Proc. Natl. Acad. Sci. USA* **77**:1437.

Wahli, W., Dawid, I. B., Wyler, T., Weber, R., and Ryffel, G. U., 1980, Comparative analysis of the structural organization of two closely related vitellogenin genes in *X. laevis*, *Cell* **20**:107.

Wasylyk, B., Kedinger, C., Corden, J., Brison, O., and Chambon, P., 1980, Specific *in vitro* initiation of transcription on conalbumin and ovalbumin genes and comparison with adenovirus-2 early and late genes, *Nature* **285**:367.

Weeds, A. G., 1976, Light chains from slow twitch-muscle myosin, *Eur. J. Biochem.* **66**:157.

Weeds, A. G., and Lowey, S., 1971, Substructure of the myosin molecule II. The light chains of myosin, *J. Mol. Biol.* **61**:701.

Weintraub, H., and Groudine, M., 1976, Chromosomal subunits in active genes have an altered conformation, *Science* **193**:848.

Whalen, R. G., Butler-Browne, G. S., and Gros, F., 1976, Protein synthesis and actin heterogeneity in calf muscle cells in culture, *Proc. Natl. Acad. Sci. USA* **73**:2018.

Whalen, R. G., Buckingham, M. E., Goto, S., Merlie, J. P., and Gros, F., 1977, The pattern and control of protein synthesis in cultured calf muscle cells, in: *Pathogenesis of Human Muscular Dystrophies*, (L. P. Rowland, ed.), pp. 433–450, Excerpta Medica, Amsterdam and Oxford.

Whalen, R. G., Butler-Browne, G. S., and Gros, F., 1978, Identification of a novel form of myosin light chain present in embryonic muscle tissue and cultured muscle cells, *J. Mol. Biol.* **126**:415.

Whalen, R. G., Schwartz, K., Bouveret, P., Sell, S. M., and Gros, F., 1979, Contractile protein isozymes in muscle development: identification of an embryonic form of myosin heavy chain, *Proc. Natl. Acad. Sci. USA* **76**:5197.

Whalen, R. G., Sell, S. M., Butler-Browne, G. S., Schwartz, K., Bouveret, P., and Pinset-Harstrom, I., 1981, Three myosin heavy chain isozymes appear sequentially in rat muscle development, *Nature* **292**:805.

Wilson, B. W., Nieberg, P. S., Walker, C. R., Linkart, T. A., and Fry, D., 1973, Production and release of acetylcholinesterase by cultured chick embryo muscle, *Dev. Biol.* **33**:285.

Wu, G. J., 1978, Adenovirus DNA-directed transcription of 5.5S RNA *in vitro*, *Proc. Natl. Acad. Sci. USA* **75**:2175.

Wydro, R., Nguyen, H. T., Gubits, R., and Nadal-Ginard, B., 1981, Myosin heavy chain is coded by a multigene family with highly conserved domains, in: *Molecular and Cellular Control of Muscle Development*, p. 51, Sept. 8, Cold Spring Harbor Laboratory, New York.

Yablonka, Z., and Yaffe, D., 1976, Synthesis of polypeptides with the properties of myosin light chains directed by RNA extracted from muscle cultures, *Proc. Natl. Acad. Sci. USA* **73**:4599.

Yaffe, D., 1968, Retention of differentiation potentialities during prolonged cultivation of myogenic cells, *Proc. Natl. Acad. Sci. USA* **61**:477.

Yaffe, D., and Dym, H., 1973, Gene expression during differentiation of contractile muscle fibers, *Cold Spring Harbor Symp. Quant. Biol.* **37**:543.

Yaffe, D., and Feldman, M., 1965, The formation of hybrid multinucleated muscle fibers from myoblasts of different genetic origins, *Dev. Biol.* **11**:300.

Young, R. A., Hagenbachle, O., and Schibler, U., 1981, A single mouse α amylase gene specifies two different tissue specific mRNAs, *Cell* **23**:451.

Zengel, J. M., and Epstein, H. F., 1980, Muscle development in *Caenorhabditis elegans:* a molecular genetic approach, in: *Nematodes as Biological Models*, pp. 73–126, Academic Press, New York.

Zimmer, W. E., Jr., and Schwartz, R. J., 1982, Amplification of chicken actin genes during myogenesis, in: *Gene Amplification* (Schimke, R. T., ed.), pp. 137–146, Cold Spring Harbor, New York.

8

Role of Microtubules and Centrioles in Growth Regulation of Mammalian Cells

Robert W. Tucker

1. Introduction

Recent studies have shown that hormonal growth factors can change the cytoskeleton (Tucker *et al.*, 1979a; Tucker, 1980) and that changes in the cytoskeleton can affect the efficiency with which exogenous growth factors induce DNA synthesis (Teng *et al.*, 1977; Vasiliev *et al.*, 1971; Friedkin *et al.*, 1979; Otto *et al.*, 1979; Crossin and Carney, 1981a; Tucker, 1980; Baker, 1976b). The role of microtubules and centrioles in these interactions between growth factors and the cell has not yet been defined. However, one hypothesis, the surface modulation theory (Yahara and Edelman, 1979), which explains how microtubules might affect the role of growth factors, has been widely accepted. This theory assumes that growth factor receptors are connected to cytoplasmic microtubules and actin filaments, and that unhooking receptors from the cytoskeleton results in increased receptor motility and enhanced clustering or capping of receptors. This change of receptor position in the plasma membrane in turn either decreases or increases stimulation of DNA synthesis (McClain and Edelman, 1980; McClain *et al.*, 1977). While this hypothesis is attractive because it explains many diverse pieces of data in the literature, there is no definitive evidence that microtubules in fact directly affect the mobility of surface receptors. Moreover, recent studies (DeMey *et al.*, 1978; Osborn and Weber, 1977; Tucker *et al.*, 1978; Eichhorn and Peterkofsky, 1979) have not supported previous evidence (Brinkley *et al.*, 1975;

Robert W. Tucker • The Johns Hopkins Oncology Center, Baltimore, Maryland 21205 .

Fuller and Brinkley, 1976; Edelman and Yahara, 1976; Rubin and Warren, 1979; Zimmer *et al.*, 1980; Fonte and Porter, 1974) for different distributions of microtubules in cells (neoplastic and nonneoplastic) that respond differently to growth factors. The main problem is that most of the studies have been only correlative. Since there are many functions of the cell which involve changes in cytoplasmic microtubules and cytoskeleton, it has not been possible to prove that there are changes in the cytoskeleton specifically causing growth. For example, changes in cell motility often accompany DNA synthesis in lymphocytes stimulated by LPS (lipopolysaccharide) (Bhalla *et al.*, 1979), and motility changes themselves most certainly include drastic alterations in the cytoskeleton.

These same problems of correlative studies will also be apparent in the present discussion. However, my main purpose here is to use changes in various parts of the cytoskeleton to suggest how growth factors produce cell growth. I will discuss changes in microtubules, centrioles, and the primary cilium during the cell cycle in both neoplastic and nonneoplastic cells. A more detailed discussion of the roles of microtubules (mitotic and cytoplasmic), primary cilium, and centriole separation in the initiation of DNA synthesis stimulated by growth factors will follow. I will then discuss one way of interpreting how the cytoskeleton affects growth by considering the effect of calcium on microtubule structures.

1.1. Growth Factors and DNA Synthesis

Hormonal growth factors must interact with receptors on cell surfaces in order to start a sequence of events which eventually leads to the initiation of DNA synthesis. It has not yet been established whether internalization of growth factors and metabolism of their receptors actually produce growth stimulation (Herschman and Perdue, 1981). However, epidermal growth factor (EGF) has been shown to stimulate growth even when down-regulation or metabolism of its receptors does not occur (Wolfe *et al.*, 1980). Moreover, clustering of EGF receptors can be blocked (Maxfield *et al.*, 1979) without preventing growth stimulation. It is impossible to exclude that a few molecules of clustered or internalized growth factor are enough to stimulate growth, but in the remainder of my discussion, I will assume that growth stimulation depends primarily on events at the surface of the cell. I will begin the discussion of growth stimulation by distinguishing between early and late mitogenic events which are produced by different classes of polypeptide growth factors.

1.1.1. Early Mitogenic Events (Competence)

Platelet derived growth factor (PDGF) and fibroblast growth factor (FGF) have both been shown to induce early mitogenic events ("competence") in quiescent fibroblasts (3T3 cells) (Pledger *et al.*, 1977; Ross *et al.*, 1974; Ross *et al.*, 1978; Kohler and Lipton, 1974; Balk *et al.*, 1973; Vogel *et al.*, 1978). Once these early mitogenic events have occurred, the cells are then primed to respond to a second class of growth factors (e.g., plasma) which produce later mitogenic events ("progression") (Fig. 1). The first reports (Pledger *et al.*, 1978) suggested that competence could be induced for up to 24h by a brief exposure to growth factors. However, recently Dicker and Rozengurt (1981) suggested that the early mitogenic events can persist after a transient exposure to a growth factor only because the growth factor itself persists on the cell. These authors studied a tumor promotor (TPA; phorbol-12-13-dibutyrate) that was easily washed from the cell and then extrapolated the results to polypeptide growth factors (PDGF, FGF). However, TPA does not completely substitute for PDGF in growth promotion (Franz *et al.*, 1979), and may act to potentiate rather than mimic polypeptide growth factors. The remaining question is whether the specific proteins induced by competence factors (Pledger *et al.*, 1981; Smith and Stiles, 1981) are sufficient for competence and whether the maintenance of competence requires the continual presence of growth factors on the surface of the cell.

These early mitogenic events are particularly interesting because non-neoplastic fibroblasts require competence growth factors (e.g., PDGF, FGF), whereas neoplastic fibroblasts are able to proliferate without such factors (Stiles *et al.*, 1979; Scher *et al.*, 1978). Thus, cellular events induced by polypeptides such as PDGF and FGF must be closely related to growth regulatory events which are different in normal and neoplastic cells.

1.1.2. Late Mitogenic Events (Progression)

Only quiescent cells that have been briefly (4 h) exposed to FGF or PDGF can initiate DNA synthesis 12–14 h after exposure to an adequate concentra-

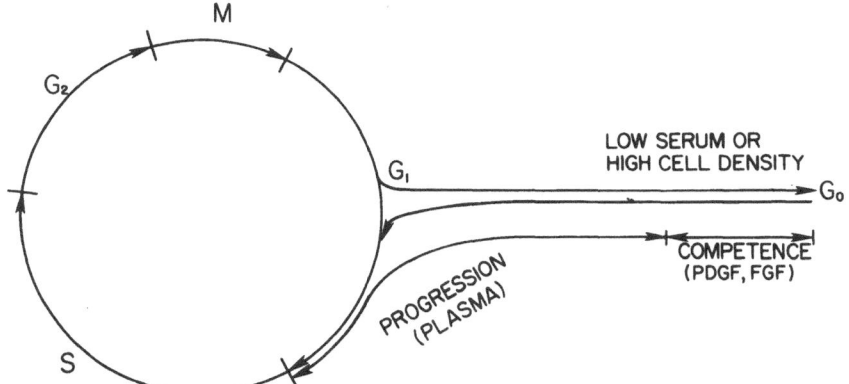

Figure 1. DNA synthesis cycle. M, Mitosis. G_1, first gap period. S, DNA synthesis. G_2, second gap period. G_0, Quiescence. Nonneoplastic cells leave G_1 period if deprived of serum growth factors or suitable substratum for attachment. Cells leaving G_0 undergo early mitogenic events ("competence," produced by PDGF or FGF) which prepare cells to enter late mitogenic events ("progression", produced by plasma).

tion of plasma (5% PPP—platelet–poor plasma) (Pledger *et al.*, 1977; Pledger *et al.*, 1978). The late mitogenic events (progression) in plasma require the continuous presence of the appropriate growth factors. Cells placed in plasma and then placed in PDGF do not enter DNA synthesis, and neither PDGF nor plasma alone will produce DNA synthesis (Fig. 1). While the molecular mechanism for synergy between plasma and PDGF in inducing DNA synthesis is not known, the two factors probably stimulate different but complementary protein and RNA synthesis events. In addition, as I will discuss later, there are also different events induced in the centriole cycle by the two classes of growth factors. It has also recently been suggested that at least one factor in plasma, somatomedin C, can stimulate DNA synthesis because competent cells have acquired receptors for somatomedin C (Clemmons *et al.*, 1980). Thus, synergy between different growth factors may be at the level of receptor formation in the plasma membrane.

1.2. Cell Cycle (DNA Synthesis)

The point in the cell cycle where these factors exert their cumulative effect on growth is in the G_1 period. Cells deprived of either plasma or PDGF will stop in G_1 and not enter DNA synthesis until the missing factor is supplied (Pledger *et al.*, 1977; Yen and Riddle, 1979) (Fig. 1). Nonneoplastic cells will stop in G_1 if placed in low serum or at high cell density (Todaro *et al.*, 1966). In contrast, neoplastic cells (at least virally transformed cells) usually do not stop at any one particular point in the cell cycle, but continue growing and eventually die at various points in the cell cycle (Pardee, 1974). Thus, the stopping point of nonneoplastic cells in G_1 represents a stable, viable state which some neoplastic cells can not enter. A corollary of these facts is that a cell (nonneoplastic or neoplastic) which has initiated DNA synthesis is committed to continue through the cell cycle and produce two daughter cells at mitosis. Thus, this "restriction point" (Pardee, 1974) in G_1 is a growth regulatory step at which cells must commit themselves either to continue growing or to become quiescent. Quiescence (G_0) may or may not be a state which differs fundamentally from that of cycling cells (Baserga, 1978), but operationally it can be distinguished from other G_1 states by the length of time it takes quiescent cells to enter DNA synthesis. Mitotic 3T3 cells will enter DNA synthesis in 5 h, while quiescent 3T3 cells stimulated with serum require 12–14 h before they start DNA synthesis.

1.2.1. G_1 Arrests

Most nonneoplastic cells in culture will arrest in G_1 when placed in physiologically deficient growth conditions, such as low amounts of growth factor or high cell density. Cells starved for amino acids such as isoleucine stop in G_1 (Yen and Pardee, 1978a; Baker, 1976a) in a different state (closer to DNA synthesis) than that induced by physiological deprivations. G_2 arrests in epidermal cells have been documented *in vivo*, but nonetheless, many epidermal cells lines in culture arrest in G_1 (Tucker, 1981). The significance of G_2 arrests is unknown and has not been extensively studied because of the lack of suitable cell culture models. G_1 arrests are not artifacts of tissue culture since many epithelial cells *in vivo* are in G_1 (G_0) arrested states. For example, uterine parenchymal cells are in G_1 until they are treated with estrogens (Tachi *et al.*, 1974), and liver cells have G_1 DNA content until they are stimulated by hepatectomy (Bucher and McGowan, 1979). A G_1 (G_0) arrest for nonneoplastic cells has the effect of preventing unnecessary duplication of the cell's genetic material under conditions (high cell density or low serum) in which the cell may not be able to form two daughter cells. Moreover, the existence of G_1 cell cycle arrests are ubiquitous throughout the animal kingdom; even yeast (Hartwell, 1978a; Hartwell, 1978b) and ciliates (Wolfe, 1973) can be stopped in G_1 under restrictive growth conditions. In order for mammalian cells not to enter quiescence, the cells must be exposed to serum factors during most of the G_1 period of the cell cycle. Thus, stimulated quiescent or postmitotic cells exposed to serum for only 1-2 h are not able to progress into DNA synthesis. In contrast, removal of serum from the same cells 1 or 2 h before initiation of DNA synthesis does not prevent DNA synthesis (Yen and Pardee, 1978b; Pledger *et al.*, 1978).

1.2.2. Neoplastic vs. Nonneoplastic Cells

As already mentioned, G_1 arrest may also identify growth regulatory differences between neoplastic and nonneoplastic cells. Nonneoplastic cells are able to utilize a stable stopping point in G_1, whereas neoplastic cells continue to grow despite restrictive growth conditions (Pardee, 1974). We assume that some of the events that nonneoplastic cells undergo when stimulated out of quiescence are also the events which neoplastic cells must bypass as they continue growing despite low amounts of serum or high cell density. Indeed, as we shall see, hormonal growth factors produce centriole changes which are bypassed when the same cells are stimulated by oncogenic viruses (Tucker *et al.*, 1979b).

It has been tempting to attribute some of these growth regulatory differences between normal and neoplastic cells to differences in their cytoarchitecture. Indeed, the prevalent view in recent years has been that neoplastic cells have a disorganized or diminished cytoplasmic complex of microtubules which increases the mobility of surface membrane receptors and allows the neoplastic cell to bypass certain normal growth requirements. The evidence for microtubule changes in neoplastic cells is controversial (see Introduction). It is, of course, impossible to disprove that some subtle change in the organization of cytoplasmic microtubules in neoplastic cells might be sufficient to change their response to growth factors. However, there is no convincing evidence that such a subpopulation of abnormal microtubules exists in neoplastic cells, let alone any idea of how they might be involved in growth control. Nonetheless, the idea that disorganized or diminished cytoplasmic microtubules might influence cell growth has received increased attention recently because a number of laboratories have reported that agents which depolymerize microtubules can potentiate growth factors (Friedkin *et al.,* 1980; Otto *et al.,* 1981; Crossin and Carney, 1981a; Tucker, 1980; Selden *et al.,* 1981). Thus, it has been proposed that neoplastic cells continue growing in restrictive growth conditions because their decreased or altered cytoplasmic microtubules increase their responsiveness to growth factors.

1.3. Centriole Cycle

During each cell cycle both centrioles duplicate, daughter centrioles mature and centriole pairs move to a mitotic pole in order to organize a mitotic spindle which participates in the separation of duplicated chromosomes to two daughter cells (Fig. 2). Details of the timing of these events with respect to the DNA synthesis cycle remain controversial. Nonetheless, it has been clear that centriole duplication and maturation are important morphologic and synthetic events in the centriole cycle which must be somehow correlated with events in the DNA synthesis cycle.

1.3.1. Centriole Duplication

There is conflicting evidence about whether centriole duplication occurs near the G_1/S border (Robbins *et al.,* 1968; Tucker *et al.,* 1981b; Snyder and

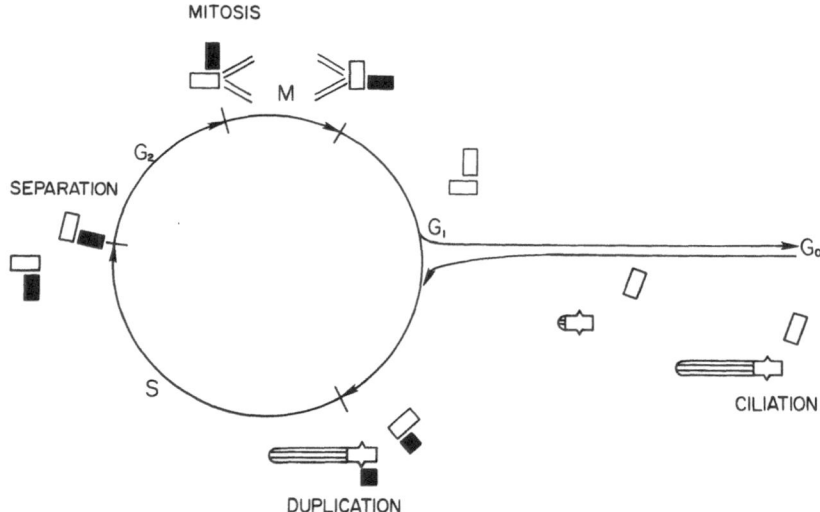

Figure 2.   Centriole cycle. G_1, Daughter centriole in pair of orthagonal centrioles continues to mature (increase length) during G_1, and parent centriole becomes capable of forming a primary cilium. G_0, Unduplicated centrioles, both mature in length, no longer orthagonal, one forming both primary cilium and basal feet. Growth factors producing "competence" cause further centriole separation and transient ciliary shortening. G_1/S, Near start of DNA synthesis, centrioles are capable of forming a procentriole. S, Procentrioles continue to mature. G_2/M, Centriole pairs separate to form spindle. Somewhere between S and M cilium is resorbed.

Liskay, 1978) or sometime during S phase (Alvey, personal communication; McGill *et al.,* 1976; Vorobjev and Chentsov, 1982). Part of the difficulty has been that cell cycle variability during G_1 has made the timing of events near the G_1/S border relatively imprecise. Centrioles duplicate by forming a rudimentary centriole, a procentriole, at right angles to the parent centriole and situated at the proximal end associated with most of the pericentriolar material. The orthogonality of parent and daughter centriole is maintained throughout most of the centriole cycle and is thought to change only when the daughter centriole starts to become a autonomous unit in the next cell cycle (Stubblefield and Brinkley, 1967). The centriole is composed of nine triplets of microtubules arranged in the periphery of a cylinder and held together by connections whose functions are unknown, but which may be involved in forming the structures associated with the proximal (pericentriole material and procentriole) and distal parts of the centriole (Albrecht-Buehler and Bushnell, 1980; Brinkley and Stubblefield, 1970; Vorobjev and Chentsov, 1980). The centriole is also associated along its lengths at various intervals with basal feet structures which are also connected with microtubular organizing material (Fig. 2). It has been suggested that the procentriole template contains RNA and protein which determine the detailed structure of the centriole (Hartman, 1975). In fact, RNA synthesis is essential for centriole duplication (McGill, *et al.,* 1976; Deutch and Shumway, 1973; Went, 1977b). A suggestion that RNA viruses (Went, 1977a) play some role in the initiation of centriole duplication has not been verified. On the other hand, one study showed that drugs which interfere with RNA metabolism can also inhibit centriole duplication (DeFoor and Stubblefield, 1974), but such high drug concentrations were used that separation of RNA and protein synthesis effects was not possible. In fact, other studies (Phillips and Rattner, 1976; Deutch and Shumway, 1973) have shown that protein synthesis is necessary for successful initiation of centriole duplication. Centriole duplication can occur if DNA synthesis is initiated, but not completed. In fact, in the presence of agents such as Ara-C that allow initiation of DNA synthesis but prevent its completion, centriole pairs mature but do not separate to form the mitotic spindle (Rattner and Phillips, 1973). Recent evidence also suggests that a nucleus is somehow able to direct the initiation of centriole duplication (Zorn *et al.,* 1979) and that, without a nucleus, centriole duplication does not occur (Kuriyama and Borisy, 1981). A nucleus may be necessary for procentriole formation, but centriole maturation continues in the absence of DNA synthesis progression. It is not known whether DNA synthesis can occur in the absence of a centriole in mammalian cells, or whether centriole duplication can occur without at least the initiation of DNA synthesis.

1.3.2. Centrosome Maturation

In sea urchins, centrioles not only start to duplicate in telophase before DNA synthesis initiation, but also continue to increase in length after mitosis (Mazia *et al.,* 1960; Bucher and Mazia, 1960). In a fertilized egg, a centriole seems capable of initiating duplication of its daughter while still maturing in length. Rattner and Philips (1973) noted in Hela cells a similar delay in maturation in the length of the daughter centriole until the next G_1 period of the cell cycle. At the same time, centriole duplication did not start until G_1/S or S Phase. Thus, in mammalian cells centrioles appear to finish their physical maturation before forming other centrioles, and centriole maturation overlaps more than one DNA synthesis cycle. This point has been recently emphasized in an elegant electron microscopic study of the centriole cycle in epithelial cells (Vorobjev and Chentsov, 1978a,b), and forms part of the basis of a cell kinetic theory which invokes parallel centriole and DNA synthesis cycles to explain the control of the cell cycle (Brooks *et al.*, 1980).

In addition to increases in centriole length which can be documented by electron microscopy, the capacity of pericentriolar material to initiate the polymerization of microtubules is another step in centriole or centrosome maturation. In mammalian cells, isolated centrosomes from mitotic cells nucleate more microtubules than do interphase centrosomes (Gould and Borisy, 1977; Kuriyama and Borisy, 1981). In sea urchins a similar maturation of centrioles results in formation of multiple mitotic poles in cells which have been arrested in mitosis either by drugs which depolymerize microtubules or by mechanical agents which do not allow the mitotic spindle to function (Sluder, 1978). In situations in which the mitotic cycle is delayed, mature centrioles are able to split and form mitotic poles which by electron microscopy appear to contain only one centriole (Mazia *et al.*, 1981). Thus, under special conditions, even the functional maturation of the centriole or centrosome can occur independently of the duplication cycle of the centriole. The function of the pericentriolar material, which contains RNA and protein, can be inhibited by RNase (Heidemann *et al.*, 1977) and by agents which interfere with RNA metabolism (McGill *et al.*, 1976; Peterson and Berns, 1978). How the RNA dependent events associated with procentriole formation and nucleation of cytoplasmic microtubules are coordinated has not been studied.

1.4. Primary Cilium

In addition to cytoplasmic and mitotic microtubules arising from the pericentriolar material at the proximal end of the centriole, a cilium is formed at the distal end of the centriole (Fig. 3). The nine microtubule triplets in the centriole become nine doublets of microtubules in a primary cilium. Originally this "9 + 0" arrangement of microtubule doublets (no central doublet of microtubules) was thought to be a precursor to the "9 + 2" arrangement found in mature cilia and flagella. Later work (Sorokin, 1968; Sorokin, 1962) showed that primary cilia are formed by an interphase centriole inside the cell near the nucleus, rather than at the free surface of the cell where "9 + 2" cilia arise. We also now know that the primary cilium remains "9 + 0" and does not change its microtubule structure during the cell cycle (Tucker *et al.*, 1979a; Rieder *et al.*, 1979; Albrecht-Buehler and Bushnell, 1980; Wheatley, 1967, 1969). The primary cilium can be found in virtually every cell (except perhaps

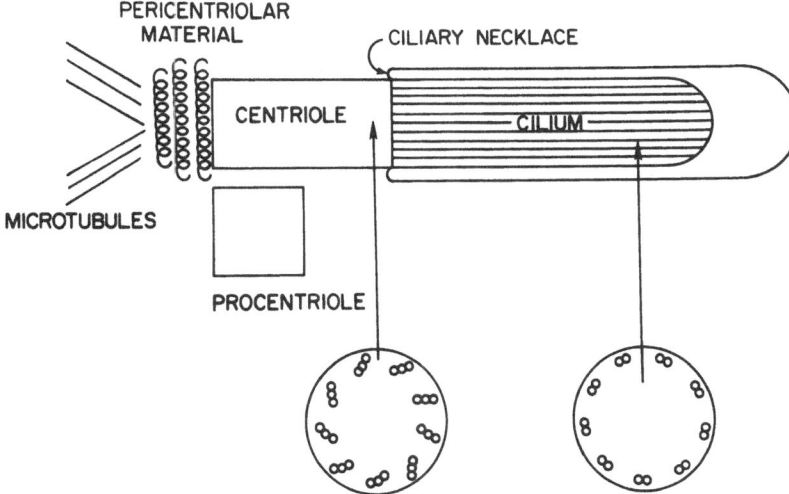

Figure 3. Centrosome. *Proximal end* of centriole associated with pericentriolar material which nucleates microtubule polymerization. Procentriole is also formed at the proximal end of the centriole. The *distal end* forms a primary ("9 + 0") cilium consisting of microtubule doublets arising out of centriole microtubule triplets. Cliniary necklace marks the junction of the ciliary and vacuole (plasma) membranes.

haematopoietic) in a multicellular organism (Tucker, 1982; Barnes, 1961; Biscoe and Stehbens, 1966; Lin and Chen, 1969; Grillo and Palay, 1962; Fritz-Niggli and Suda, 1972; Latta *et al.*, 1961; Meier-Vismara *et al.*, 1979; Przybylski, 1971; Scherft and Daems, 1967; Tanuma and Ohata, 1978; Wilsman, 1978; Dahl, 1963; Fujita, 1963; Bjorkman, 1962; Martinek *et al.*, 1973; Pathak *et al.*, 1978; McCarron and Anderson, 1973; Jirsova, 1972; Gallagher, 1980; Renard *et al.*, 1976; Flood and Totland, 1977; Quattropani, 1973; Crisp and Browning, 1968), from annelids (Gardiner and Rieger, 1980) to man (Tucker, unpublished observation; Perry *et al.*, 1979; Myklebust *et al.*, 1977; Vegge, 1963; Sturrock, 1975). Such a ubiquitous distribution argues against simply a vestigial function for the primary cilium (Tucker and Pardee, 1981).

1.4.1. Sensory Function

The function of the primary cilium is not known, but, at least in some differentiated cells, the primary cilium can be associated with sensory functions. Indeed, in the rods and cones of the mammalian eye the primary cilium has been modified to serve a primary sensory function. Many workers have now documented that modifications of both primary and motile cilia can act as mechanical receptors in insects (Wiederhold, 1976; Stommel *et al.*, 1980). It has been suggested that a bend in the primary cilium changes membrane properties near the base of the cilium, thereby producing changes in ionic fluxes through the membrane (Moran and Rowley, 1975; Moran and Varela, 1971). The resulting change in membrane potential is a signal to the organism that pressure has been applied to that position. Ionic changes have been suggested to occur through the primary cilia in mammalian cells. Freeze fracture studies of the ciliary necklace (Fig. 3) of primary cilia have revealed intramembranous particles which are very similar to those in the sarcoplasmic reticulum, a structure involved with the uptake and release of calcium. These studies suggest that the primary cilium may help mediate calcium flux changes in mammalian cells (Gilula and Satir, 1972; Satir *et al.*, 1976). However, in some receptor systems a primary cilium can be removed without changing the ultimate sensory function of the apparatus and in other systems, taking away only the ciliary microtubules can disrupt sensory function (Moran *et al.*, 1976). Thus, primary cilia are involved in sensory structures, but their precise role in sensory function has not been defined. It is even more unclear what sensory role, if any, the primary cilium could be serving in mammalian cells.

1.4.2. Quiescence and Differentiation

Ciliation was originally found associated with nondividing cells (Zimmerman, 1898). Some early studies suggested that formation of a cilium by a centriole acts to inhibit mitosis and throw the cell out of the cell cycle into quiescence or differentiation (Lenhossek, 1898; Henneguy, 1898; Rash *et al.*, 1969; Dingemans, 1969). It was thought that the centriole could not simultaneously form both a mitotic spindle and a cilium or flagellum. However, there are many algae in which the centriole forms both a flagellum and a mitotic spindle (Floyd, 1978), and it has been long recognized that centrioles participating in meiosis can also form flagella (Roth *et al.*, 1966; Henneguy, 1898). Certainly there is no structural reason why the pericentriolar material at the proximal end of the centriole cannot form mitotic or meoitic microtubules while the distal end forms a cilium or flagellum. The idea of ciliation being an inhibitor of mitosis also was questioned when it was found that rapidly dividing tissues have a high proportion of cells with a primary cilium (Fonte *et al.*, 1971). At the same time, immunofluorescence studies showed that most quiescent cells had primary cilia (Archer and Wheatley, 1971; Albrecht-Buehler, 1977b; Mori *et al.*, 1979; Tucker *et al.*, 1979a). Only recent studies on the cell cycle distribution of ciliation have resolved some of these difficulties.

2. Mitotic Microtubules

2.1. Mitotic Arrest

While the function of mitotic microtubules in the complicated orchestration of chromosome movements has not been completely defined, it is clear that mitotic microtubules are necessary for completion of mitosis. If mitotic microtubules are depolymerized any time before the end of anaphase movement, mitosis is aborted (Inoue', 1952; Inoue' and Sato, 1967). Similarly, if microtubules are depolymerized in interphase cells, the mitotic spindle does not even form. Interestingly enough, in this latter situation, a "C-metaphase" arrest occurs because centriole pairs do not move to the poles in order to organize the mitotic spindle (Brinkley *et al.,* 1967). Since there is also often a prominent band of microtubules between the two separating centriole pairs, it has also been suggested that microtubules somehow facilitate the separation of the centriole pairs (Rattner and Berns, 1976). However, recent laser irradiation studies of fungi have shown that disruption of a band of microtubules between incipient daughter nuclei can in fact accelerate the separation of the nuclei at anaphase–telophase (Aist and Berns, 1981). Thus, our information about the role of microtubules in the separation of centriole pairs appears conflicting: microtubules are necessary for the onset of centriole pair separation near prophase, but a subset of microtubules seem to retard the speed of separation of analgous structures in fungi. These studies were done with different techniques on different mitotic phases in different cells, so more experimental work is necessary before centriole separation can be more clearly understood. It may be that microtubules do not directly pull apart the centriole pairs, but affect some cellular event which is necessary for their separation.

2.2. Mitotic Cycle

As implied above, depolymerization of microtubules affects the timing of various aspects of the mitotic cycle. The mitotic cycle includes metaphase orientation of chromosomes and anaphase movement of chromosomes to the poles. Timing of events during the mitotic cycle appears to depend on the structural integrity of microtubules. For example, taking away microtubules for a period of time will cause the subsequent acceleration of later events in order to minimize the total amount of time involved in mitosis (Hamilton and Snyder, 1981). In preparation for anaphase movement, there seem to be some critical events which depend on the presence of polymerized microtubules in the mitotic spindle (Sluder, 1978; Sluder, 1979). In contrast, the initial separation of chromosomes at the onset of anaphase seems to be independent of a complete set of mitotic microtubules, since removal of kinetochore-pole microtubules by laser beam irradiation does not prevent the chromosome pair attached to the irradiated kinetochore from separating at the onset of anaphase (McNeill and Berns, 1981). At the same time, the structural integrity of mitotic microtubules does seem necessary for specifying the plane of cytokinesis perpendicular to the plane of the mitotic spindle (Rappaport, 1971; Rappaport and Rappaport, 1974). Another membrane change which may be related to cytokinesis, but is independent of microtubules, are rhythmic surface contractions observed in amphibian eggs (Kirschner *et al.,* 1980; Hara *et al.,* 1980). Thus, although the details are far from worked out, it is clear that mitotic microtubules affect the timing and spatial location of some but not all membrane and organelle movements during cell division.

3. Cytoplasmic Microtubules

3.1. Disruption of Microtubules

As mentioned in the Introduction, there are now many studies showing that agents which depolymerize microtubules (e.g., colchicine, nocodazole) can potentiate the initiation of DNA synthesis induced by growth factors. One of these studies in fact suggested that depolymerization of microtubules is itself a growth stimulus (Crossin and Carney, 1981a). However, in this latter study, colchicine actually potentiated growth already occurring in unstimulated cultures, rather than directly stimulating the cells in the absence of growth factor. We have also found that treating cells with agents which are capble of depolymerizing microtubules only potentiates, and does not substitute for growth factors. Preliminary evidence indicates that a disruption of microtubules increases growth effects produced by plasma and not those related to platelet derived growth factor or fibroblast growth factor (Tucker, 1980). Thus, disruption of microtubules may mimic early mitogenic events related to the establishment of a competent state (which thereby increases DNA synthesis produced by plasma) by either directly producing that state itself, or greatly potentiating attached growth factors. However, our studies and those in the literature have not shown conclusively that this potentiating effect is definitely related to the depolymerization of microtubules. Most of the studies have used drug concentration which have been shown to depolymerize microtubules in other systems (Otto *et al.*, 1979; Friedkin *et al.*, 1979; Friedkin *et al.*, 1980), and a more recent study using immunofluorescence with anti-tubulin antibody to monitor depolymerization of microtubules suggests but does not prove that actual depolymerization of microtubules is a prerequisite for growth potentiation (Crossins and Carney, 1981a). Some studies have also documented growth factor potentiation even after the depolymerizing agent has been removed from the cell (Otto *et al.*, 1981). However, in these cases it has not been documented by either immunofluorescence microscopy or electron microscopy that microtubules repolymerized during the subsequent addition of growth factors.

Depolymerization of microtubules might potentiate growth factors by increasing the number of occupied receptors (Friedkin *et al.*, 1980). This theory assumes that intact cytoplasmic microtubules are necessary for internalization (down-regulation) of receptors and subsequent decrease in the number of surface receptors. A decrease in cytoplasmic microtubules might then result in an increased number of occupied receptors on the surface membrane. There is evidence that the amount of EGF associated with plasma membranes decreases more slowly in the absence of polymerized microtubules (Brown *et al.*, 1980). However, there is currently no independent proof that this change in occupancy of receptors in fact causes changes in growth produced by microtubular depolymerizing agents. It is equally possible that potentiation of growth factors by microtubular depolymerization might occur at a step after the occupancy of the growth factor receptor. For example, receptor clustering might influence growth. In fact, the surface modulation theory (Yahara and Edelman, 1979; Edelman, 1976) suggests that unhooking membrane receptors from cytoplasmic microtubules would allow increased mobility of receptors and subsequent clustering. Treatment of fibroblasts with agents which depolymerize microtubules can increase capping of Con A receptors (Ukena *et al.*, 1974) and, in other studies, increased growth. However, there is no evidence that both microtubular depolymerization and growth have occurred in the same cells, nor is there any dose response showing that only when microtubules are depolymerized does increased capping or growth occur. In lymphocytes, colchicine inhibits growth stimulation, and releases Con A's inhibition of I_GG capping (Wang *et al.*, 1975). However, capping of I_GG is on B cells which are not activated by Con A (Betel and Martijnse, 1976), and colchicine is probably allowing Con A-treated lymphocytes to increase their cellular motility (Unanue and Karnovsky, 1974). Capping may actually reflect a balance between aggregation in the plasma membrane and interiorization, with capping predominant in lymphocytes, and endocytosis dominant in fibroblasts (Storrie and Edelson, 1977). In neither cell type is it clear whether capping or endocytosis of receptors simply removes used receptors from the plasma membrane, or is an integral part of growth stimulation.

Another approach to this problem has been to look at changes in the cytoskeleton which accompany capping. It is possible that binding of gamma globulin or Con A to its receptor may actually induce changes in the cytoskeleton which can mimic depolymerization changes induced by drugs. Fluorescence studies on lymphocytes which have been exposed to Con A show that cells which have caps tend to have a decreased microtubule complex as revealed by indirect immunfluorescence with antitubulin antibody (Yahara and Kakimoto-Sameshima, 1978). However, such antibody stains for microtubules in lymphocytes are controversial, and indeed, one group has claimed that capping of lymphocytes does not involve a change in cytoplasmic microtubules (Rogers and Brown, 1979). Part of the experimental problem in the latter case is the use of microtubular stabilizing buffers during the fixation process which may potentiate the polymerization of microtubules. The microtubular networks under the sites of capping may be a very dynamic network (Albertini and Anderson, 1977) and in fact may be able to repolymerize during the fixation process in a polymerization buffer. Alternatively, experiments which do not use stabilizing buffers may not preserve cytoplasmic microtubules in a capped cell. Studies have also shown that there are aggregations of actin and myosin near the capping area (Braun *et al.*, 1978a; Braun *et al.*, 1978b) which may be explained by a reorganization of the cytoskeleton induced by calcium whose submembranous release is triggered by the occupancy of a receptor on the surface membrane (Hoover *et al.*, 1981; Klausner *et al.*, 1980). At the present time it is not clear that these particular changes in actin and myosin can be induced by growth factor stimulation of fibroblasts. There are also no obvious changes in cytoplasmic microtubules in quiescent cells stimulated with growth factors, but we will see later that changes in another microtubular system (primary cilium) do reflect changes induced by growth factors (Tucker, 1980).

3.2. Stabilization of Microtubules

If the growth response of cells normally depends on the depolymerization of some microtubules in the cell, then prevention of microtubule depolymerization should inhibit cell growth in response to growth factors. A recent study takes this point of view and shows that taxol, an agent which potentiates the polymerization of microtubules (Schiff *et al.*, 1979; Schiff *et al.*, 1980), does indeed inhibit the growth of some cell lines *in vitro* (Crossin and Carney, 1981b). However, in this latter study a dose response was not done and we do not know to what extent the growth-inhibitory response is due specifically to the effect of taxol on the polymerization of microtubules. In our own studies, we have found that taxol can increase the stability of microtubule systems without affecting the response of DNA synthesis to a maximal growth stimulus (Tucker, 1980). More studies should be done on taxol's effect on the cell's response to different concentrations of growth factors. Since high doses of taxol can be toxic to cells, it is especially important to distinguish a reversible effect on polymerization of microtubules from general cellular toxicity. This is a promising area of research, especially if one can find other drugs which stabilize microtubules. In the future, correlating changes in subpopulations of microtubules with cell growth might give us a clearer picture of the necessary or sufficient changes in microtubule patterns in the cell required for increased susceptibility of cells to growth factors.

4. Primary Cilium

4.1. Light Microscopic Observations

Many authors have shown that a primary cilium is present in quiescent cells, but the relationship of ciliation to the rest of the cell cycle has been controversial (see Introduction). Lloyd (1979) has suggested that a loss of a primary cilim may be necessary in order to enter mitosis and may reflect a primitive growth switch which enables non-growing cells to divide. In our own work we studied changes in the primary cilium in quiescent 3T3 cells which had been stimulated by various growth factors (Tucker *et al.*, 1979b; Tucker, 1980). Within 2–4 h after stimulation, cells transiently lost the primary cilium as detected by immunofluorescence using antitubulin antibody; the primary cilium reappeared again within 6–8 h after growth factor stimulation. This decrease in our ability to see the primary cilium by light microscopy only occurred when competence growth factors were used (Tucker *et al.*, 1979b). Factors such as plasma did not produce any change in the primary cilium. Competent cells whose cilia had already gone through this transient change were able to enter DNA synthesis in the presence of plasma. As the cells entered DNA synthesis and subsequent mitosis, the primary cilia disappeared and were not seen again until the next cell cycle (Tucker *et al.*, 1979a; Tucker *et al.*, 1979b). This late ciliary change probably reflects resorption of the primary cilium which has been documented by electron microscopy to occur near mitosis in at least one cell line (PTK$_1$) (Rieder *et al.*, 1979). This data indicates that cells probably acquire the capacity to form a primary cilium sometime during G$_1$ and lose the cilium sometime after the initiation of DNA synthesis (Tucker, 1980). The timing of ciliation explains why both quiescent and rapidly dividing cells can have a primary cilium. Cells in late G$_1$ are capable of forming a primary cilium and do so before or after they become quiescent. Similarly, cells which do not become quiescent also form a primary cilium in G$_1$ or S. In other words, instead of inhibiting DNA synthesis, formation of the primary cilium marks a part of the cell cycle during which cells can become committed to growth or quiescence.

Exactly which phases of the cell cycle are associated with the formation of a primary cilium in exponentially growing cells is not clear. Mitotic cells in metaphase are not ciliated (Wheatley, 1969; Reider *et al.*, 1979), while cilia have been seen in S phase cells (Tucker, 1980; Pendergast and Lockwood, personal communication). The major uncertainty concerns whether a primary cilium is made during G$_1$ or S in exponentially growing cells. Direct attempts to measure ciliation in G$_1$ (mitotic shake) cells were ambiguous (Archer and Wheatley, 1971), and extrapolations from our own results with quiescent cells may not be justified. It is also not known whether 3T3 cells have a remnant of a primary cilium near prophase, as has been shown for PTK$_1$ cells (Rieder *et al.*, 1979). Thus, at the present time we can only estimate that 3T3 cells have primary cilia in perhaps late G$_1$, all or part of S, and perhaps G$_2$ up to prophase.

Changes in ciliation may reflect early mitogenic events which are necessary before a cell can enter into DNA synthesis. A concentration of growth factor which produces DNA synthesis is also capable of producing a transient ciliary change (Tucker *et al.*, 1979b). In individual cells we have preliminary evidence that each cell must lose or shorten its primary cilium in order to eventually enter DNA synthesis. Moreover, these early ciliary events seem to differ in nonneoplastic and neoplastic cells (Fig. 4). For example, when SV40 virus was added to quiescent 3T3 cells, DNA synthesis occurred without an accompanying "deciliation" of the primary cilium (Tucker *et al.*, 1979b). Because the early "deciliation" changes correlated with important growth events, we examined these stimulated quiescent cells in the electron microscope to further define the structural changes in the centriole and its associated structures.

4.2. Electron Microscopic Observations

Random electron microscopy sections done in an early study (Tucker *et al.*, 1979a) revealed that single longitudinal profiles of centrioles without an attached primary cilium were more abundant after serum or growth factor

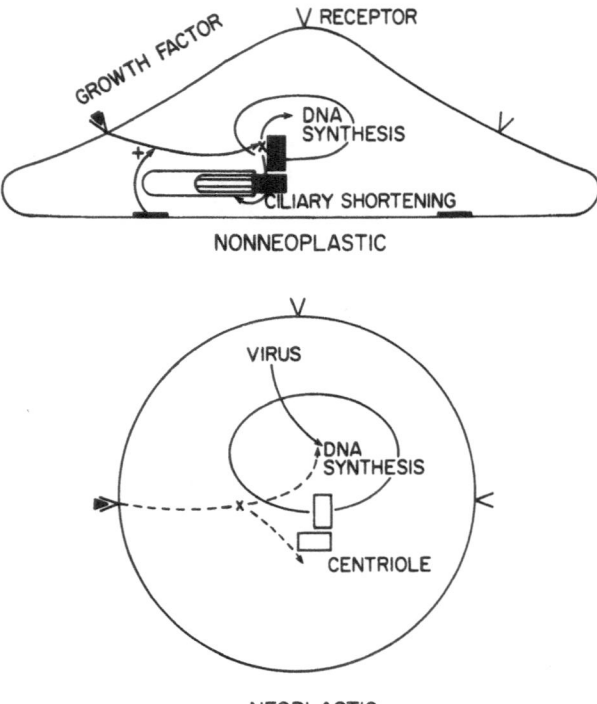

Figure 4. Nonneoplastic and neoplastic cells. *Nonneoplastic* cells need substratum attachment and/or spreading to facilitate (+) pathway by which growth factors induces both ciliary shortening and steps toward DNA synthesis. A transient increase in intracellular calcium (x) may produce both of these effects. In some *neoplastic* cells, the growth factor pathway which depends on substratum attachment is bypassed by cellular factors (e.g., virus or oncogene) which directly stimulate DNA synthesis.

stimulation of quiescent 3T3 cells than before stimulation. However, subsequent analysis by serial thin section electron microscopy revealed that parent and daughter centrioles separated from each other after growth factor stimulation; thus, some of the single centrioles in random EM sections never had a cilium, and their lack of ciliation can not be used to imply deciliation. Instead of deciliating after growth stimulation, the primary cilium shortens dramatically (two to four times shorter) so that it cannot be easily visualized using a light microscope. In fact, in 2–4 h, over 50% of the cells have shortened primary cilia as observed by electron microscopy and 50–70% of cells have primary cilia undetectable by light microscopy. The ciliated centriole maintains its membrane attachment at the ciliary necklace and continues to form transitional fibers, basal feet and striated rootlets. The only structural change is a shortening of microtubules in the primary cilium. Thus, we now know that "deciliation" reflects a depolymerization of ciliary microtubules (without any appreciable involvement of cytoplasmic microtubules) and is one structural change in the cytoskeleton which is produced by growth factors involved specifically in the growth of nonneoplastic fibroblasts.

4.3. Relationship between Ciliary and Cytoplasmic Microtubules

In studying the effects of various drugs on the polymerization of ciliary microtubules, we have found some surprising differences and interactions between ciliary and cytoplasmic microtubules. Agents such as nocodazole and colchicine depolymerize cytoplasmic microtubules, but not ciliary microtubules. In contrast, as already mentioned, growth factors can induce the depolymerization of ciliary microtubules, while leaving cytoplasmic microtubules intact. Taxol (10^{-5} M) is able to potentiate the polymerization of both cytoplasmic and ciliary microtubules in 3T3 cells (Tucker, 1980). However, after growth factor stimulation, low doses of taxol (10^{-6} M or 10^{-7} M) actually prevent repolymerization of ciliary microtubules. The most likely explanation for these results depends on the preferential effect of taxol on cytoplasmic microtubules. Depolymerized ciliary microtubules induced by growth factors probably enter the cellular tubulin dimer pool and are in equilibrium with both ciliary and cytoplasmic microtubules. Ordinarily, tubulin dimers can be reinserted into the primary cilium, but low concentrations of taxol must increase the polymerization of cytoplasmic more than ciliary microtubules. The resulting incorporation of tubulin dimers into cytoplasmic rather than ciliary microtubules prevents reciliation of stimulated cells. Not only do these results show that ciliary and cytoplasmic microtubules can be in equilibrium under special conditions, but we now have agents which can selectively affect the polymerization of ciliary microtubules. The use of such drugs should enable investigators to ask detailed questions about the relationships between ciliary changes and DNA synthesis in individual cells.

5. Centriole Separation

5.1. Light Microscopic Observations

We and others (Sherline and Mascardo, 1982b; Brooks, personal communication) have recently found that agents (nocadazole) which depolymerize microtubules also cause increased separation of microtubule organizing centers (MTOCs). In addition, MTOCs that are stained with an antibody to microtubule associated proteins (MAPs) appear to separate following stimulation of the cells with growth factors (Sherline and Mascardo, 1982a). Sherline has also found that this MTOC separation is produced by calcium ionophore and prevented by taxol, thereby implicating calcium-sensitive microtubules in the MTOC separation induced by growth factors (Sherline and Mascardo, 1982b). Thus, both ciliary shortening and MTOC separation are cytoskeletal changes probably caused by depolymerization of calcium-sensitive microtubules induced by growth factors (see Section 7.1). The time course of the MTOC separation has not been studied, but it has been found that both normal and neoplastic cells seem to respond in similar ways. Since it occurs in both ciliated (3T3) and unciliated (Hela) cells, the separation of MTOCs do not require the presence of a primary cilium.

5.2. Electron Microscopic Observations

In a recent analysis of serial EM sections of 3T3 cells treated with serum, we found that parent and daughter centrioles separated (up to 20 μm) following stimulation of quiescent 3T3 cells (Jayaraman and Tucker, 1981). Electron microscopy has documented that nocodazole also induces a physical separation of the parent and daughter centrioles (Tucker, unpublished observations). In quiescent 3T3 cells, the centriole pair forming the primary cilium is in an indentation of the nucleus near the substratum to which the cell is attached. After stimulation of quiescent cells by growth factors, parent and daughter centrioles separate from each other. However, not enough information is presently available to quantitate the extent and direction of centriole movements which can be induced by growth factors. Nor has it yet been documented which growth factors can induce these kinds of separations. Such extensive centriole separation normally only occurs near the time of prophase separation of the centriole pairs (Rattner and Berns, 1976), and occurs, therefore, only after centriole duplication. A small separation (a fraction of a micron) of parent and daughter centriole occurs near the time of centriole duplication, but, at least in stimulated quiescent 3T3 cells, centriole duplication occurs more than 12 h after serum stimulation. Thus, the centriole separation which has been documented in these more recent studies reflects the separation of parent and daughter centrioles and not that of one mature pair with respect to another mature pair of centrioles. Whether the separation of unduplicated centrioles has any particular function in the cell cycle, or whether it reflects important cellular events has not been established.

6. Relationships between Centriole and DNA Synthesis Cycles

Implicit in previous discussions has been the fact that centriole and DNA synthesis cycles must be coordinated for normal cell growth and population doubling. The centrioles must be duplicated and positioned at the mitotic poles by an appropriate time in the cell cycle in order to help separate the duplicated genetic material to the two daughter cells. It obviously would make no teleologic sense for the cell to duplicate its DNA if its centrioles were also not committed to their duplication and movements so that mitosis could be accomplished. Indeed, a series of cell cycle mutations in yeast have revealed that the spindle plaque, which is an analogue of the centrosome (centriole plus associated structures), must duplicate and separate before the initiation of DNA synthesis can occur (Hartwell, 1978a). In fact, there is a mutation in the structure of the spindle plaque which appears capable of stopping yeast in the cell cycle (Hartwell, 1978b). Mating hormone and limited nutritional conditions can also stop yeast in the cell cycle with G_1 DNA content (Byers and Goetsch, 1975). When mating hormone is removed, or nutrition supplied, spindle plaque separation and duplication occur before the initiation of DNA synthesis. Hartwell (1978a) has postulated that the centrosome analogue in yeast, the spindle plaque, integrates many cellular phenomena such as bud formation, initiation of DNA synthesis and morphologic changes in the spindle plaque itself. Definitive proof for similar responsibilities of mammalian centrioles has not been found, but there are clues that centriole events are related in important ways to DNA synthesis in mammalian cells.

6.1. Parallel Pathways

Since centriole duplication and maturation can occur regardless of whether S phase is allowed to progress (Rattner and Phillips, 1973), these centriole events must be independent of, but concurrent with, DNA synthesis. In this sense, centriole and DNA synthesis must be on parallel pathways. Thus, once DNA synthesis is initiated, the cell has either committed itself to, or has actually initiated, centriole duplication, and centriole maturation progresses regardless of whether DNA synthesis itself continues. This is also compatible with recent evidence from electron microscopy autoradiography of PTK_2 cells which demonstrates that cells entering S phase start to make a recognizable procentriole (Tucker *et al.*, 1981). This close temporal association between DNA synthesis initiation and centriole formation has also been suggested by an earlier study (Robbins *et al.*, 1968), but has not been verified in recent studies in which centriole duplication appears to be in the middle of S phase (McGill *et al.*, 1976; Alvey, personal communication; Vorobjev and Chentsov, 1982) (see Section 1.3.1). Unfortunately, all electron microscopy studies suffer from the fact that centriole duplication can only be marked by a morphological event, the growth of a large enough procentriole to be recognized in the electron microscope. The time at which the procentriole is being seen in any electron microscopy study may have been long after the actual initiation of procentriole formation. The most compelling data about the temporal relationship of centriole formation and the initiation of DNA synthesis comes from drug studies which show that cells blocked just after the initiation of DNA synthesis are not blocked from forming new centrioles (Rattner and Phillips, 1973). This data suggests that either the centriole or the nucleus has become committed to forming procentrioles by the time the cell has initiated DNA synthesis (Fig. 2). Either the actual initiation of DNA synthesis releases a signal from the nucleus which then induces a cytoplasmic change resulting in the formation of procentrioles, or a change in a procentriole occurs first and influences the initiation of DNA synthesis. This latter possibility has a counterpart in yeast, as we have already discussed. A third possibility, which would be analogous to ciliary shortening and competence (Tucker, 1980), is that both centriole and DNA replication are produced by a third process.

We have also found that changes in ciliation occur in parallel to the DNA synthesis cycle. By using taxol, a drug which potentiates the polymerization of microtubules, we have been able to prevent shortening of the primary cilium without inhibiting the initiation of DNA synthesis (Tucker, 1980). This simple experiment shows that ciliary shortening does not have to occur in order for cells to eventually enter DNA synthesis in the presence of plasma. Instead, ciliary shortening is independent of and occurs at the same time as competence for DNA synthesis; ciliary shortening must reflect and not cause events that are necessary in order for cells to enter DNA synthesis (Fig. 5). Growth factors must produce a cellular event (X) which then produces at least two effects: ciliary shortening and initiation of DNA synthesis. These two events must be related to a common factor (X) since only doses of growth factors which are capable of inducing ciliary shortening are also capable of inducing competence and eventual DNA synthesis in the presence of plasma (Tucker *et al.*, 1979b). The common factor, X, may be related to calcium fluxes (see Section 7.1).

Whether there are parallel (independent and concurrent) changes occurring in centriole and DNA synthesis cycles at mitosis is not yet clear. One of the first centriole events suggested to be related to the DNA synthesis cycle was a change in ciliation associated with mitosis. Early workers (Henneguy, 1898; Lenhossek, 1898) suggested that a cell in which a centriole near the nucleus formed a cilium (now known to be a primary cilium) was committed to a nondividing, differentiated state. As already mentioned, we now know that ciliation does not prevent cell growth, but it is true that many organisms (Lloyd, 1979) resorb their cilia or flagella ("9 + 2"), and other microtubular structures before mitosis (deTerra, 1978; Tucker, 1970; Zeuthen and Williams, 1967). Although cultured mammalian cells also lose their primary ("9 + 0") cilia by the time a metaphase mitotic spindle forms (Rieder et al.,

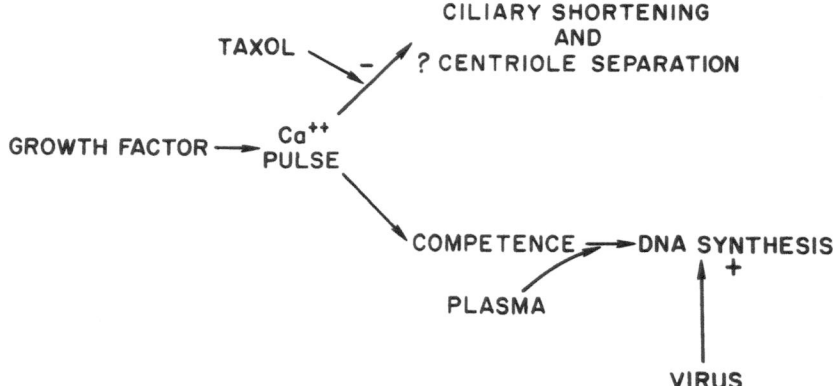

Figure 5. Growth factor produces a cellular event (e.g., calcium flux) which produces both ciliary shortening and competence for DNA synthesis. Taxol inhibits ciliary shortening without inhibiting DNA synthesis, and DNA viruses can stimulate DNA synthesis without inducing ciliary changes.

1979; Wheatley, 1971), it is not yet clear whether cilia resorption and metaphase formation are independent events. There have also not been enough studies in different cell lines to define the temporal relationship of primary cilium resorption to other mitotic events, such as nuclear membrane breakdown or centriole pair separation. The timing of centriole pair separation with respect to nuclear membrane breakdown also seems to be variable (Rattner and Berns, 1976). It is possible that both centriole pair separation and ciliary resorption are related to forces (e.g., changes in pericentriolar function) which reorganize cytoplasmic microtubules from an interphase to a mitotic configuration.

6.2. Coordinated Changes

The temporal coordination of centriole and DNA synthesis cycles can be insured if related events in the two cycles are both stimulated by a particular exogenous agent, such as a growth factor. We have found that growth factors can induce a coordinated response in centriole and DNA synthesis cycles in both fibroblast and epidermal cells. Any hormonal growth factor which causes early mitogenic events (competence) also induces ciliary shortening. Competence and ciliary shortening can only occur independently of one another if ciliary shortening is prevented by taxol, or if DNA synthesis is stimulated directly by SV40 or polyoma virus (Figs. 4 and 5). Thus, one way in which the temporal sequence of centriole and DNA synthesis cycles may be kept in register is to have related parts of both cycles responsive to the same exogenous stimulus (Fig. 5). It remains to be proven whether other parallel cycle events (e.g., possibly cilia resorption and mitosis; centriole duplication and initiation of DNA synthesis) are also examples of coordinated control.

6.3. Overlapping Cycles

DNA synthesis and centriole cycles in the parent cell also overlap with those in the daughter cells, and provide another level of coordination of these cycles. Relationships between centriole cycles in successive cell generations have become clearer with our increased understanding of the different roles of parent and daughter centrioles. Only the ciliated centriole forms specialized structures such as basal feet (Tucker, 1982; Albrecht-Buehler and Bushnell, 1980; Vorobjev and Chentsov, 1980) (Fig. 2). Moreover, it appears that the centriole which forms both basal feet and the primary cilium is probably the parent centriole (Jirsova, 1972; Vorobjev and Chentsov, 1982; Rieder and Borisy, 1981). Ciliation of the parent centriole results in ciliation of the same centriole during each successive cell cycle; ciliation of the daughter centriole would mean that a centriole would ciliate only during the first cell cycle after it was formed. These sequences of centriole ciliation are related to the fact that a centriole is a daughter for only one cycle, while a parent centriole becomes a parent again each successive cell cycle.

These results emphasize the delayed maturation of centrioles during the

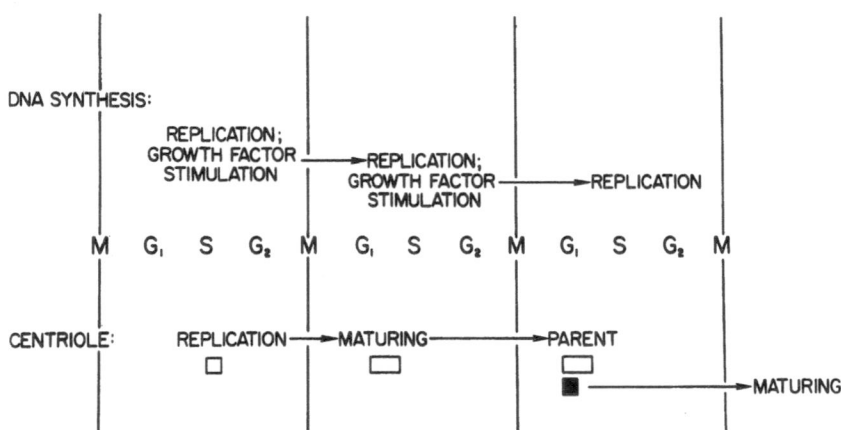

Figure 6. Overlap of centriole and DNA synthesis cycles. In the DNA synthesis cycle, growth factor stimulation in one cycle makes cells competent for DNA synthesis in the next cycle. In the centriole cycle, a daughter centriole is formed in one cell, matures in the next, and then becomes a parent (capable of forming a procentriole and primary cilium). Centriole and DNA synthesis cycles are in parallel (concurrent and independent) with each other, and are related from one cell generation to another.

centriole cycle. Not until the G_1 period of the next cell cycle does the centriole reach its full length, and acquire a capacity, at least in mammalian cells, to form a procentriole. Furthermore, if only parent centrioles form a primary cilium, then it is not until the second cell cycle after its formation that a centriole can form a cilium. Thus, the process of centriole maturation probably overlaps part of three DNA synthesis cycles (Fig. 6). Presumably, there are also differences in the architecture of the centrioles associated with these maturation changes. The DNA synthesis cycle may also not be complete within one cell generation (Fig. 6). One study (Scher *et al.*, 1979) has in fact suggested that exposure of 3T3 cells to PDGF in S or G_2 is enough to stimulate competence for DNA synthesis in the *next* cell cycle.

There may be multiple cycles, some of which are parallel and overlapping, some of which may be relatively independent of others. For example, in *Xenopus* eggs (Kirschner *et al.*, 1980; Hara *et al.*, 1980), oscillations of surface membrane contractions which correlate with cytokinesis occur independently of either the centriole or the nucleus. There are probably many events which must be coordinated so that the cell's final response to growth stimulation is the formation of two nearly identical daughter cells. Presumably, the points of coincidence of these multiple parallel pathways are important in preserving their coordination during the cell cycle, and perhaps in coordinating events in daughter cells. Indeed, sister cells have more similar morphologies (Soloman, 1979; Albrecht-Buehler, 1977a), migration paths (Albrecht-Buehler, 1977b), and growth kinetics (Brooks *et al.*, 1979) than would be expected on chance alone. Whether the centriole lineage from one cell generation to another explains these similarities has not yet been defined.

7. Role of Calcium

Since some centriole changes appear to be closely coordinated with those in the DNA synthesis cycle, events in the centriole cycle can be studied to yield clues about what may be triggering events in the DNA synthesis cycle. In fact, changes in centriolar structures may reflect important growth regulatory events which distinguish nonneoplastic cells from neoplastic cells. I propose that one such regulatory event is a change in cellular calcium.

7.1. Calcium and DNA Synthesis

It has long been appreciated that neoplastic and normal cells have different dependencies on external calcium for growth (Balk, 1971). In medium with a low calcium concentration, virally transformed cells will often continue to grow, whereas nonneoplastic cells will stop growing in G_1. In fact, these growth differences have been shown to be a very good phenotypic marker for neoplastic transformation in culture (Swierenga *et al.*, 1978; Boynton and Whitfield, 1976; Boynton *et al.*, 1977). Not only do nonneoplastic cells stop growing without exogenous calcium, but quiescent cells also undergo changes in calcium and calmodulin (calcium-binding protein) when treated with growth factors. Serum decreases the cellular content of calmodulin, a calcium regulatory protein (Chafouleas and Means, personal communication), perhaps analogous to changes in calmodulin localization produced by hormones (Harper *et al.*, 1981; Conn *et al.*, 1981). Serum also causes a decrease in the EGTA-sensitive component of cellular calcium and a subsequent increase in cellular calcium 6–8 h after the addition of serum (Tupper *et al.*, 1978; Eilam and Szydel, 1981). This initial decrease in EGTA-sensitive calcium has been presumed to be calcium associated with the surface membrane. However, it is also possible that washes with EGTA [ethyleneglycol-bis-(β-amino-ethyl ether) N_1N-tetraacetic acid] and even PBS (phosphate-buffered saline) mobilize some part of the internal stores of cellular calcium. Prolonged washing with EGTA can in fact completely abolish the chlorotetracycline fluorescence of membrane associated calcium (Tucker, Klausner and Henkart, unpublished observations). Since mechanical stimulation of the surface of L cells in the absence of exogenous calcium has been shown to mimic the effect of an intracellular injection of calcium (Henkart and Nelson, 1979), stimulation of the cell membrane may release internal stores of calcium. Thus, prolonged washing may itself stimulate a release of calcium and change the cellular distribution of calcium.

By binding exogenous calcium with EGTA, one can show that cells have to have external calcium in order to go through particular cellular events (Dulbecco and Elkington, 1975). Some stimulated quiescent cells need calcium both early (within hours after serum stimulation) (Tupper *et al.*, 1980) and late, just before DNA synthesis (Whitfield *et al.*, 1974; Boynton *et al.*, 1977). It is tempting to associate these two periods of calcium requirement with the different classes of cellular events induced by PDGR (or FGF) and plasma. Exactly how calcium might be involved in stimulating events leading to DNA synthesis has not been extensively studied. It is known that calcium can bind to calmodulin and that calmodulin antagonists can inhibit cell growth (Hidaka *et al.*, 1981; Chafouleas *et al.*, 1982). Calcium—calmodulin can activate a number of cellular processes, one of which might be a protein kinase (Wasserman and Smith, 1981) that could conceivably produce changes in phosphorylated proteins such as that reportedly associated with early mitogenic events (Pledger *et al.*, 1981; Kletzien *et al.*, 1977). The concentration of exogenous calcium has also been shown to affect RNA metabolism in hormonally responsive cells (White *et al.*, 1981) and thus calcium could also conceivably explain changes in RNA metabolism associated with the induction of competence in quiescent fibroblasts (Smith and Stiles, 1981). Since these early mitogenic events are bypassed by virus (Tucker *et al.*, 1979b), and since virally transformed cells need lower exogenous concentrations of calcium for growth than do nonneoplastic cells (Balk *et al.*, 1973; Swierenga *et al.*, 1978), virally transformed cells may be relatively independent of a calcium-dependent step necessary for nonneoplastic cell growth (Fig. 5).

If a calcium flux through the plasma membrane or calcium released from internal stores triggers early or late mitogenic events, one would expect those mitogenic events to be sensitive to calcium inhibitors. We have found that a calcium flux inhibitor, verapamil, reversibly inhibits late but not early mitogenic events. On the other hand, we have evidence that doses of verapamil which inhibit late events also can stimulate early mitogenic events in quiescent fibroblasts. These preliminary results imply that calcium may play different roles in early and late mitogenic events in stimulated quiescent cells.

7.2. Calcium and Centriole Changes

The close association between ciliary shortening and early mitogenic events has lead us to consider what cellular changes produced by growth factors in the cell could induce depolymerization of ciliary microtubules. A change in the free intracellular concentration of calcium is an obvious candidate because of the documented role of free calcium in mediating the depolymerization of mictotubules both *in vitro* (Weisenberg, 1972) and *in vivo* (Kiehart, 1981). Indeed, we can produce both ciliary shortening and the associated mitogenic events with calcium ionophore (A23187) (Tucker, 1979, 1980). In addition, it has been shown in 3T3 cells that cytoplasmic microtubules, and not ciliary microtubules, have microtubule associated proteins (Connolly and Kalnins, 1978) whose counterparts in cytoskeletal preparations of BSC-1 cells make microtubules stable to the depolymerization effects of calcium (Schliwa *et al.*, 1981). Thus, if growth factors do produce an intracellular increase in calcium at a certain place in the cell, we would not expect cytoplasmic microtubules to be affected nearly as much as ciliary microtubules. In fact, growth factors do not alter cytoplasmic microtubules, but do produce ciliary shortening due to a depolymerization of ciliary microtubules.

It remains to be shown whether such putative changes in cellular calcium reflected in ciliary shortening actually produce some of the important mitogenic events which we have been discussing. In addition, there is still a question about whether putative cellular calcium changes are produced by an increased transmembrane flux of calcium, or by an internal release of calcium. As a working hypothesis we are postulating that early mitogenic events and ciliary shortening are associated with a spatially restricted release of intracellular calcium stores. Such a hypothesis is consistent with verapamil's induction of early mitogenic events, since verapamil has been postulated to release internal stores of calcium in muscle preparations (Publicover and Duncan, 1979). In mammalian cells there are subcisternal membrane appositions of surface membranes (Henkart *et al.*, 1976) resembling the sarcoplasmic reticulum that mediates the release of internal stores of calcium in response to a membrane potential change in muscle. Fibroblasts may have retained part of this specialized membrane system in order to release internal stores of calcium in response to surface membrane events. At the present time, not much is known about changes in the compartmentation of calcium in cells stimulated with growth factors.

7.3. Calcium-Sensitive Microtubules

There are at least two calcium-sensitive microtubule changes which are produced by the interaction of growth factors with the surface membrane. One is ciliary shortening which we have documented both by electron microscopy and indirect immunofluorescence in cells treated with a particular class of growth factors; another is a separation of MTOCs which has been documented by light microscopy in two different cells (Tucker, 1982; Sherline and Mascardo, 1982a) and which EM has shown to be a separation of the parent and daughter centrioles (Tucker, 1982). Because both of these morphological events can be produced by calcium ionophore and inhibited by taxol, these cellular changes may depend on calcium-sensitive microtubules which are depolymerized by an increase in intracellular calcium localized at the centrosome.

Cold-stable microtubules, at least in rat brain, are a class of microtubules which appear to be exquisitely sensitive to calcium only in the presence of calmodulin (Job *et al.*, 1981; Margolis and Rauch, 1981). This property distinguishes them from cold-labile microtubules that are the most easily isolated form of microtubules and have been most extensively studied over the past years. Cold-labile microtubules are sensitive to low concentrations of calcium in the absence of calmodulin, and can be made more sensitive to calcium only at very high doses of calmodulin (Marcum *et al.*, 1978). Cold stable microtubules in rat brain are stabilized by dephosphorylation of a specific 64,000 molecular weight protein associated with microtubules (Margolis and Rauch, 1981). These cold-stable microtubules are also drug stable and, as already mentioned, exquisitely calcium sensitive. In mammalian cells the analogous cold and drug-stable microtubules are the ciliary microtubules which we have proposed are also sensitive to calcium. The role of associated proteins in determining the stability of microtubules in mammalian cells has not been defined. Conceivably, there is a balance between microtubule associated proteins (HMW), and phosphorylation of associated proteins which determine whether cytoplasmic microtubules *in vivo* are more sensitive to exogenous calcium in the presence of calmodulin.

The functions of calcium-sensitive microtubules in the kinetechore-pole fibers in the mitotic spindle have been recently studied. Calmodulin has been found to be associated with these fibers (DeMey *et al.*, 1980) and it has been postulated that calcium released in the mitotic spindle can combine with calmodulin, cause depolymerization of kinetechore-pole microtubules and produce anaphase movement of chromosomes (DeMey *et al.*, 1980). In contrast, it is not known what function the calcium-sensitive microtubules associated with the primary cilium and centriole separation might have in the normal functioning of the cell.

8. Cell Shape

One obvious physiological condition in which cytoskeletal changes produce changes in growth are those associated with cell spreading (O'Neill *et al.*, 1979; Dulbecco, 1970; Dulbecco and Ellington, 1973; Stoker *et al.*, 1968). Recently it has been shown that nonneoplastic cells at high cell density probably stop growing because they are unable to spread onto the substratum (Folkman and Moscoma, 1978). Nonneoplastic cells placed on a polymer [Poly (Hema)] which prevents their spreading are less stimulated to grow by a given serum concentration as compared to the same cells on a plastic substratum (Tucker *et al.*, 1981). In fact, cell spreading and growth factors interact in specific ways to influence the growth of cells in culture (Fig. 7).

8.1. Cell Shape and DNA Synthesis

When cultured cells are not allowed to spread out sufficiently, they stop in G_1 before DNA synthesis is initiated. This stopping point appears to be very similar to that induced by serum starvation and further supports the contention that rounding up of cells prevents them from effectively utilizing exogenous growth factors. The mechanism of this effect is not known, but in at least one study, poorly spread glial cells stopped responding to EGF despite the fact that the number of receptors for EGF was unchanged (Westermark, 1977; Westermark, 1978). It is not known whether cells need to be attached to substrata and spread in order to go through both the early and late mitogenic events leading to DNA synthesis. The requirement for cell spreading in order to respond to growth factors is lost or modified (Folkman and Tucker, 1980; Tucker *et al.*, 1981a) when cells become neoplastic. The hallmark of many neoplastic cells is their capacity to grow suspended in methylcellulose or agarose, or poorly spread on poly (HEMA). It is not known whether changes

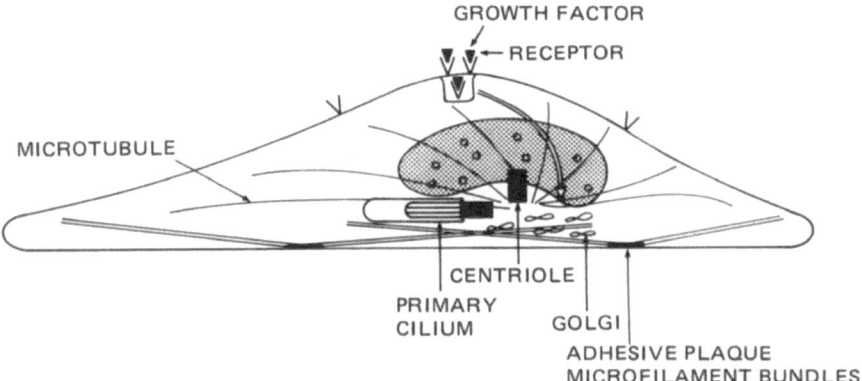

Figure 7. Cytoskeleton and DNA synthesis. In nonneoplastic cells, cell spreading is associated with the formation of adhesive plaques plus microfilament bundles and with the processing of growth factor information (growth factor and receptor) which produces coordinated early mitogenic response (ciliary shortening and competence) and perhaps later growth events (centriole duplication and DNA synthesis initiation).

in microtubules, microfilaments or intermediate filaments associated with cell spreading influence the cell's response to growth factors.

8.2. Cell Shape and Centriole Changes

Cells made quiescent either at high cell density or in low serum form a primary cilium from one of an unduplicated pair of centrioles. It has not yet been established whether centrioles are also ciliated in cells rounded up on nonadhesive polymers [poly (HEMA)] or in suspension, although we have preliminary evidence that many cells in suspension do have primary cilia (Rieder, Wheatley, and Tucker, unpublished observations). Whether such rounded cells can respond with ciliary shortening or centriole separation when stimulated with growth factors is unknown. On the one hand, cellular changes (e.g., calcium flux) usually responsible for changes in calcium-sensitive structure might occur, but the cell may be unable to respond. Alternatively, initial cellular changes in response to growth factors may not occur in rounded cells. Studying the sequence of cellular events leading to changes in calcium-sensitive microtubules in poorly spread cells may suggest why rounded cells are resistant to growth stimulation.

9. Summary

9.1. Coordination of Centriole and DNA Synthesis Cycles

There are at least four points at which the DNA synthesis and centriole cycles are closely related: ciliary shortening and competence; centriole duplication and initiation of DNA synthesis; centriole pair separation and termination of DNA synthesis; cilia resorption and mitosis. At least one of these (ciliary shortening and competence) is also an example of parallel events which are both stimulated by the same exogenous agent (PDGF, FGF). All of these temporally related events are also possible places of coordination which could insure that the centriole pairs are in position to separate the duplicated genetic material to the two daughter cells at mitosis. Another level of coordination involves continuity between events in parent and daughter cells. Centriole maturation and possibly growth factor induction of competence are started in one cell generation and finished in another. The similarities of morphologies, migration paths and growth kinetics in sister cells may depend on these morphologic and functional relationships which span cell generations.

9.2. Calcium-Sensitive Microtubules

In the coordinated stimulation of ciliary shortening and centriole separation by PDGF or FGF, changes in calcium-sensitive microtubules play an important role. These morphological changes in the centriole may indicate how the associated mitogenic changes (competence) are produced by growth factors. For example, depolymerization of specific microtubule populations in the centrosome (ciliary microtubules) may indicate that regulatory cellular events (e.g., calcium fluxes) are occurring in the vicinity of the primary cilium. It is also possible that the cellular events related to calcium and specifically necessary for the growth of nonneoplastic cells may require a primary cilium.

9.3. Calcium and Growth Control

Depolymerization of calcium-sensitive microtubules and stimulation of nonneoplastic cells to leave quiescence are probably both produced by an increase in free intracellular calcium. These same calcium-sensitive changes in microtubules are bypassed in nonneoplastic cells stimulated by virus. The actual changes in calcium during different phases of the cell cycle, especially G_1, need to be more clearly defined. There is already preliminary evidence that early and late mitogenic events differ in their sensitivity to calcium flux and calmodulin inhibitors. These facts suggest that different growth events during the cell cycle may be related to calcium fluxes that are fundamentally different.

9.4. Microtubules and Centrioles in Growth Control

Microtubules in the primary cilium and perhaps those associated with centriole separation can be depolymerized in response to growth factors without a concomitant depolymerization of cytoplasmic microtubules. Centriole duplication occurs near the time of initiation of DNA synthesis. At present there is no evidence that any of these changes in microtubules and centrioles are regulatory, either triggering important cellular events leading to DNA synthesis, or determining the kinetics of response. However, at the very least, changes in centrioles and microtubules closely reflect events that are growth regulatory and different in nonneoplastic and neoplastic cells. In this sense, a study of cellular events which are responsible for ciliary shortening and centriole separation may bring us to a clearer understanding of how neoplastic cells continue to grow under growth conditions which restrict the proliferation of nonneoplastic cells.

References

Aist, J. R., and Berns, M. W., 1981, Mechanics of chromosome separation during mitosis in fusarium (Fungi imperfecti): new evidence from ultrastructural and laser microbeam experiments, *J. Cell Biol.* **91:**446.

Albertini, D. F., and Anderson, E., 1977, Microtubule and microfilament rearrangement during capping of Concanavalin A receptors on cultured ovarian granulosa cells, *J. Cell Biol.* **73:**111.

Albrecht-Buehler, G., 1977a, Daughter 3T3 cells: are they mirror images of each other? *J. Cell Biol.* **72:**595.

Albrecht-Buehler, G., 1977b, Phagokinetic tracks of 3T3 cells: parallels between the orientation of track segments and of cellular structures which contain actin or tubulin, *Cell* **12:**333.

Albrecht-Buehler, G., and Bushnell, A., 1980, The ultrastructure of primary cilia in quiescent 3T3 cells, *Exp. Cell Res.* **126:**427.

Archer, F. L. and Wheatley, D. N., 1971, Cilia in cell-cultured fibroblasts. II. Incidence in mitotic and postmitotic BHK 21/c 12 fibroblasts, *J. Anat.* **109:**277.

Baker, M. E., 1976a, Arrest of C1300 neuroblastoma cells by limiting serum or isoleucine: implications for growth control in malignant cells, *Biochem. Biophys. Res. Commun.* **68:**1059.

Baker, M. E., 1976b, Colchicine inhibits mitogenesis in C1300 neuroblastoma cells that have been arrested in G_0, *Nature (London)* **262:**785.

Balk, S. D., 1971, Calcium as a regulator of the proliferation of normal, but not of transformed, chicken fibroblasts in a plasma-containing medium, *Proc. Natl. Acad. Sci. USA* **68:**271.

Balk, S. D., Whitfield, J. G., Youdale, T., and Braun, A. C., 1973, Roles of calcium, serum, plasma and folic acid in the control of proliferation of normal and Rous sarcoma virus-infected chicken fibroblasts, *Proc. Natl. Acad. Sci. USA* **70:**675.

Barnes, B. G., 1961, Ciliated secretory cells in the pars distalis of the mouse hypophysis, *J. Ultrastructure Res.* **5:**453.

Baserga, R., 1978, Resting cells and the G_1 phase of the cell cycle, *J. Cell Physiol.* **95:**377.

Bershadsky, A. D., and Gelfand, V. I., 1981, ATP-dependent regulation of cytoplasmic microtubule disassembly, *Proc. Natl. Acad. Sci. USA* **78:**3610.

Betel, I., and Martijnse, J., 1976, Drugs that disrupt microtubuli do not inhibit lymphocyte activation, *Nature (London)* **261:**318.

Bhalla, D. K., Braun, J. and Karnovsky, M. J., 1979, Lymphocyte surface and cytoplasmic changes associated with translational motility and spontaneous capping of Ig, *J. Cell Sci.* **39:**137.

Biscoe, T. J. and Stehbens, W. E., 1966, Ultrastructure of the carotid body, *J. Cell Biol.* **30:**563.

Bjorkman, N., 1962, A study of the ultrastructure of the granulosa cells of the rat ovary, *Acta. Anat.* **51:**125.

Boynton, A. L., and Whitfield, J. F., 1976, Different calcium requirements for proliferation of conditionally and unconditionally tumorogenic mouse cells, *Proc. Natl. Acad. Sci. USA.* **73:**1651.

Boynton, A. L., Whitfield, J. F. Issacs, R. J., and Tremblay, R., 1977, The control of human W1-38 cell proliferation by extracellular calcium and its elimination by SV40 virus-induced proliferative transformation, *J. Cell Physiol.* **92:**241.

Braun, J., Fujiwara, K., Pollard, T. D. and Unanue, E. R., 1978a, Two distinct mechanisms for redistribution of lymphocyte surface macromolecules I. Relationship to cytoplasmic myosin, *J. Cell Biol.* **79**:409.

Braun, J., Fujiwara, K., Pollard, T. D. and Unanue, E. R., 1978b, Two distinct mechanisms for redistribution of lymphocyte surface macromolecules II. Contrasting effects of local anesthetics and a calcium ionophore, *J. Cell Biol* **79**:419.

Brinkley, B. R., Fuller, G. M. and Highfield, 1975, Cytoplasmic microtubules in normal and transformed cells in culture: analysis by tubulin antibody immunofluorescence, *Proc. Natl. Acad. Sci. USA.* **73**:4981.

Brinkley, B. R., and Stubblefield, E., 1970, Ultrastructure and interaction of the kinetochore and centriole in mitosis and meiosis, in: *Advances in Cell Biology* (D. M. Prescott, L. Goldstein, E. McConkey, eds.) pp. 119–185, Appleton-Century-Crofts, New York.

Brinkley, B. R., Stubblefield, E., and Hsu, T. C., 1967, The effects of colcemid inhibition and reversal on the fine structure of the mitotic apparatus of Chinese Hamster cells in vitro, *J. Ultrastruct. Res.* **19**:1.

Brooks, R. F., Bennett, D. C., and Smith, J. A., 1980, Mammalian cell cycles need two random transitions, *Cell* **19**:493.

Brown, K. D., Friedkin, M., and Rozengurt, E., 1980, Colchicine inhibits epidermal growth factor degradation in 3T3 cells, *Proc. Natl. Acad. Sci. USA* **77**:480.

Bucher, N. L. R., and Mazia, D., 1960, Deoxyribonucleic acid synthesis in relation to duplication of centers in dividing eggs of the sea urchin, Stronglylocentrotus Purpuratus, *J. Biophys. Biochem. Cytol.* **7**:651

Bucher, N. L. R. and McGowan, J. A., 1979, Regeneration: regulatory mechanisms, in: *Liver and Biliary Disease* (E. Wright, K. G. M. M. Alberti, S. Karran, G. Millward-Sadler, eds.), pp. 210–227, W. B. Saunders, New York.

Byers, B. and Goetsch, L., 1975, Behavior of spindles and spindle plaques in the cell cycle and conjugation of Saccharyomyces Cervisiae, *J. Bacteriol.* **124**:511.

Chafouleas, J. G., Bolton, W. E., Hidaka, H., Boyd, A. E., and Means, A. R., 1982, Calmodulin and the cell cycle: involvement in regulation of cell-cycle progression, *Cell* **28**:41.

Clemmons, D. R., Van Wyk, J. J. and Pledger, W. J., 1980, Sequential addition of platelet factor and plasma to Balb/c 3T3 fibroblast cultures stimulates Somatomedin-C binding early in the cell cycle, *Proc. Natl. Acad. Sci. USA* **77**:6644.

Conn, P. M., Chafouleas, J. G., Rogers, D. and Means, A. R., 1981, Gonadotroprin releasing hormone stimulates calmodulin redistribution in rat pituitary, *Nature (London)*, **292**:264.

Connolly, J. A., and Kalnins, V. I., 1980 Immunofluorescent localization of microtubule-associated proteins in vivo, in *Microtubules and Microtubule Inhibitors* (M. DeBrabander and J. De-Mey, eds.) pp. 175–189, Elsevier/North-Holland Biomedical Press, Amsterdam.

Crisp, T. M., and Browning H. C., 1968, The fine structure of copora lutea in ovarian transplants of mice following luteotrophin stimulation, *Am. J. Anat.* **122**;169.

Crossin, K. L., and Carney, D. H., 1981a, Evidence that microtubule depolymerization early in the cell cycle is sufficient to initiate DNA synthesis, *Cell* **23**:61.

Crossin, K. L., and Carney, D. H., 1981b, Microtubule stabilization by taxol inhibits initiation of DNA synthesis by thrombin and by epidermal growth factor, *Cell* **27**:341.

Dahl, H. A., 1963, Fine structure of cilia in the rat cerebral cortex, *Z. Zellforsch* **60**:369.

DeFoor, P. H. and Stubblefield, E., 1974, Effects of actinomycin D, amethopterin, and 5-fluor-2'-deoxyuridine on procentriole formation in Chinese Hamster fibroblasts in culture, *Exp. Cell Res.* **85**:136.

DeMey, J., Janiau, M., DeBrabander, M., Moens, W. and Geuens, G., 1978, Evidence for unaltered structure and *in vivo* assembly of microtubules in transformed cells, *Proc. Natl. Acad. Sci. USA.* **75**:1339.

DeMey, J., Moereman, M., Geuens, E., Nuydens, R., Van Belle, H., and DeBrabander, M., 1980, Immunocytochemical evidence for the association of calmodulin with microtubules of the mitotic apparatus, in: *Microtubules and Microtubule Inhibitors* (M. DeBrabander, J. DeMey, eds.) pp. 227–243, Elsesier/North-Holland, Biomedical Press, Amsterdam.

deTerra, N., 1978, Control of cell division in the ciliate stentor, in: *Cell Reproduction: In honor of Daniel Mazia* (E. D. Prescott and C. F. Fox, eds.), pp. 525–537, Academic Press, New York.

Deutch, A. A., and Shumway, L. K., 1973, Centriole behavior in chloramphenicol-treated eggs of the sand dollar, Dendraster Excentricies, *Protoplasma* **76**:387.

Dicker, P., and Rozengurt, E., 1981, Stimulation of DNA synthesis by transient exposure of cell cultures to TPA or polypeptide mitogens: Induction of competence or incomplete removal? *J. Cell. Physiol.* **109**:99.

Dingemans, K. P., 1969, The relation between cilia and mitosis in the mouse adenohypophysis, *J. Cell Biol.* **43**:361.

Dulbecco, R., 1970, Topoinhibition and serum requirement of transformed and untransformed cells, *Nature (London)* **227**:802.

Dulbecco, R. and Elkington, J., 1973, Conditions limiting multiplication of fibroblastic and epithelial cells in dense culture, *Nature (London)* **246**:197.

Dulbecco, R., and Elkington, J., 1975, Induction of growth in resting fibroblastic cell cultures by Ca^{++}, *Proc. Natl. Acad. Sci. USA.* **72**:1584.

Edelman, G. M., 1976, Surface modulation in cell recognition and cell growth, *Science* **192**:218.

Edelman, G. M., and Yahara, I., 1976, Temperature-sensitive changes in surface modulating assemblies of fibroblasts transformed by mutants of Rous sarcoma virus, *Proc Natl. Acad. Sci. USA* **73**:2047.

Eichhorn, J. H., and Peterkofsky, B., 1979, Quantitative biochemical analysis of microtubule content in normal and transformed 3T3 cells, *J. Cell Biol.* **82**:572.

Eilam, Y. and Szydel, N., 1981, Calcium transport and cellular distribution in quiescent and serum stimulated primary cultures of bone cells and skin fibroblasts, *J. Cell. Physiol.* **106**:225.

Flood, P. R., and Totland, G. K., 1977, Substructure of solitary cilia in mouse kidney, *Cell Tissue Res.* **183**:281.

Floyd, G. L., Mitosis and cytokinesis in Asteromonas-Gracilis, a wall-less green monad, *J. Phycol.* **14**:440.

Folkman, J., and Moscoma, A., 1978, Role of cell shape in growth control, *Nature (London)* **278**:345.

Folkman, J., and Tucker, R. W., 1980, Cell configuration, substratum and growth control, in: *The Cell Surface: Mediator of Developmental Processes* (S. Subtelny and N. K. Wessells, eds.) pp. 259–275, Academic Press, New York.

Fonte, V., and Porter, K., 1974, "Topographic changes associated with the viral transformation of normal cells to tumorigenicity," Eighth International Congress on Electron Microscopy, Cambridge, 2: pp 334–335.

Fonte, V. G., Searles, R. L., and Hilfer, R. S., 1971, The relationship of cilia with cell division and differentiation, *J. Cell Biol.* **49**:226.

Frantz, C. N., Stiles, C. D., and Scher, C. D., 1979, The tumor-promotor 12-0-tetradecanoyl-phorbol-13-acetate enhances the proliferative response of Balb/c 3T3 cells to hormonal growth factors *J. Cell. Physiol.* **100**:413.

Friedkin, M., Legg, A., and Rozengurt, E., 1979, Antitubulin agents enhance the stimulation of DNA synthesis by polypeptide growth factors in 3T3 mouse fibroblasts, *Proc. Natl. Acad. Sci. USA.* **76**:3909.

Friedkin, M., Legg, A., and Rozengurt, E., 1980, Enhancement of DNA synthesis by colchicine in 3T3 mouse fibroblasts stimulated with growth factors, *Exp. Cell Res.* **129**:23.

Fritz-Niggli, H., and Suda, T., 1972, Formation and significance of the centriole: A study and new interpretation of the meiosis in Drosophilia, *Cytobiologie* **5**:12.

Fujita, H. E., 1963, Studies on the thyroid gland of domestic fowl, with special reference to the mode of secretion and the occurence of a central flagellum in the follicular cell, *Z. Zellforsch* **60**:615.

Fuller, G. M., and Brinkley, B. R., 1976, Structure and control of assembly of cytoplasmic microtubules in normal and transformed cells, *J. Supramol. Struct.* **5**:497.

Gallagher, B. C., 1980, Primary cilia of the corneal endothelium, *Amer. J. Anat.* **159**:475.

Gardiner, S. L., and Rieger, R. M., 1980, Rudimentary cilia in muscle cells of annelids and echinoderms, *Cell Tissue Res.* **213**:247.

Gilua, N. B., and Satir, P., 1972, The ciliary necklace: a ciliary membrane specialization, *J. Cell Biol.* **53**:191.

Gould, R. R. and Borisy, G. G. 1977, The pericentriolar material in Chinese Hamster ovary cells nucleates mictrotubule formation, *J. Cell Biol.* **73**:601.

Grillo, M. A., and Palay, S. L., 1962, Ciliated Schwann cells in the autonomic nervous system of the adult rat, *J. Cell Biol.* **52**:430.

Hamilton, B. T., and Snyder, J. A., 1981, Control of mitotic cycle timing in the absence of spindle microtubules, *J. Cell Biol.* **91**:312a.

Hara, K., Tydeman, P., and Kirschner, M., 1980, A cytoplasmic clock with the same period as the division cycle of Xenopus eggs, *Proc. Natl. Acad. Sci. USA* **77**:462.

Harper, J. F., Wallace, R. W., Cheung, W. Y., and Steiner, A. L., 1981, ACTH-stimulated changes in the immunocytochemical localization of cyclic nucleotides, protein kinases, and calmodulin, *Adv. Cyc. Nucl. Res.* **14**:581.

Hartman, H., 1975, The centriole and the cell, *J. Theor. Biol.* **51**:501.

Hartwell, L. H., 1978a, Cell division from a genetic perspective, *J. Cell Biol.* **77**:627.

Hartwell, L. H., 1978b, Fourth European Cell Cycle Workshop, April 17–19. Bern *Nature (London)* **273**:594.

Heidemann, S. R., Sander, G., and Kirschner, M. W., 1977, Evidence for a functional role of RNA in centrioles, *Cell* **10**:337.

Henkart, M. P., Landis, M. D., and Reese, T. S., 1976, Similarity of functions between plasma membranes and endoplasmic reticulum in muscle and neurons, *J. Cell Biol.* **70**:338.

Henkart, M. P. and Nelson, P. G., 1979, Evidence for an intracellular calcium store releasable by surface stimulation in fibroblasts (L cells), *J. Gen. Physiol.* **73**:655.

Henneguy, L. F., 1898, Sur le rapports des cils vibratiles avec les centrosomes, *Arch. Anat. Microsc. Morphol. Exp.* **1**:481.

Herschman, H. R. and Perdue, J. F., 1981, Cell surface receptors: workshop report, in: *Control of Cellular Division and Development Part A* (D. Cunningham, E. Goldwasser, J. Watson, and C. Fred Fox, eds.) pp. 534–547 Alan R. Liss, New York.

Hidaka, H., Sasaki, Y., Tanaka, T., Endo, T., Ohno, S., Fujii, Y., and Nagata, T., 1981, N-(6-Aminohexyl)-5-chloro-1-naphthalenesulfonamide, a calmodulin antagonist, inhibits cell proliferation, *Proc. Natl. Acad. Sci. USA.* **78**:4354.

Hoover, R. L., Fujiwara, K., Klausner, R. O., Bhalla, D. K., Tucker, R., and Karnovsky, M. J., 1981, Effect of free fatty acids on the organization of cytoskeletal elements in lymphocytes, *Mol. Cell. Biol.* **1**:939.

Inoue', S., 1952, Effect of colchicine on the microscopic and submicroscopic structure of the mitotic spindle, *Exp. Cell Res. Suppl.* **2**:305.

Inoue', S. and Sato, H., 1967, Cell motility by labile association of molecules. The nature of mitotic spindle fibers and their role in chromosome movement, *J. Gen. Physiol. Suppl.* **50**:259.

Jayaraman, S., and Tucker, R. W., 1981, Changes in ciliation of the centriole in 3T3 cells, *J. Cell Biol.* **91**:47a.

Jirsova, Z., 1972, Formation of cilia in tubal epithelium, *Folia Morphol.* **20**:70.

Job, D., Fischer, E. H., and Margolis, R. L., 1981, Rapid disassembly of cold-stable microtubules by calmodulin, *Proc. Natl. Acad. Sci. USA* **78**:4679.

Kiehart, D. P., 1981, Studies on the in vivo sensitivity of spindle microtubules to calcium ions and evidence for a vesicular calcium-sequestering system, *J. Cell Biol.* **88**:604.

Kirschner, M., Gerhart, J. C., Hara, K. and Ubbels, G. A., 1980, Initiation of the cell cycle and establishment of bilateral symmetry in Xenopus eggs, in: *The Cell Surface: Mediator of Developmental Processes* (S. Subtelny and N. K. Wessells, eds.) pp. 187–215, Academic Press, New York.

Klausner, R. D., Bhalla, D. K., Dragsten, P., Hoover, R. L., and Karnovsky, M. J., 1980, Model for capping derived from inhibition of surface receptor capping by free fatty acids, *Proc. Natl. Acad. Sci. USA* **77**:437.

Kletzien, R. F., Miller, M. R., and Pardee, A. B., 1977, Unique cytoplasmic phosphoproteins are associated with cell growth arrest, *Nature (London)* **270**:57.

Kohler, N., and Lipton, A., 1974, Platelets as a source of fibroblast growth-promoting activity, *Exp. Cell Res.* **87**:297.

Kuriyama, R. and Borisy, G. G., 1981, Centriole cycle in Chinese Hamster ovary cells as determined by whole-mount electron microscopy, *J. Cell Biol.* **91**:814.

Latta, H., Maunsbach, A. B. and Madden, S. C., 1961, Cilia in different segments of the rat nephron, *J. Biophys. Biochem. Cytol.* **11**:248.

Lenhossek, M. Von, 1898, Iber flimmerzellen, *Verh. Anat. Ges. Kiel,* **12**:106.

Lin, H. -S., and Chen, I. -L., 1969, Development of the ciliary complex of microtubules in the cells of rat subcommissural organ, *Z. Zellforsch* **96**:186.

Lloyd, C., 1979, Primitive model for cell cycle control, *Nature (London)* **280**:631.

Marcum, J. M., Dedman, J. R., Brinkley, B. R. and Means, A. R., 1978, Control of microtubule assembly-dissassembly by calcium-dependent regulator protein, *Proc. Natl. Acad. Sci. USA.* **75**:3771.

Margolis, R. L., and Rauch, C. T., 1981, Characterization of rat brain crude extract microtubule assembly: correlation of cold stability with the phosphorylation state of a microtubule-associated 64K protein, *Biochemistry* **20**:4451.

Martinek, J., Kraus, R. and Jirsova, Z. 1973, A contribution to the solitary ciliogenesis, *Folia Morphol.* **21**:265.

Maxfield, F. R., Davies, P. J. A., Klempner, L., Willingham, M. C., and Pastan, I., 1979, Epidermal growth factor stimulation of DNA synthesis is potentiated by compounds that inhibit its clustering in coated pits *Proc. Nat. Acad. Sci. USA* **76**:5731.

Mazia, D., Harris, P. J., and Bibring, T., 1960, The multiplicity of the mitotic centers and the time-course of their duplication and separation, *J. Biophys. Biochem. Cytol.* **7**:1.

Mazia, D., Paweletz, N., Sluder, G., and Finze, E-M., 1981, Cooperation of kinetochores and pole in the establishment of monopolar mitotic apparatus, *Proc. Nat. Acad. Sci. USA* **78**:377.

McCarron, L. K. and Anderson, E., 1973, A cytological study of the postnatal development of the rabbit oviduct epithelium, *Biol. Reprod.* **8**:11.

McClain, D. A., D'Eustachio, P., and Edelman, G. M., 1977, Role of surface modulating assemblies in growth control of normal and transformed fibroblasts, *Proc. Nat. Acad. Sci. USA* **74**:666.

McClain, D. A., and Edelman, G. M., 1980, Density-dependent stimulation and inhibition of cell growth by agents that disrupt microtubules, *Proc. Nat. Acad. Sci. USA* **77**:2748.

McGill, M., Highfield, D. P., Monahan, T. M., and Brinkley, B. R., 1976, Effects of nucleic acid specific dyes on centrioles of mammalian cells, *J. Ultrastruct. Res.* **57**:43

McNeill, P. A., and Berns, M. W., 1981, Chromosome behavior after laser microirradiation of a single kinetochore in mitotic PTK$_2$ cells, *J. Cell Biol.* **88**:543.

Meier-Vismara, E., Walker, N., and Vogel, A., 1979, Single cilia in the articular cartilage of the cat, *Exp. Cell Biol.* **47**:161.

Moran, D. T., Rowley, J. C., Zill, S. N., and Varela, F. G., 1976, The mechanism of sensory transduction in a mechanoreceptor, *J. Cell Biol.* **71**:832.

Moran, D. T., and Rowley, J. C., 1975, The fine structure of the cockroach subgenual organ, *Tissue and Cell* **7**:91.

Moran, D. T., and Varela, F. G., 1971, Microtubules and sensory induction, *Proc. Nat. Acad. Sci. USA* **68**:757.

Mori, Y., Akedo, H., Tanigaki, Y., Tanaka, K., and Okada, M., 1979, Ciliogenesis in tissue-cultured cells by the increased density of cell population, *Exp. Cell Res.* **120**:435.

Myklebust, R., Engedal, H., Saetersdal, T. S. and Ulstein, M., 1977, Primary 9 + 0 cilia in the embryonic and the adult human heart, *Anat. Embryol.* **151**:127.

O'Neill, C. H., Riddle, P. H. and Jordan, P. W., 1979, The relation between surface area and anchorage dependence of growth in hamster and mouse fibroblasts, *Cell* **16**:909.

Osborn, M. and Weber, K., 1977, The display of microtubules in transformed cells, *Cell* **12**:561.

Otto, A. M., Ulrich, M. O., Zumbe', A., and Jimenez deAsua, L., 1981, Microtubule-disrupting agents affect two different events regulating the initiation of DNA synthesis in Swiss 3T3 cells, *Proc. Natl. Acad. Sci. USA* **78**:3063.

Otto, A. M., Zumbe', A., Gibson, L., Kubler, A. M. and Jimenez de Asua, L., 1979, Cytoskeleton-disrupting drugs enhance effect of growth factors and hormones on initiation of DNA synthesis, *Proc. Natl. Acad. Sci. USA* **76**:6435.

Pardee, A. B., 1974, A restriction point for control of normal animal cell proliferation, *Proc. Natl. Aced. Sci. USA.* **71**:1286.

Pathak, R. K., Bajapi, V. K., Shipstone, A. C., Chandra, H., and Karkun, J. N., 1978, Observation on the mucosa of the fallopian tube of immature rhesus monkey Macaca Mulatta, *Ind. J. Exp. Biol.* **16**:1244.

Perry, M. M., Tassin, J., and Courtois, Y., 1979, A comparison of human lens epithelial cells in situ and in vitro in relation to aging: an ultrastructural study, *Exp. Eye Res.* **28**:327.

Peterson, J. B., and Berns, M. W., 1978, Evidence for centriolar region RNA functioning in spindle formation in dividing PTK$_2$ cells, *J. Cell Sci.* **34**:289.

Phillips, S. G., and Rattner, J. B., 1976, Dependence of centriole formation on protein synthesis, *J. Cell Biol.* **70**:9.

Pledger, W. J., Hart, C. A., Locatell, K. L., and Scher, C. D., 1981, Platelet-derived growth factor-modulated proteins: constitutive synthesis by a transformed cell line, *Proc. Natl. Acad. Sci. USA.* **78**:4358.

Pledger, W. J., Stiles, C. D., Antoniades, H. N., and Scher, C. D., 1977, Induction of DNA synthesis in Balb/c 3T3 cells by serum components: reevaluation of the commitment process, *Proc. Natl. Acad. Sci. USA* **74**:4481.

Pledger, W. J., Stiles, C. D., Antoniades, H. N., and Scher, C. D., 1978, An ordered sequence of events is required before Balb/c 3T3 cells become committed to DNA synthesis, *Proc. Natl. Acad. Sci. USA* **75**:2839.

Przybylski, R. J., 1971, Occurrence of centrioles during skeletal and cardiac myogenesis, *J. Cell Biol.* **48**:214.

Publicover, S. J., and Duncan, C. J., 1979, The action of verapamil on the rate of spontaneous release of transmitter at the frog neuromuscular junction, *Euro. J. Pharm.* **54**:119.

Quattropani, S. L., 1973, Morphogenesis of the ovarian interstital tissue in the neonatal mouse, *Anat. Rec.* **177**:569.

Rappaport, R., 1971, Cytokinesis in animal cells, *Int. Rev. Cytol.* **31**:169.

Rappaport, R., and Rappaport, B. N., 1974, Establishment of cleavage furrows by the mitotic spindle, *J. Exp. Zool.* **189**:189.

Rash, J. E., Shay, J. W., and Biesele, J. J., 1969, Cilia in cardiac differentiation, *J. Ultrastruct. Res.* **29**: 470.

Rattner, J. B., and Berns, M. W., 1976, Distribution of microtubules during centriole separation in rat kangaroo (Potorous) cells, *Cytobios,* **15**:37.

Rattner, J. B., and Phillips, S. G., 1973, Independence of centriole formation and DNA synthesis, *J. Cell Biol.* **57**:359.

Renard, G. M., Hirsch, M., Galle, P., and Pouliquen, Y., 1976, Les cellules ciliees de l'endothelium' corneen, *Arch. Ophtalmol.* **36**:59.

Rieder, C. L., and Borisy, G. G., 1981, Assymetric distribution and cell-cycle changes of pericentriolar material, *J. Cell Biol.* **91**:319a.

Rieder, C. L., Jensen, C. G., and Jensen, L. C. W., 1979, The resorption of primary cilia during mitosis in a vertebrate (PTK$_1$) cell line, *J. Ultrastruct. Res.* **68**:173.

Robbins, E., Jentzsch, G., and Micali, A., 1968, The centriole cycle in synchronized HeLa cells, *J. Cell Biol.* **36**:329.

Rogers, K. A., and Brown, D. L., 1979, Colchicine effects on microtubules and capping of I$_G$G in lymphocytes, *J. Cell Biol.* **83**:337a.

Ross, R., Glomset, J., Kariya, B., and Harker, L., 1974, A platelet-dependent serum factor that stimulates the proliferation of arterial smooth muscle cells in vitro, *Proc. Natl. Acad. Sci. USA* **71**:1207.

Ross, R., Nist, C., Kariya, B. Rivest, M. J., Raines, E., and Callis, J., 1978, Physiological quiescence in plasma-derived serum: influence of platelet-derived growth factor on cell growth in culture, *J. Cell. Physiol.* **97**:497.

Roth, L. E., Wilson, H. J., and Chakraborty, J., 1966, Anaphase structure in mitotic cells typified by spindle elongation, *J. Ultrastruct. Res.* **14**:460.

Rubin, R. W., and Warren, R. H., 1979, Organization of tubulin in normal and transformed rat kidney cells, *J. Cell Biol.* **82**:103.

Satir, B., Sale, W. S., and Satir, P., 1976, Membrane renewal after dibucaine deciliation of tetrahymena, *Exp. Cell Res.* **97**:83

Scher, C. D., Stone, M. E., and Stiles, C. D., 1979, Platelet-derived growth factor prevents G_0 growth arrest, *Nature (London)* **281**:390

Scher, C. D., Pledger, W. J., Martin, P., Antoniades, H. N., and Stiles, C. D., 1978, Transforming viruses directly reduce the cellular growth requirement for a platelet derived growth factor, *J. Cell Physiol.* **97**:371.

Scherft, J. P., and Daems, W. T., 1967, Single cilia in chondrocytes, *J. Ultrastruct. Res.* **19**:546.

Schiff, P. B., Fant, J., and Horwitz, S. G., 1979, Promotion of microtubule assembly in vitro by taxol, *Nature (London)* **277**:665.

Schiff, P. B., and Horwitz, S. B., 1980, Taxol stabilizes microtubules in mouse fibroblast cells, *Proc. Natl. Acad. Sci. USA.* **77**:1561.

Schliwa, M., Euteneuer, U., Bulinski, J. C., and Izant, J. G., 1981, Calcium lability of cytoplasmic microtubules and its modulation by microtubule-associated proteins, *Proc. Natl. Acad. Sci. USA.* **78**:1037.

Selden, S. C., Rabinovitch, P. S., and Schwartz, S. M., 1981, Effects of cytoskeletal disrupting agents on replication of bovine endothedium, *J. Cell. Physiol.* **108**:2190.

Sherline, P., and Mascardo, R., 1982b, Epidermal growth factor-induced centrosomal separation: mechanism and relationship to mitogenesis, *J. Cell Biol.* **95**:316.

Sherline, P., and Mascardo, R., Epidermal growth factor-induced centrosomal separation: mechanism and relationship to mitogenesis, *J. Cell Biol.* **95**:316.

Sluder, G., 1978, The reproduction of mitotic centers: new information on an old experiment, in: *Cell Reproduction* (Daniel Mazia, E. R. Dirksen, D. Prescott, and C. F. Fox, eds.) pp. 563–569, Academic Press, New York.

Sluder, G., 1979, Role of spindle microtubules in the control of cell cycle timing, *J. Cell Biol.* **80**:674.

Smith, J. C., and Stiles, C. D., 1981, Cytoplasmic transfer of the mitogenic response to platelet-derived growth factor, *Proc. Natl. Acad. Sci. USA* **78**:4363.

Snyder, J. A., and Liskay, R. M., 1978, Timing of centriole duplication in Chinese Hamster cells that have no G_1 period, *J. Cell Biol.* **79**:13a.

Soloman, F., 1979, Detailed neurite morphologies of sister neuroblastoma cells are related, *Cell* **16**:165.

Sorokin, S. P., 1962, Centrioles and the formation of rudimentary cilia by fibroblasts and smooth muscle cells, *J. Cell Biol.* **15**:363.

Sorokin, S. P., 1968, Reconstruction of centriole formation and ciliogenesis in mammalian lungs, *J. Cell Sci.* **3**:207.

Stiles, C. D., Capone, G. T., Scher, C. D., Antoniades, H. N., Van Wyk, J. J., and Pledger, W. J., 1979, Dual control of cell growth by somatomedins and platelet-derived growth factor, *Proc. Natl. Acad. Sci. USA* **76**:1279.

Stoker, M. G. P., O'Neill, C., Berryman, S., and Wasman, V., 1968, Anchorage and growth regulation in normal and virus-transformed cells, *Int. J. Cancer* **3**:683.

Stommel, E. W., Stephens, R. E., and Alkon, D. L., 1980, Motile Statocyst cilia transmit rather than directly transduce mechanical stimuli, *J. Cell Biol.* **87**:652.

Storrie, B., and Edelson, P. J., 1977, Distribution of Concanavalin A in fibroblasts: direct endocytosis versus surface capping, *Cell* **11**:707.

Stubblefield, E. and Brinkley, B. R., 1967, Architecture and function of the mammalian centriole, in *Formation and Fate of Cell Organelles* (K. B. Warren, ed.), pp. 175–218, Academic Press, New York.

Sturrock, R. R., 1975, A light and electron microscopic study of proliferation and maturation of fibrous astrocytes in the optic nerve of the human embryo, *J. Anat.* **119**:223.

Swierenga, S. H. H., Whitfield, J. F., and Karasaki, S., 1978, Loss of proliferative calcium dependence: simple in vitro indicator of tumorigenicity, *Proc. Natl. Acad. Sci. USA* **75**:6069.

Tachi, S., Tachi, C., and Lindner, H. R., 1974, Influence of ovarian hormones on formation of solitary cilia and behavior of the centrioles in uterine epithelial cells of the rat, *Biol. Reprod.* **10**:391.

Tanuma, Y., and Ohata, M., 1978, Transmission electron microscope observation of epithelial cells with single cilia in intrahepatic biliary ductules of bats, *Arch. Histol. Jpn.* **41**:367.

Teng, M., Bartholomew, J. C., and Bissell, M. J., 1977, Synergism between anti-microtubule agents and growth stimulants in enhancement of cell cycle traverse, *Nature (London)* **268**:739.

Todaro, G. J., Lazer, G. K., and Green, H., 1966, The inhibition of cell division in a contact-inhibited mammalian cell line, *J. Cell. Comp. Physiol.* **66**:325.

Tucker, J. B., 1970, Morphogenesis of a large microtubular organelle and its association with basal bodies in the ciliate Nassula, *J. Cell Sci.* **6**:385.

Tucker, R. W., 1979, Centriole deciliation associated with the early response of 3T3 cells to mitogenic stimuli, *J. Cell Biol.* **83**:12a.

Tucker, R. W., 1980, Cytoplasmic microtubules, centriole ciliation and mitogenesis in Balb/c 3T3 cells, in *Microtubules and Microtubule Inhibitors* (M. DeBrabander and J. DeMey, eds.) pp. 497–508, Elsevier/North-Holland Biomedical Press, Amsterdam.

Tucker, R. W., 1981, Effect of tumor promotors on centriole and DNA synthesis cycles in 3T3 and epidermal cells, Proceedings of the American Association for Cancer Research, **20** (March, 1981) p. 127.

Tucker, R. W., 1982, Centriole separation and ciliary shortening are both associated with early mitogenic events in Balb/c 3T3 cells, *Cell Tissue Kinet.* (in press).

Tucker, R. W., and Pardee, A. B., 1981, Primary cilia and their role in the regulation of DNA replication and mitosis, in: *Cell Growth* (C. Nicolini, ed.) pp. 365–376, Plenum Press, New York.

Tucker, R. W., Sanford, K. K., and Frankel, F. R., 1978, Tubulin and actin in paired non-neoplastic and spontaneously transformed neoplastic cell lines in vitro: fluorescent antibody studies, *Cell* **13**:629.

Tucker, R. W., Pardee, A. B., and Fujiwara, K., 1979a, Centriole ciliation is related to quiescence and DNA synthesis, *Cell* **17**:527.

Tucker, R. W., Scher, C. D., and Stiles, C. D., 1979b, Centriole deciliation associated with the early response of 3T3 cells to growth factors but not to SV40, *Cell* **18**:1065.

Tucker, R. W., Butterfield, C. E., and Folkman, J., 1981a, Interaction of serum and cell spreading affects the growth of neoplastic and non-neoplastic fibroblasts, *J. Supramol. Struct. (Proc. ICN-UCLA Symp.)* **15**:29.

Tucker, R. W., Liaw, L.-H. and Berns, M. W., 1981b, Centriole duplication in PTK$_2$ cells, *J. Cell Biol.* **91**:48a.

Tupper, J. T., Del Rosso, M., Hazelton, B., and Zorgniotti, F., 1978, Serum-stimulated changes in calcium transport and distribution in mouse 3T3 cells and their modification by dibutryl cyclic AMP, *J. Cell. Physiol.* **95**:71.

Tupper, J. T., Kaufman, L., and Bodine, P. V., 1980, Related effects of calcium and serum on the G_1 phase of the human W138 fibroblast, *J. Cell. Physiol.* **104**:97.

Ukena, T. E., Borysenko, J. Z., Karnovsky, M. J. and Berlin, R. D., 1974, Effects of colchicine, cytochalasin B, and 2-deoxyglucose on the topographical organization of surface-bound Concanavalin A in normal and transformed fibroblasts, *J. Cell Biol.* **61**:70.

Unanue, E. R. and Karnovsky, M. J., 1974, Ligand-induced movement of lymphocyte membrane macromolecules V. Capping, cell movement, and microtubular function in normal and lectin-treated lymphocytes, *J. Exp. Med.* **140**:1207.

Vasiliev, J. M., Gelfand, I. M., and Guelstein, V. I., 1971, Initiation of DNA synthesis in cell cultures by colcemid, *Proc. Natl. Acad. Sci. USA* **68**:977.

Vegge, T., 1963, Ultrastructure of normal human trabecular endothelium, *Acta. Ophthalmol. (Copen)* **41**:193.

Vogel, A., Raines, E., Kariya, B., Rivest, M. J. and Ross, R., 1978, Coordinate control of 3T3 cells proliferation by platelet-derived growth factor and plasma components, *Proc. Natl. Acad. Sci. USA.* **75**:2810.

Vorobjev, I. A. and Chentsov, Yu. S., 1978a, Centrioles in the cell cycle. I. Ultrastructure and orientation of centrioles in the second half of mitosis (English translation), *Biologicheskie Nauki* **9**:65.

Vorobjev, I. A. and Chentsov, Yu. S., 1978b, Centriole in the cell cycle. II. Interphase and first half of mitosis (English translation), *Biologicheskie. Nauki.* **9**:54.

Vorobjev, I. A. and Chentsov, Yu. S., 1980, The ultrastructure of centriole in mammalian tissue culture cells, *Cell Biol. Int. Reports* **4**:1037.

Vorobjev, I. A., and Chentsov, Yu. S., 1982, Centrioles in the cell cycle. I. Epithelial cells, *J. Cell Biol.* **98**:938.

Wang, J. L., Gunther, G. R. and Edelman, G. M., 1975, Inhibition by colchicine of the mitogenic stimulation of lymphocytes prior to the S phase, *J. Cell Biol.* **66**:128.

Wasserman, W. J. and Smith, L. D., 1981, Calmodulin triggers the resumption of meiosis in amphibian oocytes, *J. Cell Biol.* **89**:389.

Weisenberg, R. C., 1972, Microtubule formation in vitro in solutions containing low calcium concentrations, *Science* **177**:1104.

Went, H. A., 1977a, Can a reverse transcriptase be involved in centriole duplication?, *J. Theor. Biol.* **68**:95.

Went, H. A., 1977b, Centriole duplication in sand dollar eggs. I. The effects of selected reagents, *Exp. Cell Res.* **108**:63.

Westermark, B., 1977, Local starvation for epidermal growth factor cannot explain density-dependent inhibition of normal human glial cells, *Proc. Nat. Acad. Sci. USA* **74**:1619.

Westermark, B., 1978, Growth control in miniclones of human glial cells cells, *Exp. Cell Res.* **111**:295.

Wheatley, D. N., 1967, Cilia and centrioles of the rat adrenal cortex, *J. Anat.* **101**:223.

Wheatley, D. N., 1969, Cilia in cell-cultured fibroblasts. I. On their occurrence and relative frequencies in primary cultures and established cell lines, *J. Anat.* **105**:351.

Wheatley, D. N., 1971, Cilia in cell-cultured fibroblasts. III. Relationship between mitotic activity and cilium frequency in mouse 3T6 fibroblasts, *J. Anat.* **110**:367.

White, B. A., Bauerle, L. R., and Bancroft, F. C., 1981, Calcium specifically stimulates prolactin synthesis and messenger RNA sequences in GH$_3$ cells, *J. Biol. Chem.* **256**:5942.

Whitfield, J. F., MacManus, J. P., Boynton, A. L., Gillian, D. J., and Issacs, R. J., 1974, Concanavalin A and the initiation of thymic lymphoblast DNA synthesis and proliferation by a calcium-dependent increase in cyclic GMP level, *J. Cell. Physiol.* **84**:445.

Wiederhold, M. L., 1976, Mechanosensory transduction in "sensory" and "motile" cilia, *Annu. Rev. Biophys. Bioeng.* **5**:39.

Wilsman, N. J., 1978, Cilia of adult canine articular chondrocytes, *J. Ultrastruct. Res.* **64**:270.

Wolfe, J., 1973, Cell division, ciliary regeneration and cyclic AMP in a unicellular system, *J. Cell. Physiol.* **82**:39.

Wolfe, R. A., Wu, R., and Sato, G. H., 1980, Epidermal growth factor-induced down regulation does not occur in HeLa cells grown in defined medium, *Proc. Nat. Acad. Sci. USA* **77**:2735.

Yahara, I. and Edelman, G. M., 1979, Modulation of receptor redistribution by Concanavalin A, anti-mitotic agents and alterations of ph, *Nature (London)* **246**:152.

Yahara, I., and Kakimoto-Sameshima, F., 1978, Microtubule organization of lymphocytes and its modulation by patch and cap formation, *Cell* **15**:251.

Yen, A., and Pardee, A. B., 1978a, Arrested states produced by isoleucine deprivation and their relationship to the low serum produced arrested state in Swiss 3T3 cells, *Exp. Cell Res.* **114**:389.

Yen, A. and Pardee, A. B., 1978b, Exponential 3T3 cells escape in mid-G$_1$ from their high serum requirement, *Exp. Cell Res.* **116**:103.

Yen, A., and Riddle, V. G. H., 1979, Plasma and platelet associated factors act in G$_1$ to maintain proliferation and to stabilize arrested cells in a viable quiescent state, *Exp. Cell Res.* **120**:349.

Zeuthen, E., and Williams, N. E., 1967, Division-limiting morphogenetic process in Tetrahymena, in: *Nucleic Acid Metabosism, Cell Differentiation and Cancer Growth* (E. V. Cowdry and S. Seno, eds.), pp. 203–217, Pergamon Press, Oxford.

Zimmer, D. B., Turner, D., Goldstein, M. A., and Brinkley, B. R., 1980, Microtubules and transformation: A quantitative electron microscope study, *J. Cell Biol.* **87**:244a.

Zimmerman, K. W., 1898, Beitrage zur kenntniss cinigen drusen und epithelien, *Arch. Mikrosk. Anat.* **52**:552.

Zorn, G. A., Lucas, J. J., and Kates, J. R., 1979, Purification and characterization of regenerating mouse L929 karyoplasts, *Cell* **18**:659.

Index